SPECIAL RELATIVITY

SPECIAL RELATIVITY

T. M. Helliwell

HARVEY MUDD COLLEGE

UNIVERSITY SCIENCE BOOKS
Mill Valley, California

University Science Books
www.uscibooks.com

Production Manager: *Paul C. Anagnostopoulos, Windfall Software*
Manuscript Editor: *Lee Young*
Proofreader: *Rick Camp*
Illustrator: *LM Graphics*
Compositor: *Windfall Software, using ZzTEX*
Cover Design: *Genette Itoko McGrew*
Printer & Binder: *RR Donnelly*

This book is printed on acid-free paper.

Library of Congress Cataloging-in-Publication Data

Helliwell, T. M. (Thomas M.), 1936–
Special relativity / Thomas Helliwell.
p. cm.
Includes index.
ISBN 978-1-891389-61-0
1. Special relativity. I. Title.
QC173.65.H45 2009
530.11—dc22
2008055980

Printed in the United States of America
10 9 8 7 6 5 4 3 2

This book is dedicated to my wife Bonnie, my family, my colleagues, and my fellow students.

It is dedicated also to stalwart students in Toowoomba, Tübingen, Tangshan, Tangiers, Trujillo, Troll Station, Thunder Bay, and all points in between.

Brief Contents

Contents

CHAPTER 5 / Lengths 51

CHAPTER 6 / Simultaneity 63

CHAPTER 7 / Paradoxes 81

CHAPTER 8 / The Lorentz Transformation 97

CHAPTER 9 / Spacetime 113

CHAPTER 10 / Momentum 131

Preface

THE SPECIAL THEORY OF RELATIVITY is a superb place to *begin* a serious study of physics, because it illustrates the fact that nature is stranger than we could have imagined, while also being the most accessible of the twentieth-century revolutions. It is likewise a superb subject to learn at *any* level, because it is fundamental physics with many important applications, while also being surprising, counterintuitive, and fun! Anyone who knows the Pythagorean theorem and a bit of algebra can understand its fundamentals. In this book we exploit the theory's accessibility by emphasizing its physical content. Numerous illustrations and examples are discussed at the outset, while the concise mathematical description is postponed until after the reader has had the opportunity to build up some physical intuition for what is going on. The principal challenges of special relativity are in fact conceptual, not mathematical.

The plan of the book was fixed partly by experience in trying various orders of presentation, and partly by the desire to make it useful in a variety of situations. It can serve as part of an introductory first-year course or be incorporated within a second-year "modern physics" course. It can also be used in a special topics or advance placement course, or as a supplement in advanced undergraduate courses such as theoretical mechanics, electromagnetism, or particle physics.

The main narrative of the book is as follows. Chapter 1 offers a brief review of classical mechanics—students with a good high school course can skip this chapter or refer back to it as needed. Chapter 2 describes two of the important experiments leading up to the special theory. The core of the book (the physical description of what is called "relativistic kinematics") is contained in Chapters 3–6. Here very little mathematics is used, because very little is needed. It is the physical ideas and the apparently paradoxical results that can be challenging, as exemplified particularly in Chapter 7. These chapters culminate in the Lorentz transformation of Chapter 8. Thus, the mathematical outcome of Einstein's postulates is postponed until after time dilation, length contraction, and the relativity of simultaneity have already been deduced. Four-dimensional spacetime is constructed in Chapter 9; this spacetime arena then allows us to introduce four-scalars and four-vectors. Chapters 10 and 11 show that momentum and energy join together to form the components of a four-vector and that mass is a form of energy. Chapter 12 takes up several applications, including the important topic of binding energy, especially in the context of nuclear physics, and then goes on to describe many examples of particle collisions and decays and how special relativity is used to predict outcomes, and ends up with an analysis of "photon rockets." Energy and momentum transformations are central to Chapter 13, with the aberration of light, the Doppler effect, threshold energies, and colliding-beam experiments as important applications. Finally, Chapter 14 provides a brief introduction to relativistic gravitation, introducing the Principle of Equivalence, gravitational redshifts, and the effect of gravity on clocks. The Global Positioning System (GPS) is shown to be strongly influenced by relativistic effects.

In addition to the main narrative, the book has 10 appendices, one or more of which can be taken up as interest and time allow. These are partly for fun and partly for more specialized topics. They include *The Binomial Approximation,* describing the often unsung but extraordinarily useful approximation for doing real physics calculations; *The "Paradox" of Light Spheres,* providing an interesting and thorough workout in the "three rules" of special relativity; *The Appearance of Moving Objects,* distinguishing between "optical illusions" and reality; *The Twin Paradox Revisited,* examining this (in)famous paradox in greater depth than there is room for in Chapter 7; *The "Cosmic Speed Limit,"* describing causal paradoxes and why signals are not supposed to exceed the speed of light; *"Relativistic Mass" and Relativistic Forces,* debating the pros and cons of believing that mass increases with velocity (and coming down hard on one side of this issue) and going on to explore the effect of forces in special relativity; *The Ultimate Relativistic Spaceflight,* recounting a truly fantastical voyage; *Nuclear Decays, Fission, and Fusion,* illustrating the interplay between relativity and nuclear physics, particularly in fusion reactions in the Sun and laboratory, and fission and fusion in nuclear weapons; *Some Particles,* listing mass and lifetime data for the lighter fundamental particles; and (finally) *Relativity and Electromagnetism,* giving a very brief example of the close relationship between relativity and the electric and magnetic fields around a wire and their effects upon a nearby electric charge.

This book grew out of an earlier book, *Introduction to Special Relativity,* published by the author in 1966. In one incarnation or another, it has been used ever since to teach relativity to nearly all first-year students at Harvey Mudd College, whether they intend to major in physics, engineering, biology, chemistry, computer science, or mathematics. It has been exciting to introduce a subject new to nearly everyone, and this newness, together with the subject's mathematical simplicity, means that students with stronger backgrounds have little advantage over those less well prepared. Special relativity is also an excellent training ground for careful, logical thought, a useful skill to anyone, no matter what his or her primary interests may be.

A word about how much prior mathematics and physics is needed to use this text. Students familiar with elementary algebra and high-school-level classical mechanics can skip the first two chapters, and start right in with Chapter 3 (perhaps with a brief glance back at Chapter 2), and then take up Chapters 4–7, which form the physical heart of relativistic kinematics. No calculus or background in college-level physics is needed to understand this material. A beginning knowledge of differential calculus is needed in Chapter 8 and in several subsequent chapters and appendices, and some familiarity with vectors is needed in Chapters 10 and beyond. Integral calculus is used only in Appendices D and G. The binomial approximation, which is based on the Taylor expansion of differential calculus, is explained in Appendix A and used in problems throughout the book.

The author is grateful to literally thousands of Harvey Mudd College students for their enthusiasm and direct or indirect help with this project. The book really grew out of teaching one another this fascinating subject. Many faculty colleagues offered valuable suggestions and support, including especially Profs. E. Wicher, D. Petersen, P. Saeta, G. Lyzenga, J. Townsend, P. Sparks, C-Y Chen, A. Esin, and V. Sahakian. My research colleague D. A. Konkowski read the entire manuscript and offered many important suggestions. I am also grateful to Bruce Armbruster and Jane Ellis of University Science Books for their warm encouragement and personal attention, which helped make this project a pleasurable experience, and to several reviewers who offered a great number of insightful and detailed suggestions that have greatly improved the book. I thank Lee Young for his expert copyediting that reflected his knowledge of physics as well as English usage. It was a great pleasure to work with Paul Anagnostopoulos, Joe Snowden, Laurel Muller, Rick Camp, and Genette Itoko McGrew, who managed, composed, illustrated, proofread, and designed with skill, imagination, and not inconsiderable patience. Finally, I am also greatly indebted to my wife, Bonnie, for her unflagging help and support.

A rather important person in all of this was a citizen of the world who signed his papers "A. Einstein." His 10-year effort to understand the motion of light, beginning at age 16, led him to present the bulk of what is discussed here (along with a good deal else) in two of his five breathtakingly original papers of 1905. Not even Einstein could have foreseen that his ideas would be so intertwined not only with the science of physics but also with the nature of warfare, the global economy, and even the history of nations.

The author hopes the reader enjoys learning special relativity and that he or she is sometimes befuddled by it! There is little chance of truly understanding this often counterintuitive theory without getting confused and working your way out of the confusion.

Finally, here are three favorite Einstein quotes to help guide the reader through this book and beyond:

> *Common sense is the collection of prejudices acquired by age eighteen.*
>
> *A person who never made a mistake never tried anything new.*
>
> *There are only two ways to live your life. One is as though nothing is a miracle. The other is as though* everything *is a miracle.*

T. M. Helliwell
Harvey Mudd College

SPECIAL RELATIVITY

CHAPTER 1

Inertial Frames and Classical Mechanics

THE SPECIAL THEORY OF RELATIVITY is a theory about space and time. It is not primarily about mechanics, electricity, optics, the properties of materials, thermodynamics, or fundamental particles. But since space and time are the *arena* in which physical events take place, a change in our ideas about them might well have repercussions throughout the rest of physics. In fact that is what has happened. Though revolutionary, relativity grew out of what had gone before—in particular, it developed from the confrontation of Newtonian mechanics of the seventeenth and eighteenth centuries with the theories of light and electromagnetism of the nineteenth century. We'll begin in this chapter with a brief synopsis of some highlights of classical mechanics, which describes how things move and why they move that way. Then in Chapter 2 we will discuss some critical experiments on the behavior of light.

1.1 Inertial Frames and Newton's Laws

To describe precisely how things move, we first have to choose a frame of reference in which we view the motion and in which measurements can be made. Whether we are sitting in a room, standing in a laboratory, traveling in a train, or riding a merry-go-round, we can erect metersticks to form a set of three Cartesian coordinates. There is a coordinate origin, say at one corner of a room, and three coordinate axes, the x, y, and z directions, as in Fig. 1.1. We can use these coordinates to specify the position of any object in the room.

Many other reference frames are possible. Our metersticks could have a different origin, or they might be rotated relative to the original set. Observers riding in a train can erect a set of metersticks that are moving with respect to ours, which could equally well be used to measure the positions of objects.

Although any reference frame can be used to describe motion, there is a special set of frames in which the motion of objects is particularly simple. These frames are

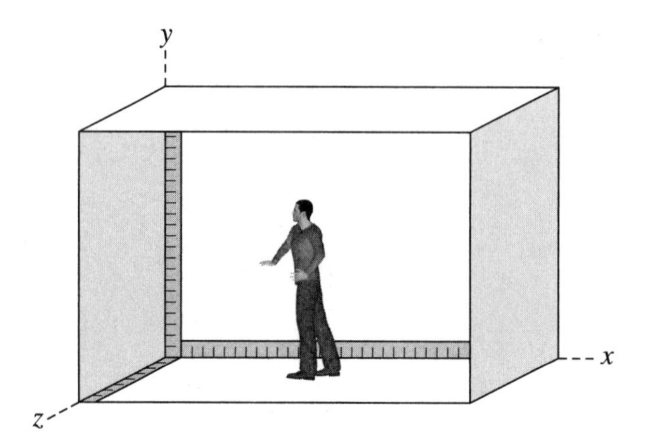

FIGURE 1.1
A frame of reference.

called *inertial frames,* because in these frames the *law of inertia* is obeyed. This law was presented by Isaac Newton in his *Principia* as the first law of motion:

> If there are no forces on an object, then if the object starts at rest it will stay at rest, or if it is initially set in motion, it will continue moving in the same direction in a straight line at constant speed.

We can use this law to test whether or not our frame is inertial. Suppose our room is inside a spaceship drifting through space. The ship is not rotating relative to the distant stars, and it has its rocket engine turned off. We remove any air inside to avoid drag forces on moving objects. Then if we set a ball at rest, it will stay put. If we toss the ball in any direction, it will keep moving in that direction with constant speed. The law of inertia is obeyed, so by definition our frame is inertial. If the ship is set spinning relative to the stars, however, our room will no longer be an inertial frame. If we toss balls in various directions, for example, we will then see that most of their paths are curved rather than straight. Of if we fire up the rocket engine and cause the ship to accelerate forward, a ball set at rest in the ship will not stay put, but will "fall" toward the rear of the ship. The law of inertia is not obeyed, so the reference frame is noninertial in either accelerating or rotating ships.

Once we have found an inertial frame, *any* frame that is neither accelerating nor rotating relative to that first frame is also inertial. The room in our drifting ship is an inertial frame, so a frame set up in any other nonrotating ship moving at constant velocity relative to ours will also be inertial. There are an infinite number of such inertial frames, moving with different speeds in different directions, all nonrotating and nonaccelerating, as shown in Fig. 1.2.

What about rooms on Earth? We stand in a laboratory, and set up metersticks to measure the position of a ball. Is our frame inertial? The first problem we face is that we can't remove all the forces on a ball, as we are supposed to do to test whether the

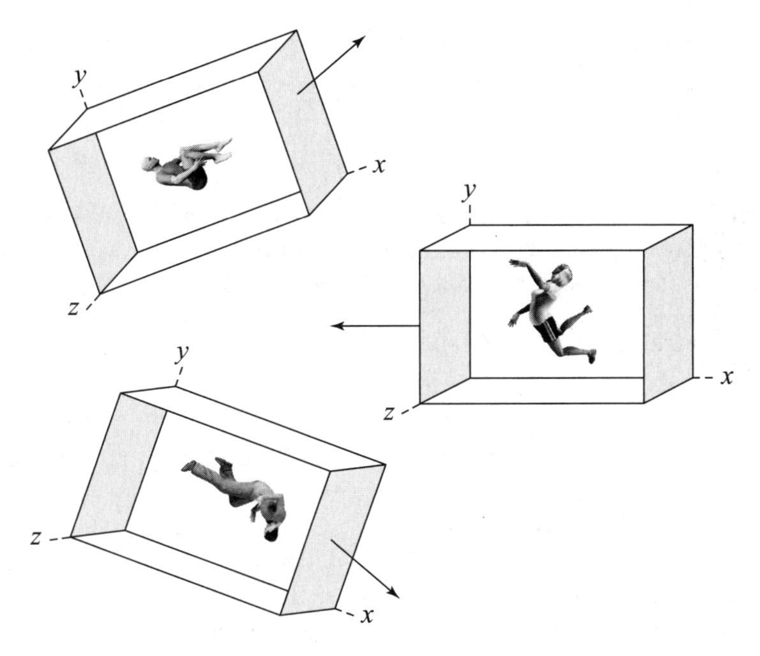

FIGURE 1.2
Various intertial frames in space.

frame is inertial: Gravity is always there—we can't eliminate it![1] A second problem is the fact that Earth rotates on its axis daily, so our lab does not fit the model of a frame that is nonrotating relative to the stars. In fact, as a result of Earth's rotation the law of inertia is not quite obeyed in our lab, because if we throw a ball horizontally, its path will bend very slightly sideways.[2]

In spite of these difficulties, as long as we include the force of gravity on objects in the lab, and the experiments we do are localized and not so precise that we can detect the effects of Earth's rotation, a local reference frame fixed on Earth's surface is close enough to being an inertial frame for most purposes.

Newton's *second* law states that for motion in a straight line the acceleration a of an object of mass m is due to the net force F exerted on it, according to the equation

---

1. It *is* actually possible to eliminate gravity on Earth! Remove the air from an elevator shaft and cut the elevator cables, so the elevator goes into free fall. Within the freely falling frame of the elevator, gravity is eliminated—nothing falls relative to the frame. Gravity is also eliminated in an Earth-orbiting spacecraft, another freely falling frame. We can even use freely falling frames as inertial frames, if we do not count gravity as a force. Einstein brilliantly exploited this fact as a step in developing his theory of gravitation, the general theory of relativity. See Chapter 14.

2. This bending, called the Coriolis effect, is small in ordinary laboratories, but very important on larger scales. Coriolis deflections occur significantly in large-scale air currents, for example, as in tornados and hurricanes, and in large-scale ocean currents.

$F = ma$. More generally, of course, objects move in all three dimensions, in which case Newton's second law becomes

$$\mathbf{F} = m\mathbf{a} \tag{1.1}$$

in terms of the *vectors* **F** and **a** (written in boldface type), each of which has a direction as well as a magnitude. That is, if a force **F** is exerted in some particular direction, Eq. 1.1 tells us that the acceleration **a** is in that same direction. The equation has the three components

$$F_x = ma_x, \quad F_y = ma_y, \quad F_z = ma_z, \tag{1.2}$$

one equation for each of the three perpendicular directions x, y, z. The force in the x direction causes acceleration in the x direction, the force in the y direction causes acceleration in the y direction, and so forth.

Forces may be due to springs, cables, friction, gravitation, or electricity, for example, and are measured in the unit *newtons*.[3] The law implies that if we remove all forces from an object, it will not accelerate: it will stay at rest if it starts at rest, and move in a straight line at constant speed if it is given an initial velocity. But that is just Newton's *first* law, so it might seem that the first law is just a special case of the second law! However, the second law is not true in all frames of reference. An observer bouncing up and down on a trampoline will see objects accelerate from her point of view, even if there is no net force on them. In fact, it is only *inertial observers,* who make their measurements in inertial reference frames, who can use Newton's second law. So the first law should not be thought of as a special case of the second law—for motion without forces—but as a means of specifying those observers for whom the second law is valid.

1.2 The Galilean Transformation

In describing motion, we not only have to know the position of an object; we also have to know at what time it has that position. To cover four measurements with a single word, we define an *event* as something that happens at an instant in time and a point in space. If a flashbulb goes off at time $t = 5$ seconds and at position $(x, y, z) = (2, 3, 1)$ meters, the event can be characterized by the four quantities (5 s, 2 m, 3 m, 1 m).

The measurement of the position of a given event obviously depends on the observer's frame of reference. Suppose, for example, there are two inertial frames moving at constant relative velocity V along their mutual x axes, as shown in Fig. 1.3. By convention, the primed frame S' (which might be a frame inside a train) moves with velocity V to the *right* as seen by people in the unprimed frame S (which might be a

3. In science we most often use *Système International* (SI) units, in which mass is measured in kilograms (kg), length in meters (m), and time in seconds (s). Therefore velocity has units m/s and acceleration has units m/s^2. Force is measured in newtons (N), where $1\text{N} = 1\,\text{kg m/s}^2$.

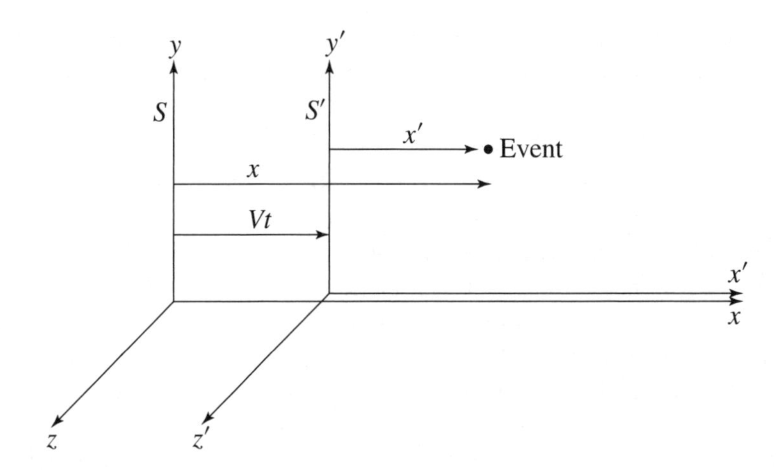

FIGURE 1.3
A primed frame S' moves to the right relative to an unprimed frame S, along their mutual x axes.

frame fixed to the ground), and the unprimed frame S moves with velocity V to the *left* as seen by people at rest in the primed frame S'. That is what we mean by relative velocity: Each frame measures the other frame to be moving at velocity V, either to the right or left.[4]

When a particular event occurs, people in S label it by (t, x, y, z), while people in S' label it (t', x', y', z'). The two sets of numbers must be related to one another, and using common sense it seems that if the same standards of length and time are used in both systems, and if all the clocks are synchronized and set to $t = t' = 0$ just when the origins of S and S' coincide, the relationships must be

$$x' = x - Vt, \;\; y' = y, \;\; z' = z, \;\; t' = t \tag{1.3}$$

for any given event. These relations, collectively called the *Galilean transformation,* simply state that an event measured in the S frame, with coordinates t, x, y, z, will be at exactly the same time and place in the S' frame, except for the x coordinate, which is corrected for the fact that the origin of the S' frame has moved away from the origin of the S frame by a distance Vt. It is possible for an event to be at positive x and negative x', which happens if the origin of the S' frame has moved past the position of the event in Fig. 1.3 before it happens.

Notice from the Galilean transformation that length *differences* and time *intervals* are the same in the two frames. If x_1 and x_2 are the x coordinates of the left and right ends of a meterstick at rest along the x axis of the S frame, for example, the length of

---

4. Physicists sometimes distinguish "speed" from "velocity" by reserving "speed" to mean how fast an object is moving, without regard to its direction of motion, while the "velocity" of the object implies a specific direction of motion as well as a speed. With this convention, we could say that a hawk flies at *speed* 5 m/s, or that it is flies at *velocity* 5 m/s *southward.* Physicists are not always consistent in this convention, however, and we do not promise to be consistent in this book.

the stick is $x_2 - x_1$. In the S' frame the meterstick is moving; its length as measured by S' observers is $x_2' - x_1'$, where they measure both ends at the same time t'. From Eqs. 1.3,

$$x_2' - x_1' = (x_2 - Vt_2) - (x_1 - Vt_1) = x_2 - x_1, \tag{1.4}$$

since $t_2 = t_1$ (because $t_2' = t_1'$). In other words, primed observers can say to their unprimed-frame friends: "You claim your meterstick is one meter long, and we have found that it is one meter long from our point of view, also." Note that *two* measurements are required in the primed frame (x_1' and x_2'), which have to be made at the *same time*—we would certainly not want to measure the position of the two ends of a moving object at *different* times. And to be sure the measurements are made at the same time, the two observers must have previously synchronized their clocks. Then if the front of the stick passes by one observer just as the back passes by the other, then the distance between them is the length of the stick.

Similarly, a time interval $t_2 - t_1$ measured by observers in the S frame is the same time interval measured by observers in the S' frame, since from the Galilean transformation, $t_2' - t_1' = t_2 - t_1$. How could this be verified experimentally? Suppose a clock C is at rest in the primed frame, while we are at rest in the unprimed frame. Clock C is therefore moving at speed V relative to us. We want to read C at two different times, so we again need *two* observers (A and B) with synchronized clocks in our frame, as shown in Fig. 1.4. As C passes A, A notices both the reading of $C(t_1')$ and also the reading of his own clock (t_1). Some time later, clock C passes by observer B, who

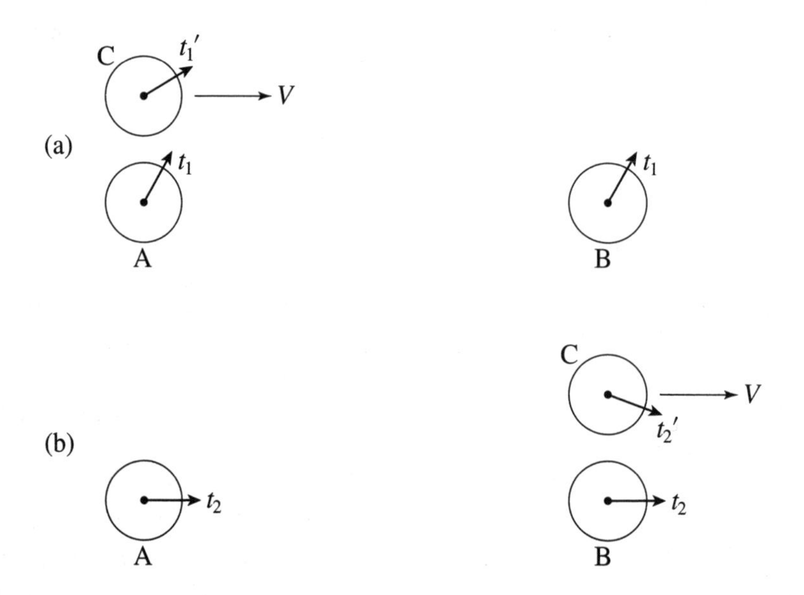

FIGURE 1.4
The time read by clock C, at rest in the primed frame, as measured by two observers (A and B) at rest in the unprimed frame. (a) The first measurement is made by A, at time t_1 on A's clock. (b) The second measurement is made by B, at time t_2 on B's clock.

notices that C reads t_2', while her own clock reads t_2. Later, A and B can compare notes, and if the Galilean transformation is correct, they will find that $t_2' - t_1' = t_2 - t_1$.

According to the Galilean transformation, space and time are entirely distinct notions, which never become confused, and which have no influence on the behavior of material bodies. People searched for laws governing the motion of objects *in* space and *in* time, without really investigating space and time themselves. According to Newton there is no reason to do so:

> Absolute space, in its own nature and without regard to anything external, always remains similar and immovable.

He also wrote,

> Absolute, true, and mathematical time, of itself, and by its own nature, flows uniformly on, without regard to anything external.

The Galilean transformation implies more than just how to find the primed coordinates in terms of the unprimed coordinates. It also shows how to find the velocity of a particle in one frame in terms of its velocity in a different frame. We have used the symbol V for the relative velocity (in the x direction) of the two frames S and S', and we will use v_x, v_y, v_z for the x, y, and z components of a particle's velocity in the S frame, and v_x', v_y', v_z' for the corresponding components of the same particle's velocity in the S' frame. By definition, $(v_x, v_y, v_z) = (dx/dt, dy/dt, dz/dt)$, so, using Eqs. 1.3 (including $dt = dt'$),

$$
\begin{aligned}
v_x' &\equiv \frac{dx'}{dt'} = \frac{d(x - Vt)}{dt} = \frac{dx}{dt} - V \equiv v_x - V \\
v_y' &\equiv \frac{dy'}{dt'} = \frac{dy}{dt} \equiv v_y \\
v_z' &\equiv \frac{dz'}{dt'} = \frac{dz}{dt} \equiv v_z.
\end{aligned}
\tag{1.5}
$$

Note that there are three different velocities involved here! There is the velocity (v_x, v_y, v_z) of a particle as measured by observers at rest in the S frame, the velocity (v_x', v_y', v_z') of the *same* particle as measured by observers at rest in the S' frame, and the relative velocity V (in the x direction) of the two frames themselves.

Equations 1.5 make sense—the velocity components in the y and z directions are the same in both frames, while the velocity components in the x direction differ by the relative velocity of the two sets of coordinates. For example, a train traveling at 100 miles/hour relative to the ground will be moving at only 40 miles/hour from your point of view, if you are riding in a car in the same direction at 60 miles/hour, as shown in Fig. 1.5. Here the ground frame is the S frame and the car frame is the S' frame, while the velocities are v_x (= 100 mi/hr, the velocity of the train relative to the ground), v_x' (= 40 mi/hr, the velocity of the train relative to the car), and V (= 60 mi/hr, the car's velocity relative to the ground.)

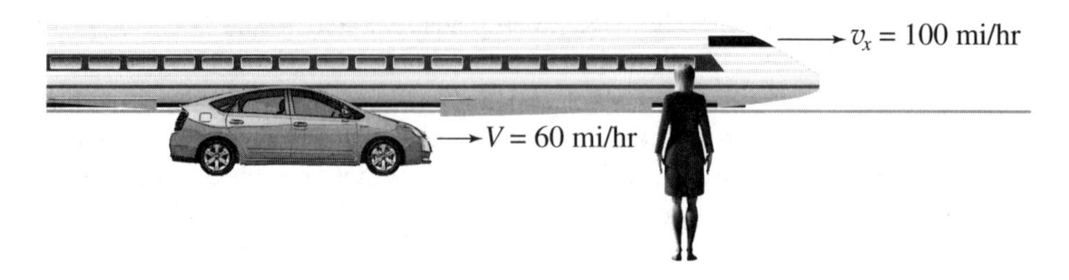

FIGURE 1.5
The velocity of a train as seen in two different reference frames, that is, by an observer on the ground and an observer in a car.

If we take a second derivative of Eqs. 1.3, we get the transformation rules for acceleration,

$$a'_x \equiv \frac{dv'_x}{dt'} = \frac{d\left(v_x - V\right)}{dt} = \frac{dv_x}{dt} \equiv a_x \tag{1.6}$$

and similarly $a'_y = a_y$ and $a'_z = a_z$. Thus the acceleration of an object is the same in all inertial frames, according to the Galilean transformation. Therefore since we take the mass of a particle to be a constant quantity, independent of frame, it follows that the force components F_x, F_y, F_z must also be the same in all inertial frames. That is because in classical physics we take Newton's second law $\mathbf{F} = m\mathbf{a}$ to be a fundamental law of nature, so that observers in any inertial frame must be able to use it. The equation $\mathbf{F} = m\mathbf{a}$ is said to be "invariant under the Galilean transformation," meaning that it is equally valid in any inertial frame if the Galilean transformation is used to translate variables from one frame to another.

In classical mechanics, any observer can use the fundamental laws: *There is no preferred frame of reference.* This requirement of invariance of fundamental laws, sometimes called the "principle of relativity," is wonderfully powerful, and we will have more to say about it when we begin discussing Einstein's relativity in Chapter 3.

1.3 Newton's Third Law and Momentum Conservation

Newton's *third* law states that "action equals reaction," which means that if one particle exerts a force on a second particle, the second particle exerts an equal but opposite force back on the first particle, as shown in Fig. 1.6. Thus if a collection of particles is considered as a unit, and if no outside forces are applied, there is no net force on the collection, because the internal forces cancel each other out. This fact is important in establishing the significance of *momentum.*

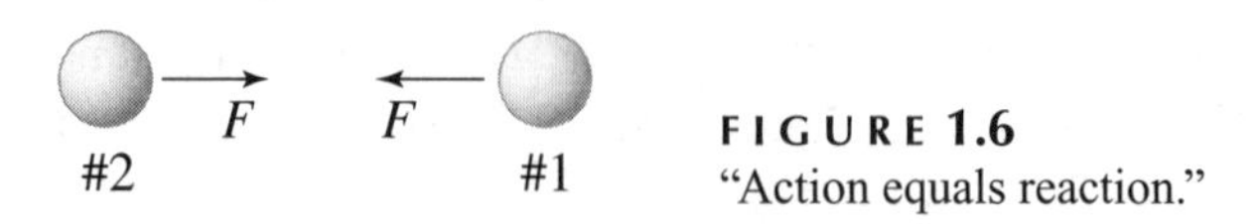

FIGURE 1.6
"Action equals reaction."

In classical mechanics, the momentum $\mathbf{p}$ of a particle is

$$\mathbf{p} = m\mathbf{v}, \tag{1.7}$$

the product of the particle's mass and its velocity.[5] The vector equation confirms that the momentum is in the same direction as the velocity; it is also short for the three component equations

$$p_x = mv_x, \;\; p_y = mv_y, \;\; p_z = mv_z. \tag{1.8}$$

Now suppose we differentiate the momentum with respect to time. The result is

$$\frac{d\mathbf{p}}{dt} = \frac{d\,(m\mathbf{v})}{dt} = m\frac{d\mathbf{v}}{dt} = m\mathbf{a}, \tag{1.9}$$

using the fact that the particle's mass does not change, and the definition $\mathbf{a} \equiv d\mathbf{v}/dt$. Therefore $\mathbf{F} = m\mathbf{a}$ can equally well be written

$$\mathbf{F} = \frac{d\mathbf{p}}{dt}, \tag{1.10}$$

so that the net force not only causes acceleration; it also causes a change in momentum. (As usual, Eq. 1.10 stands for three component equations, $F_x = dp_x/dt$, $F_y = dp_y/dt$, $F_z = dp_z/dt$.) It follows from Eq. 1.10 that if there is no net force on a particle its momentum stays constant in time, so is said to be *conserved.*

Now instead of a single particle, picture a *system* of particles, and define their total momentum $\mathbf{P}$ to be the sum of the momenta of the individual particles. That is, define

$$\mathbf{P} = \mathbf{p}_1 + \mathbf{p}_2 + \mathbf{p}_3 + \cdots \tag{1.11}$$

Similarly, define the total force $\mathbf{F}_{\text{total}}$ to be the sum of the forces on all the individual particles,

$$\mathbf{F}_{\text{total}} = \mathbf{F}_1 + \mathbf{F}_2 + \mathbf{F}_3 + \cdots. \tag{1.12}$$

It then follows that $\mathbf{F}_{\text{total}} = d\mathbf{P}/dt$, just by adding up the individual $\mathbf{F} = d\mathbf{p}/dt$ equations for all the particles. If we further split up the total force $\mathbf{F}_{\text{total}}$ into $\mathbf{F}_{\text{ext}}$ (the sum of the forces exerted by *external* agents, like Earth's gravity or air resistance) and $\mathbf{F}_{\text{int}}$ (the sum of the *internal* forces caused by members of the system themselves, like the forces of collision between two pool balls belonging to the system), then $\mathbf{F}_{\text{total}} = \mathbf{F}_{\text{internal}} + \mathbf{F}_{\text{external}} = \mathbf{F}_{\text{external}}$, because all the internal forces cancel out by Newton's third law! (When pool balls A and B collide, for example, the force of A

5. We will see in Chapter 10 that this expression is only an *approximation* to the true momentum of a particle, valid only when the velocity of the particle is much less than the velocity of light.

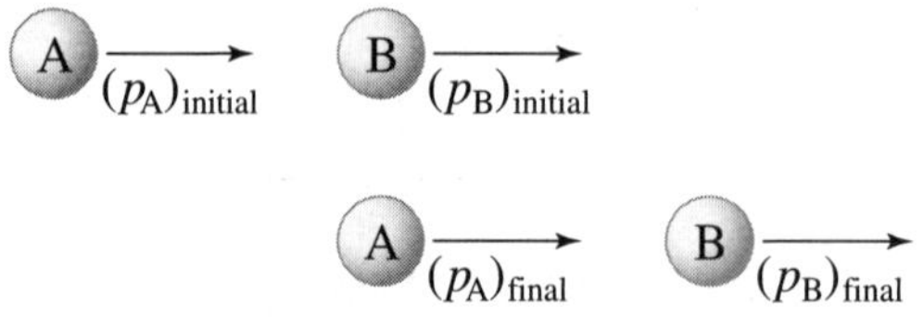

A $(p_A)_{\text{initial}}$ B $(p_B)_{\text{initial}}$

A $(p_A)_{\text{final}}$ B $(p_B)_{\text{final}}$

FIGURE 1.7
In collisions, momentum is exactly conserved if there are no external forces, and nearly conserved for moderate external forces.

on B is equal but opposite to the force of B on A, so these "internal forces" sum to zero.) Finally, we can write a grand second law for the system as a whole,

$$\mathbf{F}_{\text{ext}} = \frac{d\mathbf{P}}{dt}, \tag{1.13}$$

showing how the system as a whole moves in response to external forces.

Now the importance of momentum is clear. For if no external forces act on the collection of particles, their total momentum cannot depend upon time, so $\mathbf{P}$ is *conserved.* Individual particles in the collection may move in complicated ways, but they always move in such a way as to keep the total momentum constant.

It is interesting that in collisions, such as the collision of two balls A and B, we can assume that momentum is conserved in the collision even if there are modest external forces. Suppose, for simplicity, that the balls collide head-on, so the collision is one-dimensional. Then Eq. 1.13 becomes simply $F_{\text{ext}} = dP/dt$. For short time intervals Δt we can write this in the form $\Delta p = F_{\text{ext}}\Delta t$, so for the tiny duration Δt of the collision the change of momentum Δp is typically so very small that the momentum just after the collision is essentially the same as the momentum just after the collision. Therefore $(p_A + p_B)_{\text{initial}} = (p_A + p_B)_{\text{final}}$, as illustrated in Fig. 1.7.

Aside from the total momentum P, another quantity characterizing a collection of particles spread out along the x axis is their *center of mass* position X_{CM}. Suppose we stand in a particular reference frame and measure the distance of each particle in the collection from the origin of our coordinate system. Let the ith particle have mass m_i, and suppose its distance from the origin is x_i, as shown in Fig. 1.8. The center of mass of the collection of particles is then defined to be

$$X_{\text{CM}} = \frac{m_1 x_1 + m_2 x_2 + \cdots}{m_1 + m_2 + \cdots}, \tag{1.14}$$

which gives the point in space about which all the particles "balance" in the x direction. If the particles are atoms in a horizontal broom, for example, X_{CM} is the location just above the spot where you can balance the broom on one finger.

The velocity of the center of mass is

$$\begin{aligned} V_{\text{CM}} \equiv \frac{dX_{\text{CM}}}{dt} &= \frac{d}{dt}\left(\frac{m_1 x_1 + m_2 x_2 + \cdots}{m_1 + m_2 + \cdots}\right) \\ &= \frac{m_1(dx_1/dt) + m_2(dx_2/dt) + \cdots}{m_1 + m_2 + \cdots} \end{aligned} \tag{1.15}$$

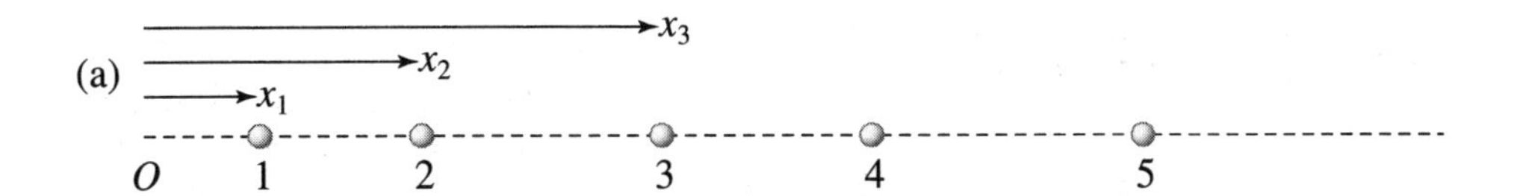

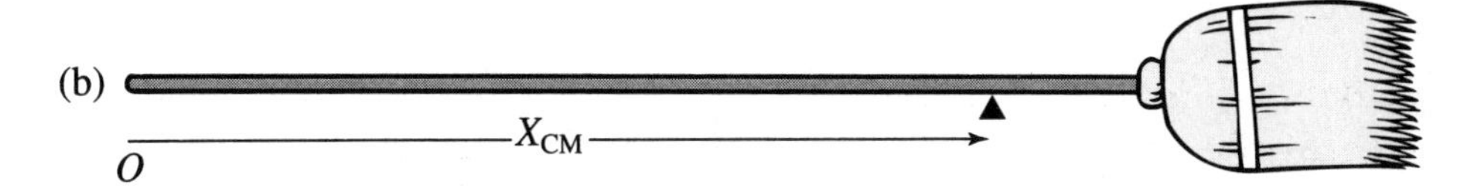

FIGURE 1.8
(a) A collection of particles with positions x_i along the x axis. (b) The center of mass X_{CM} of a broom measured from some arbitrary origin "O."

differentiating term by term and using the fact that the particle masses are constant. The velocity of particle i is $v_i = dx_i/dt$ and its momentum component is $p_i = mv_i$, so

$$V_{CM} = \frac{p_1 + p_2 + \cdots}{m_1 + m_2 + \cdots} = \frac{P}{M}, \tag{1.16}$$

where P is the total momentum of the particles and M is their total mass. So we have proven that the center of mass moves at constant velocity whenever P is conserved—that is, whenever there is no net external force.[6] In particular, if there is no external force on the particles, their center of mass stays at rest if it starts at rest.

If the total momentum of the system is zero in some particular frame of reference, then the center of mass remains fixed in that frame. This particular reference frame is called the *center of mass frame*, and can be thought of either as that frame in which the center of mass remains at rest, or in which the total momentum is zero.[7] Both are true.

1.4 Energy

Another quantity that is sometimes conserved in the classical mechanics of a single particle is its total *energy*, the sum of its *kinetic* and *potential* energies. If the particle is moving, it has a kinetic energy

$$KE = \frac{1}{2}mv^2, \tag{1.17}$$

6. If particles are spread out in all three dimensions, the ith particle is located by a *position vector* $\mathbf{r}_i = (x_i, y_i, z_i)$, with components in each of the three directions. The center of mass of all of these particles is $\mathbf{R}_{CM} = (m_1\mathbf{r}_1 + m_2\mathbf{r}_2 + \cdots)/(m_1 + m_2 + \cdots)$, and the velocity of the center of mass is $d\mathbf{R}_{CM}/dt = \mathbf{P}/M$. As usual, each vector equation is equivalent to three equations, one for each direction.

7. If we find one center of mass frame, there are actually an infinite number of others, all at rest relative to the first one, but tilted at some angle relative to it.

FIGURE 1.9
A ball moving upward.

where m is its mass and v its speed. Kinetic energy is the energy of motion, an energy that depends upon how fast the particle is moving, but not on where it is located.[8] The SI unit for kinetic energy, and for *any* kind of energy, is joules (J), where $1\ \text{J} = 1\ \text{kg m}^2/\text{s}^2$.

For each so-called *conservative force* acting on the particle, there is an associated *potential energy.* (Conservative forces include gravitational forces, spring forces, electrostatic forces, and many others.) Potential energies depend on the position of the particle, but not on its speed. The potential energy of a particle in a uniform gravitational field g, for example,[9] is $PE_{\text{grav}} = mgh$, where m is its mass and h is its altitude; similarly, the potential energy of a particle subjected to the ideal spring force $F = -kx$ is $PE_{\text{spring}} = (1/2)kx^2$, where k is the force constant of the spring and x is the spring's stretch or compression. Examples of *non*conservative forces include the surface friction between a chair and the floor, or the air resistance acting on a flying baseball. Nonconservative forces do not have associated potential energies.

Now suppose all of the forces acting on a particle are conservative—we assume there is no friction or air resistance, for example. Then the mechanical energy of the particle, defined to be the sum of its kinetic energy and all potential energies, is conserved. That is,

$$E = KE + PE_{\text{TOTAL}} = \text{constant}. \tag{1.18}$$

If a baseball is thrown vertically upward on the moon (where there is no air resistance), its kinetic energy is gradually converted into gravitational potential energy, as shown in Fig. 1.9. At the crest of its motion the energy is entirely potential, and then as the ball falls again its energy is reconverted into kinetic energy. Each kind of energy can change, but the sum stays the same.

---

8. We will show in Chapter 11 that the expression $(1/2)mv^2$ for kinetic energy is an *approximation,* a *good* approximation only for particles that move slowly compared with the speed of light.
9. The gravitational field g at some point is numerically equal to the acceleration of a particle at that point, if gravity is the only force acting on the particle.

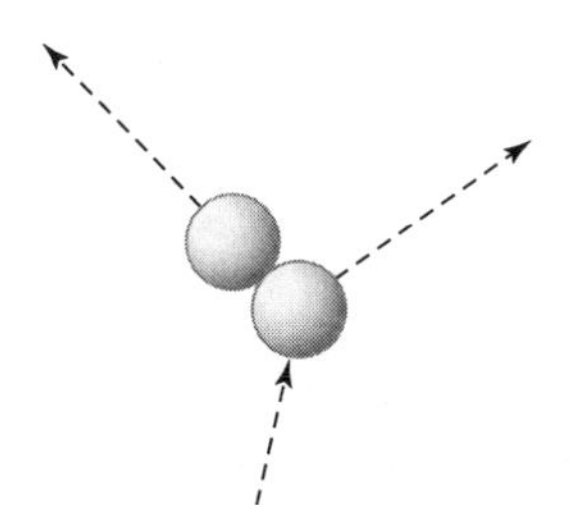

FIGURE 1.10
Colliding pool balls.

Conservation of energy also holds for a *system* of particles, as long as all forces are conservative, whether the forces are between particles in the system or are exerted from outside. In that case the kinetic energy is the sum of the kinetic energies of all particles, and the potential energy is the sum of external potential energies, plus the potential energies between particles within the system.

An especially interesting system is that of two colliding particles, such as two pool balls colliding on a pool table, as illustrated in Fig. 1.10. Suppose the balls strike one another so that they are compressed like a kind of spring and then they decompress and push each other apart so that the total kinetic energy of the balls is the same as it was before the collision. In other words, suppose that the force between the two balls is conservative. Such an idealized collision is said to be *elastic.*

In fact, this picture *is* only an idealization. *Real* collisions of pool balls are *inelastic,* because the kinetic energy after the collision is somewhat less than it was before. Energy appears not to be conserved, and in fact the sum of the kinetic and potential energies of the balls seems to *decrease* as a result of the collision. Deeper inspection shows, however, that as a result of the collision, oscillations and waves are set up within each ball, which ultimately turn into the kind of random oscillations at the molecular level that we call *heat.* That is, the large-scale (macroscopic) energy decreases, but the small-scale (microscopic) energies associated with oscillations make up the difference.

In a *completely inelastic* collision, two colliding objects stick together, moving as a unit after the collision. An example is a bullet fired into a large block of wood, where the bullet is captured by the block, as shown in Fig. 1.11. The block and bullet recoil together, moving with a velocity consistent with momentum conservation. The macroscopic kinetic energy decreases markedly, however, while the bullet and block get very hot; in fact, the metal bullet may melt in such a collision.

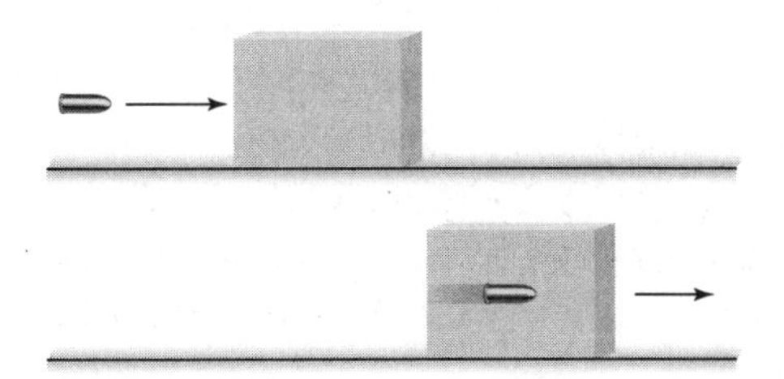

FIGURE 1.11
A completely inelastic collision.

In short, conservation of energy is universal in classical physics, if we include many types of energy besides the macroscopic kinetic and potential energies. If the energy of an isolated system seems to decrease, we look for other places it may be lurking, and we have always been able to find it.

Newton's laws, the Galilean transformation, the conservation laws, and the whole of classical mechanics provide an excellent description of the motion of objects in a wide variety of circumstances. People have used these ideas successfully for centuries. Toward the end of the nineteenth century, however, when the study of light and electromagnetism became sufficiently advanced, difficulties were uncovered. That is the topic of Chapter 2.

Sample Problems

1. A powerful laser constructed on the far side of the Moon is used to blast a small probe into space. The probe has a mass of 1 kg, and the laser manages to provide a constant force of 1000 N on it. According to classical mechanics, (a) what is the probe's acceleration? (b) How long does it take the probe to reach half the speed of light, where the speed of light is $c = 3.0 \times 10^8$ m/s?

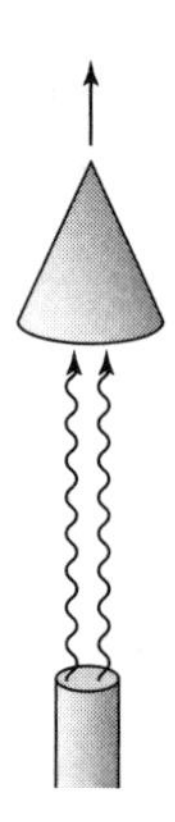

Solution: (a) Using Newton's second law, the probe's acceleration is $a_0 = F/m = 1000\text{ N}/1\text{ kg} = 1000\text{ m/s}^2$. (b) For constant acceleration a_0, the relation between velocity and time is $v = a_0 t$ if the object starts at rest. (As a check, the acceleration is $a = dv/dt$, and the time derivative of $v = a_0 t$ correctly gives $a = a_0$.) The time to reach half the speed of light is therefore

$$t = (c/2)/a_0 = \left(1.5 \times 10^8\text{ m/s}\right)/1000\text{ m/s}^2 = 1.5 \times 10^5\text{ s} = 1.74\text{ days.}$$

2. Police investigators find that two cars of equal mass crashed head-on in an icy street. Car A had an unknown initial eastward velocity v_0, while Car B had an initial westward velocity of 20 miles/hour. As a result of the collision, the investigators find that the two cars became stuck together, and moved eastward at

10 miles/hour. They ask you to determine whether Car A's initial speed exceeded the speed limit of 25 miles/hour. What would you say?

Solution: Momentum is conserved in the collision, so the initial momentum must equal the final momentum. Taking the eastward direction to be positive and the westward direction to be negative,

$$mv_0 - m(20 \text{ mi/hr}) = 2m(10 \text{ mi/hr})$$

so $v_0 = 40$ mi/hr. Car A had been traveling way over the speed limit!

3. In the preceding problem, what fraction of the initial kinetic energy of the cars was lost in the collision?

Solution: The total kinetic energies before and after the collision were $KE_0 = (1/2)mv_\text{A}^2 + (1/2)mv_\text{B}^2$ and $KE_\text{f} = (1/2)(2m)v_\text{f}^2$. The fraction lost in the collision was therefore

$$\frac{KE_0 - KE_\text{f}}{KE_0} = \frac{\left(\frac{1}{2}mv_\text{A}^2 + \frac{1}{2}mv_\text{B}^2\right) - \frac{1}{2}(2m)v_\text{f}^2}{\frac{1}{2}mv_\text{A}^2 + \frac{1}{2}mv_\text{B}^2} = \frac{\left(v_\text{A}^2 + v_\text{B}^2\right) - 2v_\text{f}^2}{v_\text{A}^2 + v_\text{B}^2}.$$

In fact, $v_\text{A} = 40$ mi/hr, $v_\text{B} = 20$ mi/hr, and $v_\text{f} = 10$ mi/hr, so the fraction of kinetic energy lost was

$$\frac{\left(v_\text{A}^2 + v_\text{B}^2\right) - 2v_\text{f}^2}{v_\text{A}^2 + v_\text{B}^2} = \frac{\left((40)^2 + (20)^2\right) - 2(10)^2}{(40)^2 + (20)^2} = \frac{1800}{2000} = \frac{9}{10}.$$

Ninety percent of the initial kinetic energy was converted into heat, sound, and other forms of energy!

Problems

1–1. A car moving along a road with initial velocity v_0 starts speeding up with constant acceleration a_0. Relative to the roadway, it has velocity $v(t) = v_0 + a_0 t$ as a function of time. Using the Galilean transformation, find its velocity $v'(t')$ measured in the frame of another car moving in the same direction at constant velocity V.

1–2. In Earth's frame of reference, two spaceships A and B approach one another with velocities v_0 and $-v_0$, respectively. Find (using Eqs. 1.5) (a) B's velocity in A's rest-frame (b) A's velocity in a frame moving at velocity $v_0/2$ to the right relative to Earth.

1–3. (a) Verify that if a particle's velocity in the x direction as a function of time is $v(t) = v_0 + a_0 t$, where v_0 is its initial velocity and a_0 is a constant, then its acceleration is a_0. (b) Then verify that if the particle has position $x(t) = x_0 + v_0 t + \frac{1}{2} a_0 t^2$, where x_0 is its initial position, then its velocity is $v(t) = v_0 + a_0 t$. (c) In a uniform gravitational field g directed downward, the force $F = mg$ on a particle of mass m is constant. At time $t = 0$, drop a ball from rest at altitude h in such a field. As functions of time, find (before it hits the ground) its acceleration a, velocity v, and height y above the ground.

1–4. Repeat Problem 1–3 (c), except suppose instead that at time $t = 0$ the ball is at altitude h and we throw it with a horizontal velocity v_0. As functions of time, find both the horizontal and vertical components of the ball's (a) acceleration $\mathbf{a}$, (b) velocity $\mathbf{v}$, (c) position $\mathbf{r} = (x, y)$ where $x = 0$ at time $t = 0$. The positive y direction is upward. *Hint:* You can use results stated in Problem 1–3 (a) and (b).

1–5. A 3 1/2 ton two-horned African rhinoceros charges at 30 m/s into a 7 ton ball of putty sitting on a frictionless sheet of ice in a perfect vacuum. Assuming the rhino gets stuck in the putty, find the final velocity of the two. Is kinetic energy conserved in the collision?

1–6. Ball A, initially at rest, is struck by ball B. Is it possible for ball A to have a larger final momentum than the initial momentum of B? If so, give an example; if not, prove why it is not possible.

1–7. In the middle of an icy intersection, a car of mass m moving north with some unknown speed crashes into a truck of mass $M = 2m$ moving west at 30 mi/hr. The investigation shows that the vehicles became locked together upon impact and slid exactly northwest. How fast was the car traveling before the collision, and how fast do they both travel just afterwards? *Hint:* Momentum is separately conserved in the east–west direction and in the north–south direction.

1–8. A system of particles is constrained to move along the x axis. Prove that the net external force on a system of particles is related to the acceleration of the center of mass of the system by $F_{\text{ext}} = Ma_{\text{CM}}$, where M is the total mass of the system and $a_{\text{CM}} \equiv d^2X_{\text{CM}}/dt^2$. This important result shows that Newton's second law can be used for a *system* of particles just as well as for a "single" particle.

1–9. The cue ball moving east strikes the equal-mass eight ball, which is initially at rest on a pool table. After the collision the cue ball moves at 2.00 m/s at an angle of 30.0° to the east, while the eight ball moves at an angle of 60.0° to the east. Find (a) the final speed of the eight ball, and (b) the original speed of the cue ball. (c) Is kinetic energy conserved in this collision?

1–10. Asteroid α, of mass m, is headed directly toward asteroid β, of mass $3m$, with velocity v_0 as measured in β's reference frame. (a) How far is α from their mutual center of mass when α and β are 100 km apart? (b) What is the velocity of their center-of-mass frame, as measured in β's frame? (c) If the asteroids merge upon colliding, how fast are they moving in the original frame of β?

1–11. A spring exerts a force $F = -kx$ on an attached ball of mass m, where k is a constant and x is the spring stretch. (a) If the ball is released from rest at $x = A$, how fast is it moving when it reaches $x = 0$? (b) If the ball starts at $x = 0$ with speed v_0, how fast is it moving when it reaches $x = -A/2$ if A is its maximum distance from the origin?

1–12. Pool ball A with initial speed v_0 collides head-on with pool ball B initially at rest. Find the final speeds of each ball, assuming their masses are equal and that the collision is *elastic*, that is, that kinetic energy is conserved as well as momentum. Then prove that in any one-dimensional elastic collision between two balls of masses m_A and m_B, the relative velocity between the two balls has the same magnitude before and after the collision.

1–13. A neutron of mass m with initial velocity v_0 in the laboratory collides head-on with a helium nucleus of mass $4m$ initially at rest. Suppose that the collision is *elastic,* so that the total kinetic energy is conserved in the collision, as well as the total momentum. (a) Find the final velocity of each particle after the collision. (b) Using the Galilean velocity transformation, find the initial and final velocities of each particle in the original rest frame of the neutron. (c) Then show that both momentum and kinetic energy are conserved in the collision, as measured by a hypothetical observer at rest in the neutron's original rest frame.

CHAPTER 2

Light and the Ether

DURING THE NINETEENTH CENTURY, as a means of trying to understand light waves, many physicists believed that the universe was filled with a substance called "ether." There were at least two excellent reasons to believe in its existence. First of all, it was thought that all waves require something to *propagate* in; water waves require water, sound requires air or some other medium, and so on. Light then must "wave" in "ether," because *something* is needed to provide the restoring force necessary to maintain oscillations. Light waves travel very well through empty space, so the ether hypothesis provided a means other than ordinary material media for supporting these oscillations.

The second reason for believing in ether was more convincing, and very hard to get around. The ether at rest defined that frame of reference in which light travels with its characteristic speed c. This means that if you happen to be moving with respect to the ether, a beam of light could move at speed either *less* than c or *more* than c, depending on whether you move with the light or against it. That is what is found with other kinds of waves. Sound moves with its characteristic speed with respect to the air. If a wind is blowing, sound travels faster than usual *in* the direction of the wind, and slower than usual *against* the direction of the wind. This is built into the Galilean transformation given in Chapter 1.

So, during the latter part of the nineteenth century, people began to try to detect the ether. Of particular interest were the questions: Is the ether at rest with respect to Earth, or is it moving? How fast is the ether wind blowing past us? We might expect offhand a seasonal variation brought about by the changing direction of Earth's velocity around the Sun. Also it would seem possible that the ether might blow more strongly on mountain tops, where the planet as a whole has less chance to impede the flow. That is why measurements were made at different times during the year, and why some of them were made on mountain tops.

2.1 The Aberration of Light

A certain effect, known as the *aberration of light,* was of importance in the ether investigation. This effect had been known ever since 1727, when the English astronomer James Bradley observed that the stars seem to perform a small annual circular motion in the sky. This apparent motion was understood to be due to the fact that the observed direction of a light ray coming from a star depends on the velocity of Earth relative to the star. The angular diameter of the circular paths is about 41 seconds of arc, which can be understood by a consideration of Fig. 2.1. Because of the motion of a telescope during the time it takes light to travel down the length of the tube, the light will appear to follow a path that is tilted with respect to the actual path.

The figure shows a star that is straight overhead being viewed through a telescope. The telescope is mounted on Earth, which is moving to the right with velocity v in the course of its orbit around the Sun. The light requires a finite time t to travel down the length of the tube, in which time it covers a distance ct, as shown in the figure. During this time the telescope itself moves to the right a distance vt, also as shown. Therefore if the light ray is to strike the eyepiece at the bottom of the tube rather than the side of the tube, the telescope has to be tilted to the *right.* The star itself then appears to be in a different position, at an angle θ to the vertical. Six months later, at a time when the star is actually again overhead, the telescope will have to be tilted to the *left* in order to see the star. More generally, as Earth orbits the Sun the telescope has to be continuously adjusted so as to point slightly along the direction of Earth's motion.

Thus as Earth orbits the Sun, the star appears to move around a small circular path. From the figure, which displays a grossly exaggerated angle θ, clearly $\tan\theta = v/c$. Using $v = 30$ km/s for the velocity of Earth in its orbit, we find that

$$\tan\theta \cong \theta = \frac{30 \times 10^3 \text{ m/s}}{3 \times 10^8 \text{ m/s}} = 10^{-4} \text{ radians}, \tag{2.1}$$

equivalent to 20.5 seconds of arc.[1] This is in excellent agreement with the observed value (of the circle's radius), as previously quoted.

We conclude from these observations that the ether cannot simply be at rest relative to Earth. If it were, the telescope could be pointed vertically, and there would be no aberration effect. The ether in Fig. 2.1 would then be moving to the right with velocity v, pulling the light ray along with it (just as a wind pulls sound along with it), so there would be no need to correct for Earth's motion by tilting the telescope. In short, if there is an ether, it must be blowing past us at an average speed of about 30 km/s! Obviously we should devise an experiment to try to confirm this!

---

1. The small-angle approximation $\tan\theta \cong \theta$ can be derived from the Taylor series given in Appendix A.

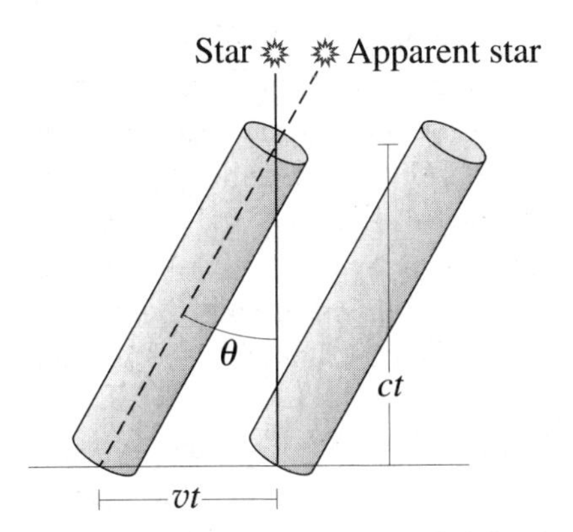

FIGURE 2.1
A telescope accepts light from a star.

2.2 The Michelson–Morley Experiment

Such an experiment was devised and performed in 1887 by the Polish American physicist Albert Michelson and the American chemist and physicist Edward Morley. By looking at the interference of light waves traveling along perpendicular paths, they hoped to measure the velocity of the ether wind. The apparatus they used was a Michelson interferometer, as it is now called, sketched in simplified form in Fig. 2.2. It contains two fully reflecting mirrors (B and C) and a half-silvered mirror (A) that reflects half and transmits half of the light incident on it.

The idea is this: Light from a source strikes A, half of it being reflected upward toward B and the other half transmitted forward to C. The light striking B is reflected back, and half of it is transmitted through A to the observer O. (The other half is reflected back to the source, and is lost from the experiment.) The light striking C is reflected back, and half of it is reflected off A to the observer O. (The other half

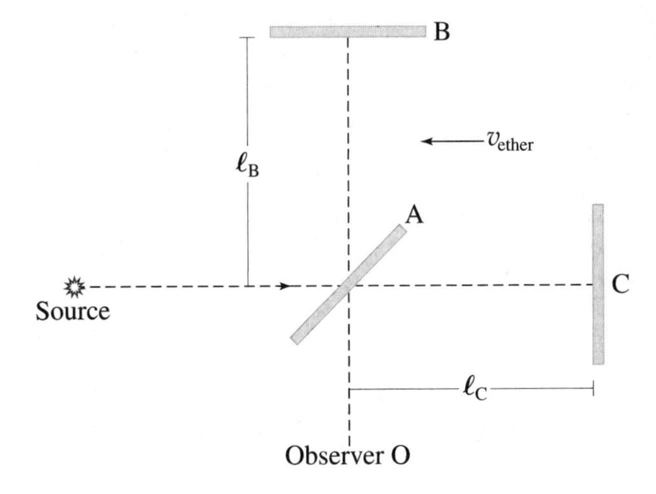

FIGURE 2.2
Outline of a Michelson interferometer.

of course is transmitted through to the source.) Altogether half the light leaving the source reaches the observer, of which half follows the path ABAO and half the path ACAO. These two paths generally have different lengths, so the peaks and valleys of the light wave reaching the observer along ACAO will not likely correspond to the peaks and valleys of the wave traveling along ABAO, so the observer will not necessarily observe constructive interference.[2] The fact that the wavelength of visible light is so small (about 6×10^{-7} m) means that there is a rapid alternation of constructive and destructive interference as the relative path lengths are changed, an indication that the interferometer is a very sensitive device.

What does the ether have to do with this experiment? Suppose we have adjusted the paths ABAO and ACAO to be of exactly equal length, and suppose the ether is sweeping past the apparatus from right to left with (unknown) velocity v, as shown in Fig. 2.2. Then in traveling from A to C, the light will have to fight upstream against the ether current, while returning from C to A it will be swept back with the current. The light traveling from A to B and back to A will be moving largely cross-current, although it will have to fight somewhat against the current or else it would be swept downsteam and never return to A. We shall see that even if the paths are of the same length, it takes longer to swim upstream and downstream than to swim cross-current, so the time intervals will be different, and the interference pattern will be shifted from that expected for equal-time paths.

Recalling that we have assumed that light moves with speed c with respect to the ether, just as sound travels with its characteristic speed with respect to air, we can calculate the time needed to traverse each path. In traveling upstream the distance ℓ from A to C, the light will travel at speed $c - v$ with respect to us, and so requires a time $\ell/(c - v)$ to arrive. Traveling downstream from C to A, the speed is $c + v$, so the time required is comparatively short, $\ell/(c + v)$. The total time for the trip ACA is therefore

$$t_{\text{ACA}} = \frac{\ell}{c - v} + \frac{\ell}{c + v} = \frac{2\ell c}{c^2 - v^2} = \frac{2\ell}{c}\frac{1}{1 - v^2/c^2}. \tag{2.2}$$

The time t_{ABA} for the cross-current trip is most easily calculated in the ether's frame of reference, as shown in Fig. 2.3. In this frame the ether is at rest, and the apparatus moves to the right with velocity v. During the time $t_{\text{ABA}}/2$ the light takes to travel from A to B, the light moves a distance $ct_{\text{ABA}}/2$, while the apparatus moves the smaller distance $vt_{\text{ABA}}/2$.

From the Pythagorean theorem, we have

$$\ell^2 + \left(\frac{vt_{\text{ABA}}}{2}\right)^2 = \left(\frac{ct_{\text{ABA}}}{2}\right)^2, \tag{2.3}$$

2. Constructive interference occurs when the peaks and valleys of one wave match the peaks and valleys of the other wave. When this occurs the waves are said to be "in phase" with one another, and the total wave has maximum amplitude. Destructive interference occurs when the peaks of one wave match the valleys of the other; the two waves then cancel one another out.

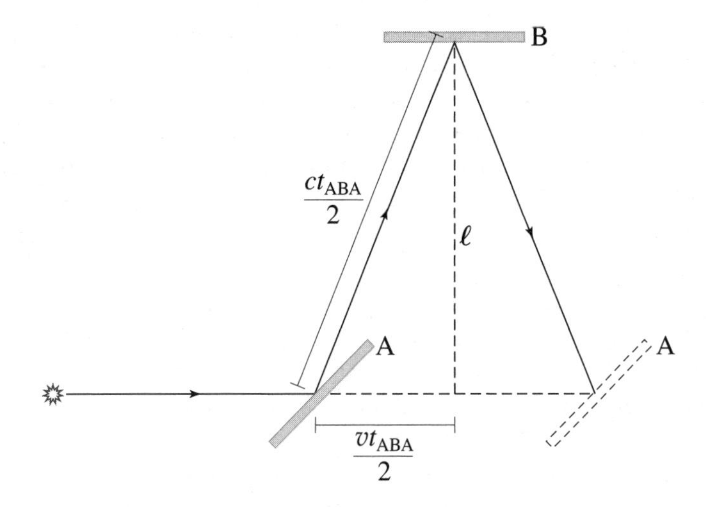

FIGURE 2.3
Ether at rest, interferometer moving to the right with speed v.

so $t_{\mathrm{ABA}}^2(c^2 - v^2) = 4\ell^2$. Therefore

$$t_{\mathrm{ABA}} = \frac{2\ell}{c}\frac{1}{\sqrt{1 - v^2/c^2}}. \tag{2.4}$$

Comparing with the upstream–downstream time t_{ACA}, we see that

$$t_{\mathrm{ACA}} = \frac{t_{\mathrm{ABA}}}{\sqrt{1 - v^2/c^2}}, \tag{2.5}$$

so it takes longer to go up and downstream than to go back and forth across the current. We do expect that $v << c$, so it takes a very sensitive device to tell the difference. The slight difference in time means that the two light rays will be out of phase with each other and consequently will produce interference patterns, even when the path lengths are exactly the same.

Now, in fact, it is not possible with a single measurement to separate the effect of different path lengths from the effect of the ether wind. A greater time spent along path ACA could be due to ether wind, as in Eq. 2.5, or because path ACA is actually longer than path ABA. Therefore, it is necessary to look at the interference fringes in one position and then rotate the entire interferometer by 90° to interchange the position of the interferometer arms with respect to the ether wind. During the second measurement the path ACA would be cross-current and path ABA would be upstream and downstream, so there would be a fringe shift between the two orientations of the interferometer if the ether wind is blowing.

Michelson and Morley's actual apparatus allowed multiple reflections to increase the path length. The optical system was mounted on a heavy sandstone slab, which was supported on a wooden float, which in turn was designed to float on a trough containing mercury. This made turning easy and smooth, and reduced the effects of vibration. The effective optical length of each arm of the interferometer was about 11 m, which would theoretically lead to a shift of 0.4 fringes when the apparatus was rotated, assuming the ether wind was about the same as the orbital speed of Earth. From very careful measurements, in July of 1887 they concluded that "if there is any displacement due to the relative motion of the earth and the luminiferous ether, this cannot be much greater than 0.01 of the distance between the fringes."

The simple conclusion from the null result of Michelson and Morley is that the ether is *not* blowing past us, at least not with a speed anywhere near the orbital speed of Earth. Perhaps whatever wind there may be far away is drastically reduced at ground level because Earth drags nearby ether along with it? But then we would be in trouble with the aberration of light. From this effect, as previously discussed, it was concluded that the ether could *not* be dragged around with Earth. An ether wind of velocity equal to Earth's orbital velocity was needed to explain the small annual circular motion of the stars.

Many other experiments contributed to the confusion surrounding the influence of the ether, such as experiments with light shining through moving water and the measurement of magnetic forces between charged capacitor plates. Naturally several explanations were advanced, some of them very interesting and clever, but none successfully explained all of the experimental results.

Sample Problems

1. A spaceship is coasting through space. The onboard astronomer observes the star Betelgeuse vertically above her, at an angle of 90° relative to the ship's longitudinal axis, as shown. The ship now accelerates straight forward, increasing its velocity by one-twentieth the speed of light. At what angle to the vertical does she now see Betelgeuse?

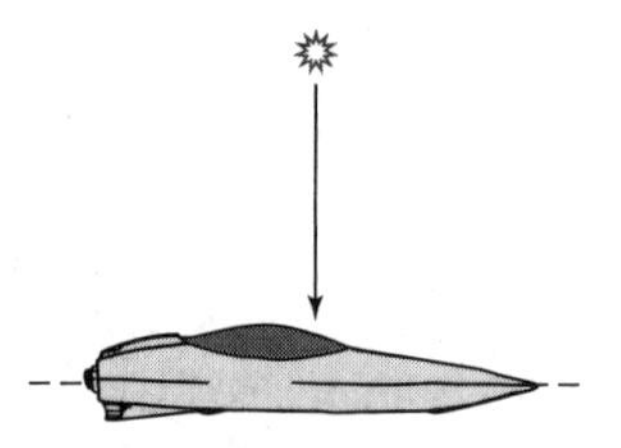

Solution: Due to the aberration of light, Betelgeuse will now appear to be shifted slightly in the forward direction, by an angle $\theta \cong v/c = 1/20 = 0.05$ radians, or about 2.9°.

2. A river flows at uniform speed $v_W = 1.00$ m/s between parallel shores a distance $D = 120$ m apart. A kayaker can paddle at 2.00 m/s relative to the water. (a) If the kayaker starts on one shore and always paddles perpendicular to the shoreline, how long does it take him to reach the opposite shore, and how far downstream is he swept? (b) If instead he wants to reach a point on the opposite shore straight across from the starting point, at what angle should he paddle relative to the stream flow direction so as to arrive in the shortest time? What is this shortest time?

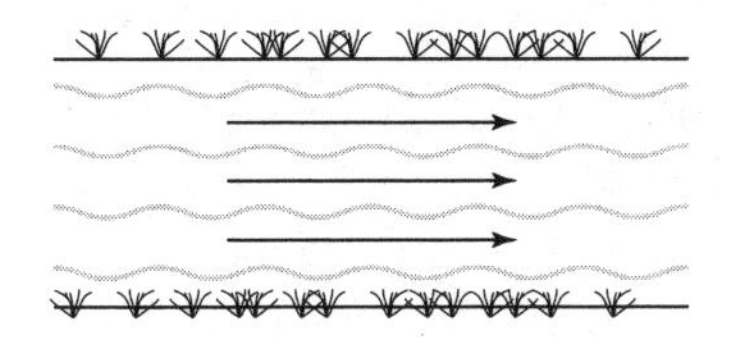

Solution: Let the unprimed frame be the shore frame, and the primed frame be the water frame. The relative velocity of the two frames is $V = 1.0$ m/s to the right, and v and v' are the speeds of the kayaker relative to the shore and water, respectively; they are related by the Galilean velocity transformation $v_x = v'_x + V$ and $v_y = v'_y$, or (in vector form) $\mathbf{v} = \mathbf{v}' + \mathbf{V}$.

(a) In this case the three velocities are related as shown, with $\mathbf{v}'$ directly toward the opposite shore, with magnitude 2.00 m/s. The time it takes him to reach the opposite shore is therefore

$$t = D'/v' = 120 \text{ m}/2.0 \text{ m/s} = 60 \text{ s},$$

and the distance he is swept downstream is

$$d = Vt = 1.00 \text{ m/s} \times 60 \text{ s} = 60 \text{ m}.$$

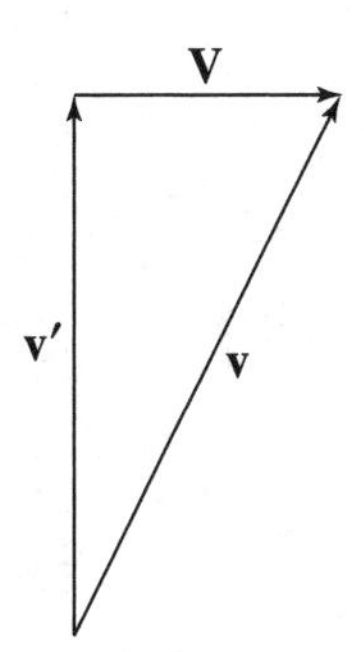

(b) In this case $\mathbf{v}$ is directed toward the opposite shore, as shown below, and $\mathbf{v}'$ has an upstream component (to cancel the flow velocity) and has magnitude 2.0 m/s. The sine of the upstream angle is

$$\sin\theta = \frac{V}{v'} = \frac{1.00}{2.00} = \frac{1}{2},$$

so that $\theta = 30^\circ$. The velocity relative to the shore is

$$v = v' \cos 30^\circ = 2.00(0.866) \text{ m/s} = 1.73 \text{ m/s},$$

so the time to cross the river is $t = D/v = 120 \text{ m}/1.73 \text{ m/s} = 69.4$ s.

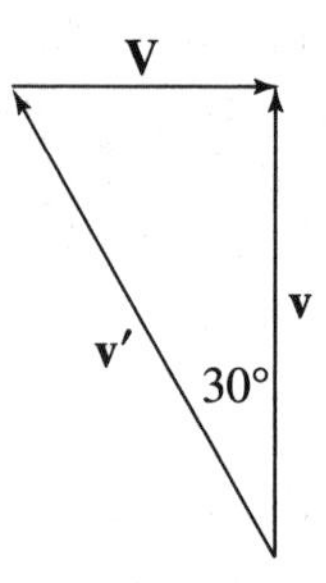

3. What could account for the failure of the Michelson–Morley experiment to detect the ether flow? Irish physicist G. F. FitzGerald and Dutch physicist H. A. Lorentz independently invented a possible explanation. They proposed that as an object passes through the ether, the object is *shortened* in the direction of the ether flow! Suppose that one arm (ACA) of the Michelson interferometer happens to be aligned with the ether flow direction. Michelson and Morley calculated the time t_{ACA} given in Eq. 2.2 for light to go back and forth by that path, and the time t_{ABA} given in Eq. 2.4 to take the other path, perpendicular to the ether flow direction. They found that $t_{\text{ACA}} > t_{\text{ABA}}$, even if the two arm lengths are identical. So the FitzGerald–Lorentz hypothesis would make these times equal by shortening the arm parallel to the flow direction, thereby reducing the time t_{ACA} to make it equal to t_{ABA}. (a) What is the necessary contraction factor? (b) According to their idea, what happens when the apparatus is rotated by 90°?

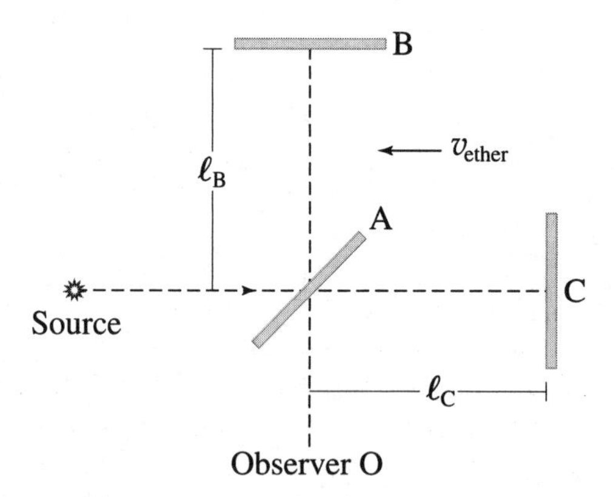

Solution: From Eqs. 2.2 and 2.4, $t_{\text{ACA}} = (2\ell/c)/(1 - v^2/c^2)$ and $t_{\text{ABA}} = (2\ell/c)/\sqrt{1 - v^2/c^2}$. (a) Those two times can be made equal by mutiplying t_{ACA} by $\sqrt{1 - v^2/c^2}$. That is, if the ACA arm of the interferometer (which is the arm parallel to the ether flow direction) is contracted by the factor $\sqrt{1 - v^2/c^2}$, then the time to execute that path will be reduced by the same factor. (b) If the apparatus is rotated by 90°, it will be the arm ABA that is now parallel to the flow direction, so *that* arm will be contracted and the times in the two arms will stay the same, resulting in no fringe shift, in agreement with observation. The FitzGerald–Lorentz contraction hypothesis therefore "explained" the failure of the Michelson–Morley experiment to detect the ether flow. In their view the ether flow was *there* all right, but the contraction effect made it impossible to detect in this experiment. (It can also be shown that contraction in the ether flow direction would explain the more general case where the flow was at some arbitrary angle to one of the interferometer arms.)

The idea of FitzGerald and Lorentz was very clever, but also somewhat ad hoc, in that it was invented to get the "right" answer in the experiment and did not emerge naturally from well-accepted theory. As we shall see in Chapter 5, there is a $\sqrt{1 - v^2/c^2}$ contraction factor in Einstein's special relativity as well, associated with what is called Lorentz contraction (or Lorentz–FitzGerald contraction); but it has nothing whatever to do with any ether flow direction!

Problems

2–1. The planet Mars has an orbit around the Sun with radius 2.28×10^8 kilometers and period 687 days. Find the angle subtended by the radius of a circle executed by a star as seen from Mars, due to light aberration.

2–2. An artificial Earth satellite completes a 26,000-mile orbit in 90 minutes. Find the angle subtended by the radius of a circle executed by a star as seen from the satellite, due to light aberration.

2–3. An airplane flies at speed $v_a = 150$ mi/hr relative to the air. A wind is blowing at velocity $v_w = 120$ mi/hr toward the east, relative to the ground. (a) If the plane flies due north according to its onboard compass, find the speed and direction (relative to north) of the plane as measured by observers on the ground. (b) If the plane flies due north according to ground observers, how fast does the plane move relative to the ground? In this case, at what angle to north must the pilot fly, according to the onboard compass?

2–4. An airplane flies at speed $v_a = 50$ m/s relative to the air. The wind is blowing at velocity $v_w = 40$ m/s toward the east, relative to the ground. (a) If the plane flies due north according to an onboard compass, find the speed and direction (relative to north) of the plane as measured by observers on the ground. (b) If the plane flies due north according to ground observers, how fast does the plane move relative to the ground? In this case, at what angle to north must the pilot fly, according to the onboard compass?

2–5. A train is traveling along a straight track at speed $v_t = 12$ m/s. Looking through a side window, a passenger sees snowflakes falling vertically relative to the train, at speed $v_s = 6$ m/s. As seen by someone standing on the ground, how fast are the flakes moving, and at what angle are they falling relative to the vertical?

2–6. An ultralight aircraft is 5 miles due west of the landing field. It can fly 25 miles per hour in stationary air. However, the wind is blowing at 25 miles per hour from the southwest at a 60° angle to the direction toward the landing field, as shown.

(a) At what angle relative to the east must the pilot aim her craft to reach the landing field? Explain. (b) How long will it take her to reach the landing field if she flies as described in part (a)?

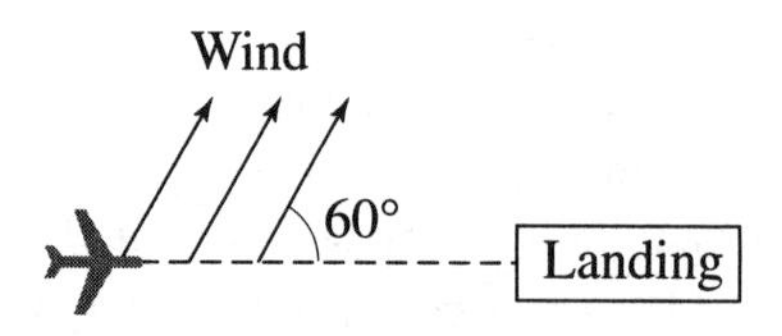

2–7. A spaceship is in the middle of a large cluster of stars, so that a passenger looking out sees an equal number of stars in all directions. Now the ship accelerates to a high velocity. Will the distribution of stars remain isotropic, or will the passenger see more stars toward the front than the rear, or more stars toward the rear than the front? Explain.

2–8. A river flows at speed $v_w = 0.50$ m/s between parallel shores a distance $D = 35$ m apart. A swimmer swims at speed $v_s = 1.00$ m/s relative to the water. (a) If the swimmer swims straight toward the opposite shore, that is, in a direction perpendicular to the shoreline as seen by the swimmer, how long does it take her to reach the opposite shore, and how far downstream is she swept? (b) If instead the swimmer wishes to reach the opposite shore at a spot straight across the stream, at what angle should she swim relative to the stream flow direction, so as to arrive in the shortest time? What is this shortest time?

2–9. A river of width D flows at uniform speed V_0. Swimmers A and B, each of whom can swim at speed v_s relative to the water, decide to have a race beginning at the same spot on the shore. Swimmer A swims downstream a distance D relative to the shore, and immediately swims back upstream to the starting point. Swimmer B swims to a point directly across from the starting point on the opposite shore, and then swims back to the starting point. Assume $v_s > V_0$. Find the total time for each swimmer. Who wins the race?

2–10. In the Michelson–Morley experiment, (a) What would v/c have to be to make t_{ACA} 1 percent larger than t_{ABA}? (b) For this value of v/c, what is t_{ACA} (in seconds) if ℓ were exactly 1 meter?

CHAPTER 3

Einstein's Postulates

AS EVIDENCED BY MANY EXPERIMENTS, including light aberration and the results of Michelson and Morley, the ether wind cannot be detected. If there is an ether, it apparently has no influence on physics. It doesn't seem to matter at all whether it is at rest or blowing past us. Therefore since we don't observe it, it seems reasonable to discard the idea that it exists. This seems easy to do, and we wonder what all the excitement was about—until we recall that then *we do not know the frame of reference in which light travels with speed c!* The ether frame was supposed to *be* that frame, but now we don't have it. The frame has vanished along with the ether.

There seems to be no way to find the frame in which light moves with speed c. Various suggestions have been put forward: for example, might it not be that light travels with speed c with respect to the *source* that emits it? The idea that this is the special frame has been contradicted by observing the light from double stars. For if at a particular time one star is approaching and the other receding from us as they orbit around one another, light would reach us from the approaching star first. A detailed analysis of this effect shows that the double star system would appear to behave in a different way from what is actually observed. So what do we do?

3.1 A Revolutionary Proposal

It is here that Albert Einstein appears on the scene. While working in the Swiss patent office in Berne, the capital of Switzerland, he published a paper in 1905 that set forth the basis of what he later called the *Special Theory of Relativity.*[1] The theory is founded on two rather innocent-sounding postulates:

1. Absolute uniform motion cannot be detected.
2. The velocity of light does not depend upon the velocity of its source.

1. "Zur Elektrodynamik bewegter Körper" (On the Electrodynamics of Moving Bodies), *Annalen der Physik* **17**, 1905. English translation in "The Principle of Relativity" by Einstein et al., Dover Publications (1952).

Neither statement seems particularly upsetting, but the combination of the two is revolutionary! The first postulate says that absolute motion in a straight line cannot be detected—meaning that there is no absolute frame with which all motion can be compared. There is no "absolute space" or "ether frame" that is at "rest." No inertial frame is fundamentally better than any other inertial frame. The fundamental laws of physics can be used by observers at rest in *any* inertial frame. The idea that no inertial frame is preferred above any other in using the laws of physics is certainly not original with Einstein, but it is a reaffirmation of the same assumption implicit in Newton's laws, as discussed in Chapter 1.

The first postulate implies that the laws of physics must look the same in any inertial frame. If they varied, one frame could be singled out as being fundamentally "better" than another (say, because of greater simplicity of the laws), and so could become the preferred frame with respect to which all velocities should be measured. We should stress that it is absolute *uniform* motion that cannot be detected, since it is usually easy to tell whether or not you are accelerating, just by watching the behavior of a pendulum or holding a spring with a mass on the end.[2] The special theory of relativity deals only with measurements made in inertial frames of reference.

Since the first postulate seems quite plausible, and has been part of physics for hundreds of years, let us proceed to the second postulate. Denying the idea that light is influenced by what the source is doing, this postulate states that the velocity of light does *not* depend on the velocity of its source. Some other things in physics behave this way, and some do not. For example, if we stand on the sidewalk watching a car go by, and somebody in the car throws a rock straight ahead, the rock's initial velocity $v_{\text{rock, us}}$ with respect to us *will* depend on the velocity of the car. In fact, as everybody knows (at least within the limits of experimental error),

$$v_{\text{rock, us}} = v_{\text{rock, car}} + v_{\text{car, us}}, \tag{3.1}$$

which is just the additive law of velocities when referred to a different frame.[3] Therefore rocks do not obey the second postulate.

---

2. As a matter of fact, if an observer with such accelerometers is accelerating at a constant rate, the behavior of these accelerometers will be the same as that of similar ones in an inertial frame at rest in a uniform gravitational field. The pendulum will swing and the spring with the mass attached will stretch or compress depending upon how it is oriented. Conversely, such an accelerometer in free fall in a uniform field will give no reading of acceleration at all. As a step in the development of the so-called General Theory of Relativity of 1915, which is a theory of gravity, Einstein postulated that in a restricted region of space no experiment can distinguish whether the apparatus is set in a uniformly accelerated frame of reference without gravity, or in an unaccelerated frame in the presence of a uniform gravitational field. This postulate is one form of the *Principle of Equivalence,* which is discussed further in Chapter 14.

3. If someone in the car throws a rock straight ahead at 10 mph relative to the car, and the car is moving at 20 mph relative to the sidewalk, then people on the sidewalk will observe that the rock is moving 30 mph relative to them. This is an example of the Galilean transformation discussed in Chapter 1.

On the other hand, consider the motion of sound in air. Sound travels at about 330 m/s with respect to the air. Therefore an observer at rest in the air will find that the measured velocity of sound has nothing to do with the velocity of the sound source through the air. Two people standing in front and in back of a moving car, at equal distances from it when it honks its horn, will hear the honk at the same time. *So as long as the velocity of the sound source is measured with respect to the air,* sound is like light in that it obeys the second postulate. The sound velocity is independent of the velocity of its source.

But suppose you measure the velocity of sound and of the sound source relative to yourself while you are *moving* with respect to the air. Then the situation is quite different. For the velocity of sound relative to you clearly does depend on how fast you are moving through the air. If you move *toward* a stationary sound source, the sound velocity relative to you will be *greater* than 330 m/s, and if you move away from a stationary sound source the sound velocity will be *less* than 330 m/s. Therefore obviously the velocity of sound does depend on the source velocity if this source velocity is measured by you while you are moving relative to the air. Sound obeys Einstein's second postulate in only one special frame of reference, in which the observer is at rest in the air.

The crucial difference between sound and light is then immediately clear. Since there is no ether (which would correspond to the air in the case of sound), light has to obey the second postulate in *all* inertial frames. Without the ether there is no preferred frame to be chosen above any other. The velocity of light cannot depend on the source velocity regardless of the reference frame of the observer.

It is this fact that produces the first surprise of relativity. Imagine a searchlight out in the middle of empty space, which sends out a continuous beam of light. Some distance away are two spaceships, one at rest with respect to the searchlight and the other moving toward it at relative velocity $c/2$, as shown in Fig. 3.1. Observers in both ships are equipped to measure the velocity of light from this searchlight. The "stationary" observers will of course measure the velocity to be its standard value, $c = 3 \times 10^8$ m/s. On the basis of intuition (for example, from the behavior of rocks and cars), we might then say that the light velocity measured by the "moving" observers will be $c + v = c + c/2 = 3/2 \times 3 \times 10^8$ m/s. But then we contradict the postulates of Einstein! For the first postulate states that the situation with the spaceship racing toward the searchlight is *exactly the same* as if the searchlight were racing toward a

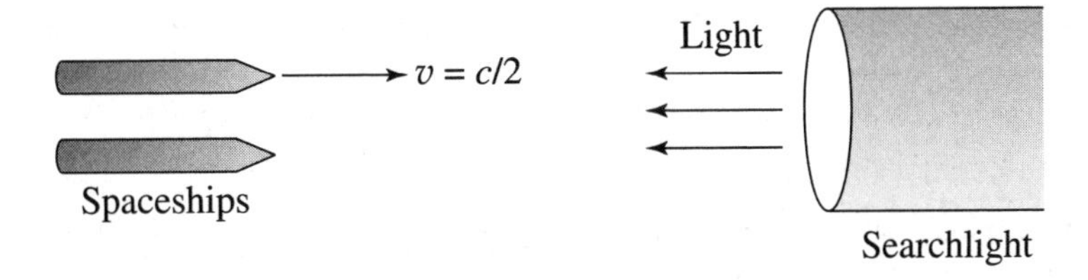

FIGURE 3.1
Observers on two different ships measure the speed of light from a searchlight

stationary spaceship (who can tell which is moving?). Then the second postulate claims that the measurement of light velocity in this latter case must give the *same* result as if the searchlight were not moving toward the spaceship. But this result would be just $v_{\text{light}} = c$! Our guess of $v_{\text{light}} = (3/2)c$ for a moving observer was wrong, and should have been $v_{\text{light}} = c$. It takes both postulates to force this conclusion. Therefore the velocity of light is independent of the observer's motion. It is the same in every inertial frame of reference. This is a revolutionary idea, unprecedented before Einstein. It took considerable nerve to write down postulates that had as a consequence that light always goes at the same velocity no matter how fast the observer is moving.

3.2 Spheres of Light

A particular consequence of the constancy of light's velocity is illustrated in the following: Imagine two sets of rectangular coordinates (inertial frames) S and S' moving with uniform relative velocity V. At a certain time, the origins of the two systems pass each other, and a bomb explodes at the point where the origins instantaneously coincide, as shown in Fig. 3.2(a). The light from the flash of the explosion will spread out in all directions, the wave front forming a sphere of radius $R = ct$. Observers in frame S note that the center of this expanding sphere is at the origin of their frame, which is where the bomb went off, as shown in Fig. 3.2(b). Meanwhile, observers in frame S' find that the center of the sphere is at the origin of *their* system of coordinates, since the light left that spot at time $t' = 0$ and spread out in all directions at the same speed. In other words, the wave front forms a sphere in *both* frames of reference, and

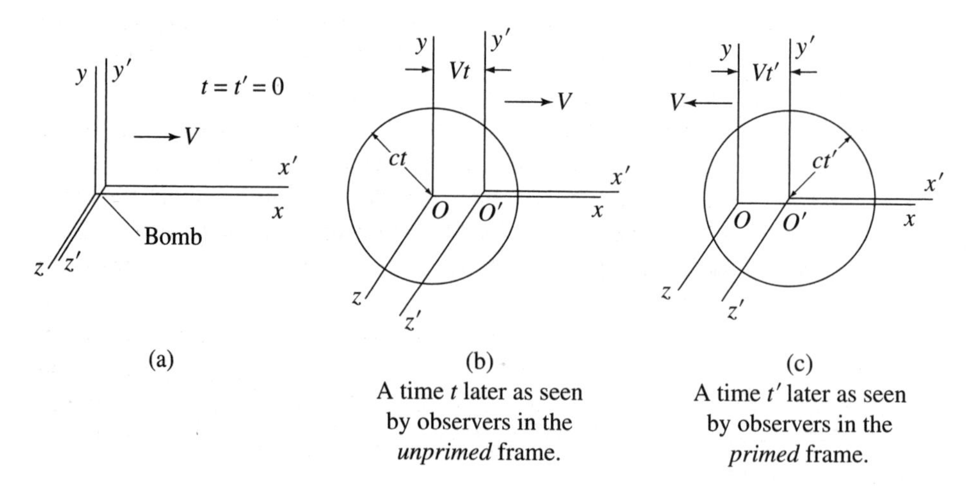

FIGURE 3.2
(a) A bomb explodes at the origins of two inertial reference frames moving relative to one another, just as the two origins coincide. (b) The resulting light sphere at time t later according to observers in S. (c) The light sphere at time t' later according to observers in S'.

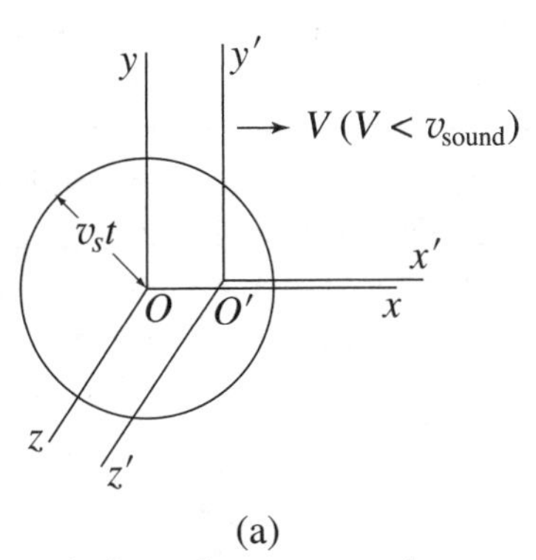

(a)
A time t later as seen by observers in the *unprimed* frame.

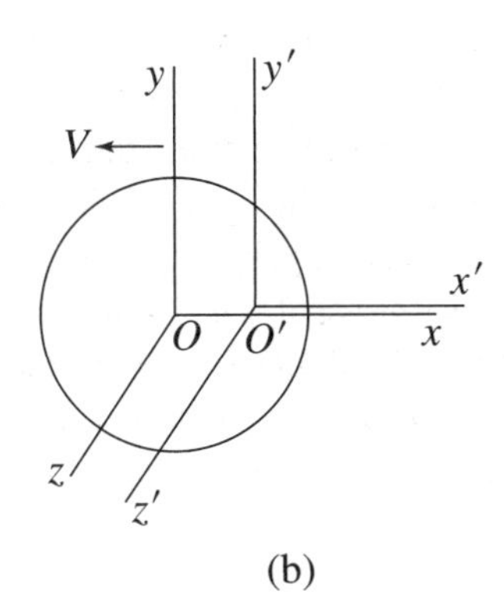

(b)
A time t' later as seen by observers in the *primed* frame.

FIGURE 3.3
A spherical *sound* wave observed from two different reference frames.

the observers in each frame claim that the center of the sphere is at their own origin of coordinates! This appears to be paradoxical, since after $t = 0$ the two origins don't coincide. Yet the conclusion is forced by the two postulates.

To actually perform this experiment, several observers are needed in each frame. A single observer can't stand back and watch a sphere of light expand. One observer should be located at the frame origin to verify that the explosion happened there at time $t = 0$. Others can be stationed here and there with synchronized clocks and a knowledge (from measurements made with metersticks) of how far they are from the origin. If each observer receives the light flash at time $t = r/c$, where r is that observer's distance from the origin, they can all compare notes afterward and be satisfied that the light spread spherically from the origin of their own frame.

So light from the explosion spreads out in such a way that observers in each frame conclude that it spreads spherically and is centered about their own origin. But how about *sound* from the explosion? Sound moves with its characteristic speed only in the frame in which the air is at rest. Only in that frame (say the unprimed frame) will the sound expand spherically with its center at the frame origin. In some other (primed) frame, the center of the expanding sound wave will always be located at the origin of the *unprimed* frame, as shown in Fig. 3.3. So as you might expect, to primed observers the center of the expanding wave will drift steadily with a velocity equal to the wind velocity felt by them due to their motion through the air. Thus again the great difference between the behavior of sound and light in different frames of reference can be traced to the absence of an ether frame for light.

We have only begun to explore the consequences of Einstein's postulates. Further investigation of time and distances will help to explain how expanding light spheres can behave in such a paradoxical fashion. From the results of the next three chapters, Appendix B will show how this apparent paradox can be understood.

Sample Problems

1. An ice-skater gliding along the ice at speed v_0 collides with a stationary skater of the same mass. They get tangled together and move off as a unit with speed v_f. Using classical mechanics, determine the following: (a) What is v_f in terms of v_0? (b) A third skater is skating along at speed $v_0/2$ in the same direction as the first skater. Find the initial and final speeds of each of the first two skaters from the point of view of the third.

Solution: (a) Momentum is conserved in the collision, because there is no appreciable horizontal force on the two-skater system. Before the collision the momentum of the moving skater is $p_0 = mv_0$ according to classical mechanics, where m is the skater's mass, and the momentum of the stationary skater is zero. Afterwards the total mass is $2m$, and the total momentum is $2mv_f$. Conservation of momentum gives

$$mv_0 = 2mv_f,$$

so the final speed of the two skaters is $v_f = v_0/2$.

(b) From the third skater's point of view, before the collision the first skater had the speed

$$v_0 - v_0/2 = v_0/2,$$

using the Galilean velocity transformation Eq. 1.3. Similarly, the "stationary" skater had speed

$$0 - v_0/2 = -v_0/2.$$

So the initial momentum in the third skater's frame is

$$mv_0/2 + m\left(-v_0/2\right) = 0.$$

The speed of the two skaters after they stick together is $v_0/2 - v_0/2 = 0$, so the final momentum is obviously zero. Therefore momentum is also conserved in the third skater's frame! This is consistent with Einstein's *first* postulate, that the fundamental laws of physics should be equally true in all inertial frames. (Conservation of momentum is a fundamental law in Newtonian mechanics for an isolated system of particles, that is, for a system on which no external forces act.) Is our result also consistent with Einstein's *second* postulate? That is a question we will postpone until Chapter 10.

2. Is the velocity of light *really* independent of the velocity of its source? That is not easy to verify experimentally, partly because the source has to be moving fast enough that we can tell whether the source velocity makes any difference. One way to achieve high source velocities is to use a high-energy accelerator to create fast-moving particles that emit light. An experiment of this type was carried out at the European accelerator facility CERN near Geneva.[4] Protons were accelerated to high energies, and fired into a target where neutral pi mesons (π^0) were created. These neutral pions then decayed into two photons ($\pi^0 \to \gamma + \gamma$), and photons are the particles associated with light. In fact, light *consists* of photons, so the velocity of photons *is* the velocity of light.[5]

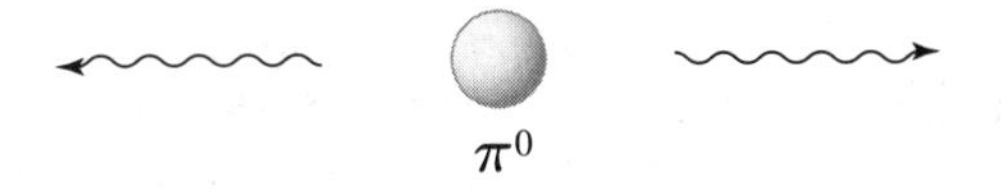

(a) The pions used in the measurements all moved in the laboratory at velocities greater than or equal to $0.99975c$. If light in fact moved at speed c relative to its *source,* how fast would forward-directed gamma-ray photons move in the *lab*?

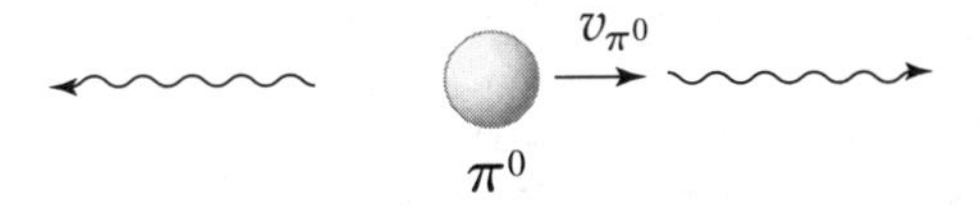

(b) The experimenters measured the time it took the photons to travel a measured distance of 31.450 ± 0.0015 meters. The time required was found to be $(1.04913 \pm 0.00013) \times 10^{-7}$ seconds. How fast did these photons travel in the lab? Compare this result with the accepted speed of light, $c = 2.99792458 \times 10^8$ m/s.

Solution: (a) If light moved at velocity c relative to its *source,* a photon moving in the forward direction (the direction of the pion's motion) at velocity c relative to the pion would move at velocity $1.99975c \sim 2c$ in the laboratory.

---

4. "Test of the second postulate of special relativity in the GeV region," by Alvager, Farley, Kjellman, and Wallin (*Physics Letters* **12**, p. 260, 1964.)

5. That light consists of discrete particles was originally suggested in another of Einstein's famous papers of 1905.

(b) The measured velocity of the gamma-ray photons in the lab was

$$v = \frac{d}{\Delta t} = \frac{(31.450 \pm 0.0015)\ \text{m}}{(1.04913 \pm 0.00013) \times 10^{-7}\ \text{s}} = (2.9977 \pm 0.0004) \times 10^{8}\ \text{m/s},$$

which (within experimental error) is equal to the accepted speed of light.[6] The experiment is consistent with Einstein's second postulate, and *not* consistent with the idea that light moves at speed c relative to the source.

Problems

3–1. Is the velocity of water waves on the surface of the ocean independent of the velocity of the source of the waves, in (a) some particular frame? (b) all frames?

3–2. Devise a way for observers in a given frame to verify experimentally that light spreads out spherically and is centered about their origin, as in the example mentioned in the chapter. Devise a scheme for measuring the shape of the sound wave front as well.

3–3. An ice-skater of mass m_1 moving at speed v_0 relative to the ice collides with a stationary skater of mass m_2. As a result of the collision they are tangled together and move off as a unit with speed v_f. (a) Using classical momentum conservation in the frame of the ice, find v_f in terms of v_0. (b) Show (using the Galilean velocity transformation of Eq. 1.5) that momentum is conserved in the collision in any other inertial reference frame as well, where the new frame moves with arbitrary velocity V relative to the ice.

3–4. From our conclusion that the speed of light is the same in all inertial frames, show that the analysis of the Michelson–Morley experiment in Chapter 2 is in error. Show also that if Einstein is correct there *should* be no fringe shift in the experiment.

3–5. The speed of light is c and the speed of sound in air is v_s. (a) A firecracker explodes in the air some distance from you, while the air is still and you are too. What are the speeds of the light and of the sound from the firecracker, in your frame of reference? (b) Another firecracker goes off, this time when there is a wind blowing at speed v_w in a direction from you toward the firecracker, when both you and the firecracker are at rest relative to the ground. What now are the speeds of light and sound from the firecracker, from your point of view? (c) A third firecracker

---

6. The error in v is found from the rms experimental errors in d and Δt by so-called propagation of error. See, for example, *An Introduction to Error Analysis* by John R. Taylor. Also, nowadays the speed of light is no longer measured, but (since it is universal, independent of reference frame) it is *defined* to be $c = 2.99792458 \times 10^{8}$ m/s. Therefore measurements involving light propagation are actually used to measure the length of the meter!

goes off; the wind is still blowing from you toward the firecracker with speed v_W, the firecracker is still at rest, but now you are running toward the firecracker at constant speed v_0. What now are the speeds of the firecracker's light and sound from your point of view?

3–6. With a powerful telescope, we observe a double-star system in which the stars have equal masses and rotate in circular orbits about their mutual center of mass halfway between them. One of the stars (α) is bright, and the other (β) is an unseen dark companion. Our line of sight passes through the orbital plane, so that once in every period α approaches head-on, and once it recedes directly away, and the same for β. Pretend that light always moves at speed c *relative to the source that emits it,* so that if v is the orbital speed of each star, light travels toward us at speed $c + v$ from α when it is headed toward us, and at speed $c - v$ when it is headed away from us, as illustrated. (a) Suppose the double-star system is a distance d from Earth. Find the time it would take light to get to us from α if the light is emitted when α is (*i*) coming directly toward us, and (*ii*) moving directly away from us. (b) Show that if T is the period of rotation of the two stars (the time to go completely around one another) then the time (as we observe it on Earth) for α to go from its position in picture (1) to its position in picture (2) is $(T/2) + (2vd)/(c^2 - v^2)$, and to go from its position in picture (2) to its position in picture (3) is $(T/2) - (2vd)/(c^2 - v^2)$. (c) Find a distance d, in terms of v, c, and the orbital period T, such that α would appear to be simultaneously both to the left and right of the center of mass point!

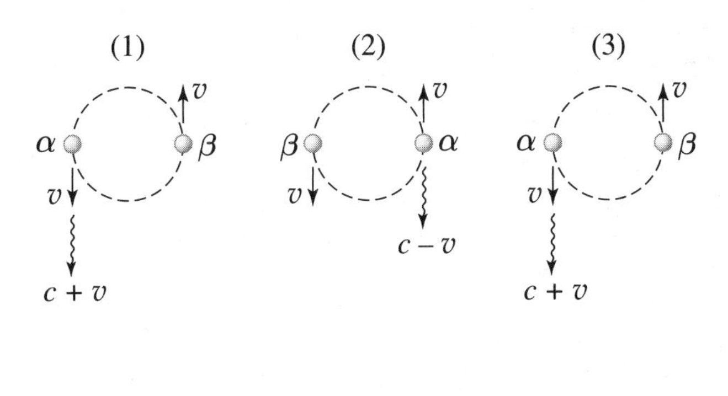

(A paper by K. Brecher explores the subject of light from binary star systems in considerable detail. The reference is "Is the speed of light independent of the velocity of the source?" *Physical Review Letters* **39**, p. 1051, 1977)

3–7. As shown in the chapter, Einstein's postulates predict that in empty space, the speed of light is the same to any inertial observer (that is, to any observer who travels in a straight line at constant speed.) However, we can show that there is a limit to how fast such an observer can travel! Consider the following thought experiment: We suddenly turn on a laser beam moving at velocity c relative to us, just as three hypothetical observers pass by us in the same direction as the beam. The first observer has velocity $c/2$, the second has velocity $0.999c$, and the third has velocity c. Give an argument why, if metersticks had different lengths in different frames, or clocks ticked at different rates in different frames, it is not impossible that the first and second observers could measure the beam to be moving at speed c relative to them, but that the third observer *could not* find that the beam is moving at speed c relative to her.

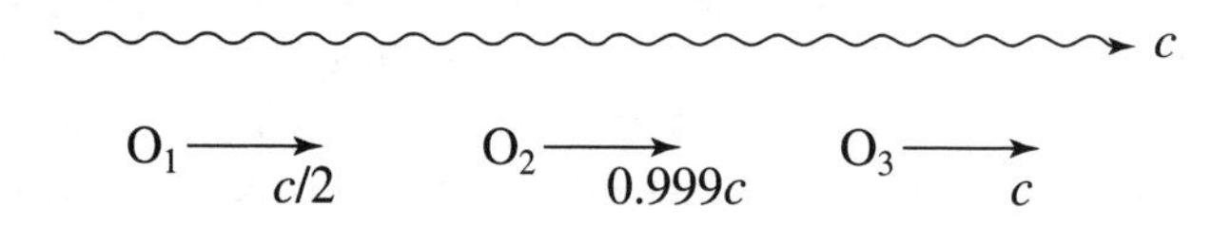

CHAPTER 4

Time Dilation

. . . until at last it came to me that time was suspect

—*A. Einstein*

THE PHENOMENON OF TIME DILATION follows from the result of Chapter 3 that light travels at the same speed in all inertial frames. The term "time dilation" means that moving clocks run slow. That is, we will show that if we were to compare the readings of moving clocks with the readings of similar clocks at rest in our own frame of reference, we would find that the moving clocks run behind in time. For example, if our clocks advance by one hour, the moving clocks might advance by only a half hour.

Consider an array of clocks at rest inside a spaceship. There are old-fashioned wind-up clocks with various wheels and springs inside, wristwatches with electronic timekeepers, battery-powered wall clocks, all kinds of clocks, as illustrated in Fig. 4.1. They are all close together. There is a beaker containing chemicals that change color from blue to red and back to blue once per second. One of the passengers has a heart that beats exactly 60 times per minute. There is even a "light clock" built as follows: A flashbulb on one side of the ship flashes a light pulse toward the opposite side of the ship, a distance D away. A mirror there reflects the pulse back toward the flashbulb. A detector beside the flashbulb causes the flashbulb to emit another pulse every time it detects the reflection of the previous pulse. The light clock therefore "ticks" with time interval $\Delta t = 2D/c$, since the light travels a distance $2D$ between successive ticks.[1] All the clocks tick at the same rate, all keeping time together.

The spaceship along with all its clocks is passing by us with speed V. We have a similar set of clocks, all ticking at the same rate as one another. And since our clocks

1. If the light clock ticks once per second, the flashbulb and mirror must be 150,000 kilometers apart! However, if the light clock's flashbulb and mirror are only 1.5 m apart, one-second ticks would correspond to every 10^8 light roundtrips.

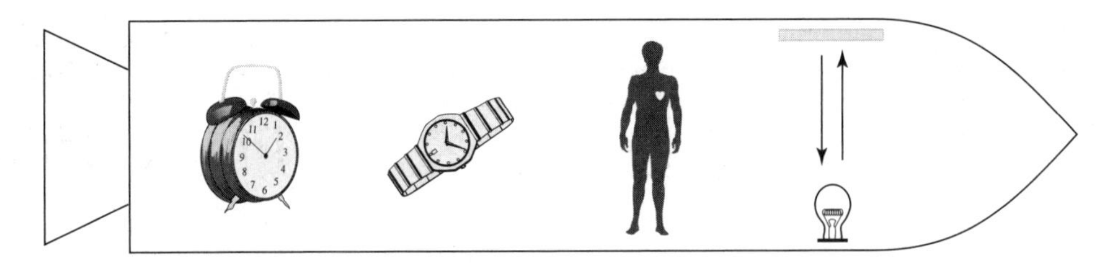

FIGURE 4.1
Various clocks, all ticking at the same rate, are mutually at rest within a spaceship.

are essentially identical to theirs, our clocks tick from our point of view at the same rate as their clocks tick from their point of view. This follows from Einstein's first postulate, that no inertial frame is fundamentally better than any other. However, if *we* observe the tick rate of *spaceship* clocks and compare it with the tick rate of our own clocks, one of the following possibilities will be true: The moving (spaceship) clocks will all tick

(a) at the same rate as our clocks,
(b) fast compared with our clocks, or
(c) slow compared with our clocks.

That is, all the moving clocks tick at the same rate as *one another* from our point of view, but they do not necessarily tick at the same rate as our own rest clocks. They might all run fast or they might all run slow compared to our own clocks.

Why is it that all moving clocks must tick at the same rate as one another from our point of view? Why couldn't moving mechanical clocks all run fast and moving electronic clocks all run slow, for example, while the moving chemical and biological clocks all run at the same rate as our rest clocks? The reason is that all the clocks, however they work, could have digital readouts that proclaim the time. All ship clocks read the same time simultaneously according to ship passengers. If a small shipboard explosion stopped all the clocks when all read 3 hours, 52 minutes and 10 seconds, then we outside the ship will also see them read the numbers 3 hours, 52 minutes, and 10 seconds, for all eternity.[2] Some clocks cannot have run slow while others ran fast and others did neither from our point of view, if the final numbers are all the same. Anyone, in any frame of reference, can verify that all the numbers are the same.

2. Recall that all shipboard clocks are placed right next to one another, so the explosion stops all clocks simultaneously.

4.1 The Light Clock

We are now going to analyze the *simplest* of all the moving clocks, to see if it runs fast, slow, or the same rate as our rest clocks. Then using the logic of the previous paragraph, we can claim that all the moving clocks, some of which may be much harder to analyze directly than the simplest clock, do the same thing as the simplest clock.

The simplest clock is the light clock. It has no wheels or springs or fancy electronic chips that determine the tick rate. It uses only the motion of light for which we can use the results of Einstein's analysis. If the moving light clock runs fast compared with our clocks, *all* the moving clocks run fast by the same amount. If the moving light clock runs slow compared with our clocks, *all* the moving clocks, however they are built, run slow by the same amount.

We have already figured out that the tick rate of the light clock on the ship, as observed by ship passengers, is $\Delta t_{\text{ship observers}} = 2D/c$, where D is the distance between the flashbulb and mirror. What is the tick rate of this ship light clock as observed by us, as the ship passes by at speed V? We assume that the light clock is oriented perpendicular to the direction of motion of the ship. The light clock as seen by ship observers is shown in Fig. 4.2(a). While the light is moving up and down, the ship, flashbulb, and mirror are all moving to the right from our point of view. Therefore in order to strike the mirror and then be reflected back to the flashbulb, the light path is angled from our point of view, as shown in Fig. 4.2(b). Note that therefore the light has to move farther in our frame than it had to move in the ship frame. The speed of light is the same in both frames, however, so the greater distance means that it requires a greater time for the clock to tick once from our point of view than it took for observers on the ship. Therefore the moving light clock runs *slow* as observed by us, when we compare its tick rate with that of our own clock that is at rest in our frame. While our clock ticks 60 times (1 minute), the onboard clock might tick only 50 times from our point of view, so that to us it runs 10 seconds slow every minute.

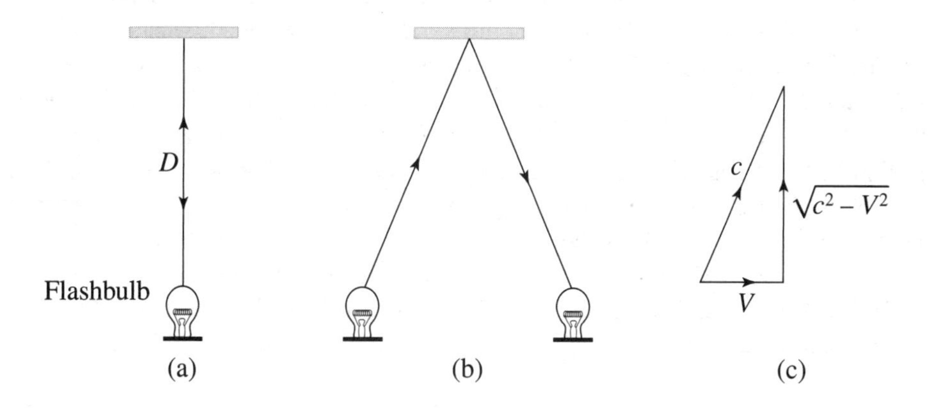

FIGURE 4.2
(a) The light clock as seen by ship observers. (b) The same light clock as seen by us. (c) The Pythagorean theorem.

How does the tick rate of the moving light clock depend upon the *speed* of the ship? The easiest way to figure this out is to consider only the *vertical* component of the light's motion. The speed of light is c. The horizontal component of the light's velocity is V (it *has* to be V so that the light pulse keeps up with the ship, the mirror, and flashbulb as they move to the right). The vertical component of the light pulse's velocity must therefore be $\sqrt{c^2 - V^2}$, using the Pythagorean theorem, as shown in Fig. 4.2(c). The total vertical distance back and forth is $2D$, so the time it takes for the light to go up and back from our point of view is[3]

$$\Delta t_{\text{between ship light clock ticks to us}} = \frac{2D}{\sqrt{c^2 - V^2}} = \frac{2D}{c\sqrt{1 - V^2/c^2}}$$
$$= \frac{\Delta t_{\text{between ticks of our own light clock}}}{\sqrt{1 - V^2/c^2}}. \tag{4.1}$$

The denominator is $\sqrt{1 - V^2/c^2} < 1$ if the ship is moving, so the time between ticks of the ship clock as we observe it is *greater* than the time between ticks of our own light clock. Therefore to us their clock is running *slow* by the factor $\sqrt{1 - V^2/c^2}$.

Suppose, for example, that just as their light clock passes our light clock, both clocks read *zero*. Then at a later time $t_{\text{stationary clock}}$ in our frame, their clock will read only

$$t_{\text{moving clock}} = t_{\text{stationary clock}}\sqrt{1 - V^2/c^2}, \tag{4.2}$$

running slow from our point of view. This effect is called *time dilation*. As explained earlier, if their *light* clock runs slow from our point of view by the factor $\sqrt{1 - V^2/c^2}$, then *all* of their clocks must run slow by this same factor. Logical consistency requires that whether they are simple or complicated, whether they are mechanical, electrical, chemical, or biological, all moving clocks run slow by the same factor $\sqrt{1 - V^2/c^2}$. If the ship is moving at speed $V = (4/5)c$, for example, then $\sqrt{1 - V^2/c^2} = \sqrt{1 - (4/5)^2} = 3/5$, so if our clocks advance by 60 minutes, their moving clocks advance by only 60 minutes $\times$ 3/5 = 36 minutes when observed from our frame. This happens for *all* moving clocks, so we say that time itself is dilated (that is, expanded). Time itself runs slow on the moving ship from our point of view; it takes longer for their clocks to advance by 1 hour than it takes our clocks to advance by 1 hour. The mechanical and electrical clocks run slow and the passengers age more slowly than we do. And the faster the ship moves the more striking the effect.

3. We have assumed that the transverse distance D is the same in both reference frames! This will be proven at the beginning of Chapter 5.

4.2 Measuring Time Dilation

It is important to point out the care required to verify time dilation. How do we *measure* the time read by a clock moving past us? We cannot just sit back and watch the clock through a telescope, because as the clock moves past us it gets closer and then farther away, so it takes varying lengths of time for light from the clock to reach our eyes. We would not see the clock reading as it is *now,* but as it was when light left the clock on its way to our eyes. That is not a reliable way to measure the actual time read by a distant clock.

Instead, suppose we have *two* clocks A and B, each at rest in our "stationary" frame of reference, as shown in Fig. 4.3. The clocks have been synchronized with one another, and observers have been stationed beside each of them.[4] As a moving clock C passes by clock A, suppose that it reads time $t = 0$, and clock A reads $t = 0$ as well, as shown in Fig. 4.3(a). These readings are noted by the stationary observer beside clock A. Later, the moving clock C passes by clock B, and the stationary observer beside B observes the reading on both clocks B and C, as shown in Fig. 4.3(b). The observer notes that clock C reads a lesser time than clock B, so concludes that the moving clock has run slow. This is what one means by the statement that moving clocks run slow. *Note that the only unambiguous way to record the time of some event is to put the clock right at the location of the event.*

Time dilation means that it would take fast-moving space travelers less time than expected to reach far-away destinations according to their own clocks. Suppose travelers want to visit Alpha Centauri, one of the nearest stars, and their ship moves at speed $(4/5)c$ relative to Earth. Alpha Centauri is a distance $d = 4$ light-years from Earth; that is, it requires 4 years for light to travel from it to us.[5] According to clocks on Earth and Alpha Centauri (we will assume they are at rest relative to one another) the spaceship takes

$$t = \frac{d}{(4/5)c} = \frac{4\,c \cdot \text{yrs}}{(4/5)c} = 5 \text{ yrs} \tag{4.3}$$

to arrive. Spaceship clocks run slow by the factor $\sqrt{1-(4/5)^2} = 3/5$, however, so the trip takes only $(3/5) \times 5$ yrs $= 3$ yrs according to spaceship clocks. All shipboard clocks, including biological clocks, run slow, so the travelers are only 3 years older when they arrive!

The faster spaceship travelers move, the more advantage there is. If they could travel at $0.999c$, it would take $4\,c \cdot \text{yrs}/0.999c = 4.004$ years to get there according to Earth clocks, but only 4.004 yr $\times \sqrt{1-(.999)^2} = 0.18$ years according to travelers' clocks. They would be only a couple of months older when they arrive! Note that the time-dilation formula fails for a hypothetical clock moving faster than the speed of light, because then the factor $\sqrt{1 - V^2/c^2}$ would be imaginary. However, it doesn't

4. There is a straightforward way to synchronize clocks, which will be described in Chapter 6.

5. Distance = speed × time, so 4 light-years can be written $4\,c \cdot$ yrs.

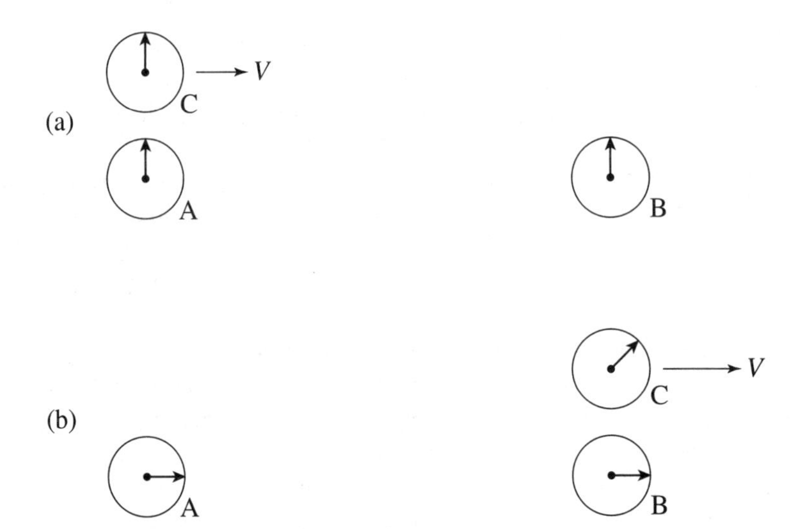

FIGURE 4.3
Two clocks A and B are placed at rest in our frame of reference, with a rest observer stationed beside each one. As a moving clock C passes by at speed V, the time it reads can be marked off by a rest observer right beside it, who can also compare the reading of the moving clock with the reading of the observer's own rest clock. *This eliminates the problem of light travel time one would have with a single rest observer who watches the moving clock through a telescope.* (a) The moving clock passes by clock A when the moving clock C and the stationary clocks all read $t = 0$. (b) Later, C passes by stationary clock B, and the rest observer beside that clock observes the reading on both clocks B and C, and finds that C reads a lesser value than B.

make sense for the light clock (or any clock, for that matter) to travel faster than light, according to Einstein's postulates. As is obvious from the light clock pictures of Fig. 4.2, in which the overall speed of light is c, neither the horizontal component nor the vertical component can exceed c. The horizontal component must be V from our point of view, so that the light will keep up with the moving spaceship, and there is also a nonzero vertical component. If the sum of the squares of the components is c, we must have $V < c$. The relative velocity between any two inertial frames must always satisfy the inequality $V < c$.

4.3 Evidence

Using Einstein's postulates, we have come to the quite incredible conclusion that moving clocks must run slow. Is there any evidence for this? When Einstein published his paper, there was no direct evidence whatever. Yet by now time dilation has been observed many times. Some of the first experiments involved the decays of unstable particles. Such particles can either be created naturally by cosmic rays hitting the upper atmosphere, or by collisions in high-energy accelerators. Since then we have observed time dilation also in the slowing of natural clocks in distant galaxies that are moving rapidly away from us, including a slowing in light intensity changes in distant supernova explosions. We have also observed time dilation in highly precise human-made clocks carried in airplanes and in artificial satellites, including those used in the Global Positioning System (GPS).[6] The results agree with Einstein's prediction.

Consider in particular the decay of particles created by cosmic rays. Cosmic rays are high-energy particles (mostly protons) from outer space that strike atomic nuclei (such as the nuclei in nitrogen atoms) in Earth's atmosphere. When a cosmic-ray proton strikes a nucleus, it can break the nucleus apart and create unstable particles. Each species of unstable particle has a characteristic average lifetime that can serve as a kind of clock whose rate can be measured as a function of the particle's velocity. For example, there is a particle called the *muon* (μ), which decays on the average in a time $T = 2.2 \times 10^{-6}\,\text{s} = 2.2\,\mu\text{s}$ as measured by clocks in the muon's frame of reference.[7] That is, when we measure the lifetimes of a large number of muons at rest in our frame of reference, their average lifetime is about 2.2 μs. But if a muon moves past us, from our point of view its clock runs slow, so to us it will last *longer* than T before it decays. This implies that if a large number of muons all move past us at some velocity v, we will find that their average lifetime is *greater* than 2.2 μs.

Muons are produced in large numbers in the upper atmosphere by the decay of particles called pi (π) mesons, or *pions,* which are themselves created in collisions of cosmic-ray protons with air molecules. If the muons actually decayed in their standard average lifetime of 2.2 μs from our point of view, they would almost all be gone before reaching Earth's surface. For example, a muon moving at nearly the speed of light would move only a distance $cT = (3 \times 10^8\,\text{m/s}) \times (2.2 \times 10^{-6}\,\text{s}) = 660\,\text{m}$, which is considerably less than the height of the atmosphere. A very large muon flux is nevertheless observed at the ground, since from our point of view they do not decay that fast. An experiment was carried out by Frisch and Smith[8] to test the time-dilation effect quantitatively. They counted the number of muons on top of Mt. Washington in New Hampshire (6265 feet in altitude) and compared this count with the number observed at sea level. Using only muons with speeds between $0.9950c$ and $0.9954c$, they found by statistical methods that these muons lasted 8.8 ± 0.8 times longer than muons at rest. Theoretically, for muons of these speeds in their detection setup, they calculated $1/\sqrt{1 - v^2/c^2} = 8.4 \pm 2$, in good agreement with experiment.

Universal time dilation is a strange consequence of the innocent-sounding postulates of Einstein. After all, the second postulate refers only to properties of light, which would seem to have little to do with the behavior of a mechanical clock or of the aging of a human body. However, as we have seen, logical consistency requires that not only moving *light* clocks but *all* moving clocks run slow. Is this the end of the consequences? Absolutely not! Physics is a house of cards. If someone removes one

---

6. See Chapter 14.
7. $1\,\mu\text{s} = 10^{-6}\,\text{s} = 1$ microsecond. Similarly, $1\,\text{ms} = 10^{-3}\,\text{s} = 1$ millisecond, $1\,\text{ns} = 10^{-9}\,\text{s} = 1$ nanosecond, and $1\,\text{ps} = 10^{-12}\,\text{s} = 1$ picosecond.
8. Frisch and Smith, *American Journal of Physics* **31**, p. 342 (1963).

traditional card (that the speed of light should depend upon the speed of the observer), another card (universal time) falls as well. And when universal time falls, there are many additional strange consequences, as we will begin to see in the next chapter.

Sample Problems

1. Barnard's star is 6.0 light-years from the Sun in the Sun's rest frame. A spaceship departs the solar system at speed $(4/5)c$ bound for the star, when both the Sun clock and spaceship clock read $t = 0$. In the rest frame of the Sun, (a) what will the Sun clock read when the spaceship arrives? (b) What will the spaceship clock read when it arrives?

Solution: Pictures of the departure and arrival are shown.

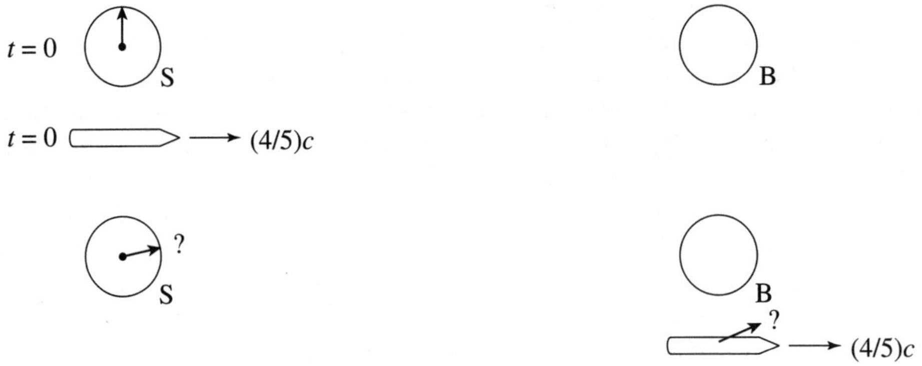

$t = 0$ S B

$t = 0$ $\longrightarrow$ $(4/5)c$

? S B ? $\longrightarrow$ $(4/5)c$

(a) Upon arrival, the Sun clock reads t = distance/speed $= 6.0\ c \cdot \text{yrs}/(4/5c) = 7.5$ yrs.

(b) During this time, the ship clock has been running slow in the Sun's frame, so it will read $\tau = t\sqrt{1 - v^2/c^2} = 7.5\text{ yrs} \cdot \sqrt{1 - (4/5)^2} = 7.5\text{ yrs} \cdot 3/5 = 4.5$ yrs.

2. Clock A is at rest in our frame of reference, and clock B is moving at speed $(3/5)c$ relative to us. Just as clock B passes clock A, both clocks read 12:00 midnight. (a) When clock A reads 5:00, five hours later, what does clock B read, as observed in our frame? (b) When clock B reads the time found in part (a), what does clock A read as observed in B's frame?

Solution: (a) In our frame, clock B runs slow by the factor $\sqrt{1 - v^2/c^2} = \sqrt{1 - (3/5)^2} = 4/5$. So since our clock A has advanced by five hours, clock B advances by only $(4/5)(5\text{ h}) = 4$ h. So in our frame, clock B reads 4:00. Pictures are shown for the important events in A's frame.

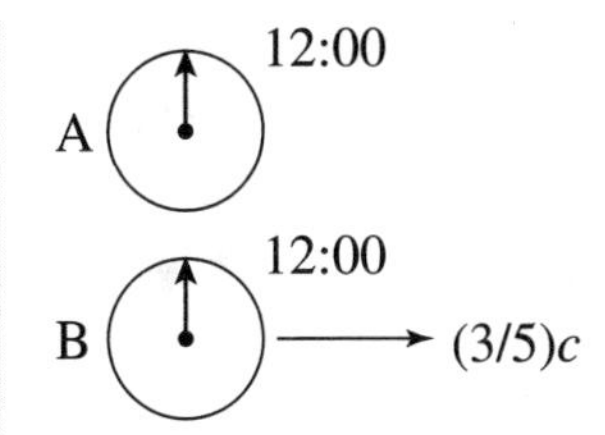

(b) In B's frame, clock A runs slow by the same factor of 4/5. So since clock B has advanced by 4 hours according to observers in B's frame, clock A advances by only $4\text{ h} \times 4/5 = 3.2\text{ h}$. Therefore observers at rest in B's frame will find that A's clock reads 3:12 when B's clock reads 4:00. Pictures are shown for the important events in B's frame.

(*Isn't there a contradiction here*? When clock B reads 4:00, we found that clock A reads 5:00 in part (a), but only 3:12 in part (b). The problem illustrates the importance of stating clearly the frame of reference in which the observations are made! A's clock runs slow in B's frame, just as B's clock runs slow in A's frame. *There is no contradiction.* When clock B reads 4:00, clock A reads 5:00 *if the observations are made by observers at rest relative to clock A.* When clock B reads 4:00, clock A reads 3:12 *if the observations are made by observers at rest relative to clock B.* All of this will become clearer in subsequent chapters, especially Chapter 7.)

3. A particular muon lives for a time $\tau = 2.0\ \mu\text{s}$ in its rest frame. If it moves in the laboratory a distance $d = 800$ m from creation to decay, how fast did it move in the lab frame, expressed as a fraction of the speed of light?

Solution: The muon's "clock" runs slow from our point of view, so in our frame it lasts the longer time $t = \tau/\sqrt{1 - v^2/c^2}$, where $\tau = 2.0\ \mu s = 2.0 \times 10^{-6}$ s. During this time, it moves a distance $d = v \cdot t = v \cdot \tau/\sqrt{1 - v^2/c^2}$, which can be rewritten in the form $d\sqrt{1 - v^2/c^2} = (v/c) \cdot c\tau$. We can solve this equation for v/c by first squaring it: This gives $d^2(1 - v^2/c^2) = (v/c)^2 \cdot (c\tau)^2$, and so

$$\frac{v}{c} = \sqrt{\frac{d^2}{d^2 + (c\tau)^2}} = \sqrt{\frac{(8.00 \times 10^2\ \text{m})^2}{(8.00 \times 10^2\ \text{m})^2 + (3.00 \times 10^8\ \text{m/s} \cdot 2.0 \times 10^{-6}\ \text{s})^2}}$$

$$= \sqrt{\frac{64}{64 + 36}} = 0.8.$$

The muon moved at speed $(4/5)c$ relative to the lab.

4. In an effort to prolong youth, you leap into your car and drive away, leaving a friend behind. If you drive round trip at the steady rate of 30 m/s for 10^4 s (about 2¾ hours) according to your friend's clock, how much less have you aged than your friend?

Solution: Your friend has aged 10^4 s, while you have aged only $10^4\ \text{s}\sqrt{1 - v^2/c^2}$ where $v = 30$ m/s. Here $v/c = (30\ \text{m/s})/(3 \times 10^8\ \text{m/s}) = 10^{-7}$ is very small compared with unity, so we can use the binomial approximation $(1 + x)^n \cong 1 + nx$ (valid for $|x| \ll 1$), as derived in Appendix A. That is, $(1 - v^2/c^2)^{1/2} \cong 1 - v^2/2c^2$ in this case. Therefore you have aged $10^4\ \text{s}\sqrt{1 - v^2/c^2} \cong 10^4\ \text{s}(1 - v^2/2c^2)$, which is less than your friend by the time

$$10^4\ \text{s}(v^2/2c^2) = 5 \times 10^{-11}\ \text{s}.$$

This does not seem worth the trouble, so you may want to try a different tack.

Problems

4–1. A clock moving at $v = (4/5)c$ passes our clock when both clocks read $t = 0$. When the moving clock reads $t = 24$ s, what do clocks in our frame read?

4–2. A clock moving at $v = (3/5)c$ reads 12:00 as it passes us. In our frame of reference, how far away will it be (in light-hours) when it next reads 1:00?

4–3. The mean lifetime of $\pi^{\pm}$ mesons is about $\tau = 26.0$ ns $= 26.0 \times 10^{-9}$ s in their own rest frame. If a beam of pions is produced that travels on the average a distance $d = 10.0$ m in the lab before decaying, how fast are they moving?

4–4. A space traveler with 40 years to live wants to see the galactic nucleus at first hand. How fast must the traveler travel? Express your answer in the form

$v/c = 1 - \epsilon$, where ϵ is a small number. [We are 8.5 kpc = 8.5×10^3 pc from the center of our galaxy. One parsec (pc) = 3.26 light-years.] *Hint:* See Appendix A.

4–5. A space traveler wants to reach the Andromeda galaxy (about 2 million light-years away) in only 2 years from the traveler's point of view. How fast must the traveler travel? Express your answer in the form $v/c = 1 - \epsilon$, where ϵ is a small number. *Hint:* See Appendix A.

4–6. Two π^+ mesons are created, one at rest in the laboratory, and one moving at $v = (4/5)c$. Each decays in 2.5×10^{-9} s in its own rest frame. Find (a) the lifetime of the moving pion as measured in the lab, and (b) the lifetime of the pion at rest in the lab as measured in the frame of the "moving" pion.

4–7. A spaceship leaves the solar system at $v = (3/5)c$, headed for the Earth-like planet Gliese 581C, 20 $c \cdot$ yrs away in the constellation Libra. Assume that the Sun and the star Gliese 581 are mutually at rest and that their clocks have been previously synchronized, both reading zero when the spaceship leaves. (a) What will the clock on Gliese 581 read when the ship arrives? (b) What will the clock on the ship read when it arrives, assuming it began at zero? (c) In the ship's frame of reference, what will the Sun's clock read when the ship arrives?

4–8. A K^+ meson is created using the Fermilab synchrotron in Batavia, Illinois. In its own rest frame the K^+ meson decays into two π mesons in a time 12.4 ns = 12.4×10^{-9} s. How fast is it traveling in the laboratory if it moves a distance $d = 10.0$ m before decaying?

4–9. A Λ particle created in a high-energy collision moved at $v = 0.99c$ in the lab and traveled 55 cm before decaying. What was its lifetime in its own rest frame?

4–10. The Sun and another star are 60 light-years apart and are at rest relative to one another. At time $t = 0$ (on both the Sun clock and the spaceship clock) a spaceship leaves the Sun at $v = (4/5)c$ headed for the star. Just as the ship arrives, a light signal from the Sun indicates that the Sun has exploded. (a) Draw a set of three pictures in the rest frame of the Sun and star for the three important events in the story, that is, at the time of the ship departure, the Sun explosion, and the ship arrival. Each picture should show the Sun, the star, and the ship. Then answer the following questions, all from the point of view of observers at rest in the Sun–star frame. (b) When the ship arrives, what does the star clock read? (c) When the ship arrives, what does the ship clock read? (d) When the ship arrives, what does the Sun clock read? (e) When the Sun explodes, what do clocks on the Sun, the star, and the ship read?

4–11. Starship A departs Earth at $v = (3/5)c$, when both its clock and clocks on Earth read $t = 0$. Starship Command on Earth knows that the captain of starship A is due to have a birthday 16 days later, according to her own onboard clocks. So at time t_1 on Earth clocks, a troupe of entertainers is sent from Earth in starship B, at speed $v = (4/5)c$, to reach starship A exactly on the captain's birthday. Ship B arrives at ship A at time t_f, according to clocks at rest in Earth's frame. (a) Draw pictures of the three critical events in the story, in the rest frame of Earth. (b) What is t_f (in days)? (c) How far (in light-days) are the ships from Earth when they meet, in Earth's frame? (d) What is t_1, in days? (e) What does B's clock read when the ships meet, assuming B read t_1 upon departure?

4–12. The International Space Agency (ISA) is designing a spaceship to reach the star Proxima Centauri, 4 light-years away, so that the onboard crew will age 4 years from departure to arrival. How fast must the ship travel?

4–13. Take two identical clocks. Set one on an ice floe at the North Pole and the other on Isla Santiago, one of the Galapagos Islands, in the Pacific Ocean, right on the equator. The clock at the North Pole is nearly inertial (neglecting the fact that the Earth orbits the Sun, the Sun orbits the galactic nucleus, etc.). A clock at rest on Isla Santiago is *not* inertial, since it circles Earth's center every 24 hours. According to time dilation, this clock should run slow compared to the clock at the North Pole. What fraction of a second does the Isla Santiago clock lose per day? Note that Earth's radius is 6400 km. *Hint:* See Appendix A.

4–14. In 1977, researchers at the European high-energy accelerator facility CERN used a muon storage ring to store both positive and negative muons, circling within the ring at a speed such that $\gamma \equiv 1/\sqrt{1 - v^2/c^2} = 29.33$. In the lab frame, the average lifetime of these muons was found to be $(64.378 \pm 0.026) \times 10^{-6}$ s. What was the average lifetime of the muons in their own rest frame, to four significant figures? (See J. Bailey et al., "Measurements of relativistic time dilatation for positive and negative muons in circular orbit," *Nature* **268**, p. 301, 28 July, 1977.)

CHAPTER 5

Lengths

AFTER FINDING that time has lost its absolute character, with clocks running at different rates when measured in different reference frames, we had better be cautious about all kinds of things. For example, might it not be true that the measurement of *distance* also depends on the observer's frame of reference?

5.1 Transverse Lengths

In looking back at the discussion of time dilation in the previous chapter, note that we actually *assumed* that distances were *not* changed! More precisely, it was assumed that distances *perpendicular* to the direction of relative motion were the same to both observers, as for example in Fig. 4.2, where it was taken for granted that the vertical distance between the flashbulb and mirror was D in both reference frames.

Fortunately, this assumption that transverse distances are unchanged is correct, as can be seen from a simple thought experiment. Two sprinters A and B are each equipped with a meterstick having a thin knife blade attached to the top end, as shown in Fig. 5.1. They run toward each other at a high relative speed, holding the sticks perpendicular to the direction of motion with the bottom end barely skimming the ground. If the sticks are really equally long, the knives should hit each other, but if one stick is longer than the other, it will be sliced off by the knife on the shorter stick. Suppose that A's stick is exactly 1 meter long to A, and that B's stick is exactly 1 meter long to B. We would like to show that in fact the knives *will* hit each other, indicating that B's stick is also 1 meter long to A and that A's stick is 1 meter long to B. This would prove that transverse lengths are unaffected by motion.

The proof follows by contradicting the other alternatives. First suppose that B's stick is shorter than 1 meter according to A. Then B's knife will slice off the top of A's stick. This fact does not depend upon who is observing it: It is definitely A's stick (and not B's) that has been cut off! The whole experiment was set up in a symmetrical way, playing no favorites between A and B, but it ends up in an unsymmetrical way, with A's stick getting cut off. This cannot happen, according to Einstein's first postulate,

FIGURE 5.1
Runners approaching one another with sticks to which knives have been attached.

because it would mean that there is an a priori reason for preferring one reference frame over the other. In such an originally symmetric experiment, with the laws of physics the same for both A and B, everything that happens to A should also happen to B.

The second alternative, that B's stick is longer than 1 meter according to A, implies that B's stick will be cut off, leading to the same contradiction. A preferred frame could again be chosen. The one possibility remaining, that the knives hit one another, is a symmetrical result showing that to either observer the metersticks have the same length. Therefore relativity agrees with our intuition that transverse lengths are unaffected by motion.

5.2 The Longitudinal Contraction of Lengths

Suppose we have a stick of length D, at rest in our frame of reference. A clock, moving to the right along the stick at speed v, reaches the left-hand end of the stick when it reads $t = 0$, as shown in Fig. 5.2(a). According to our own rest clocks, the time it takes the moving clock to reach the right-hand end of the stick is $t =$ distance/speed $= D/v$. But since the moving clock runs slow, we know that the moving clock will actually read the smaller time $t = (D/v)\sqrt{1 - v^2/c^2}$, as shown in Fig. 5.2(b).

Now look at the same experiment according to observers at rest relative to the *clock*. From their point of view the *stick* is moving, not the clock. Fig. 5.3(a) shows the stick moving to the left, just as the left-hand end reaches the clock. We already know

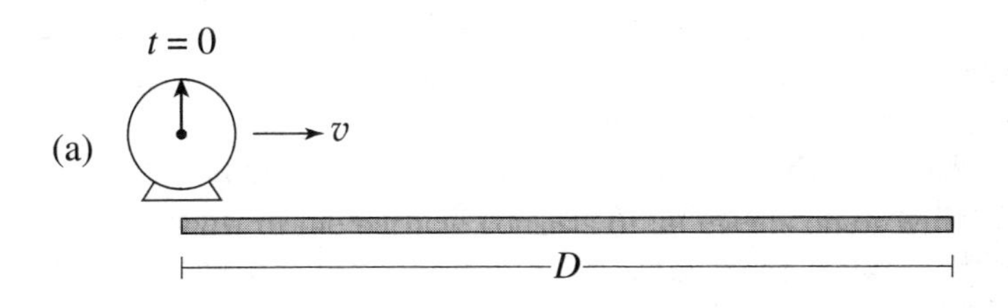

FIGURE 5.2
A moving clock passes a stick at rest in our frame of reference. (a) The clock reaches the left-hand end of the stick when it reads $t = 0$. (b) The clock reaches the right-hand end of the stick when it reads $t = (D/v)\sqrt{1 - v^2/c^2}$, where the square-root factor is due to time dilation.

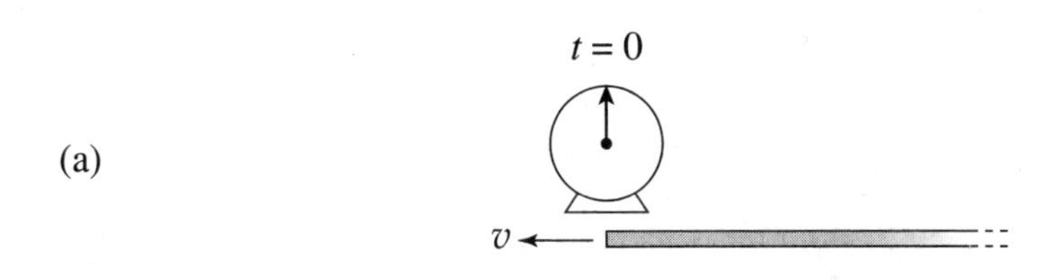

FIGURE 5.3
The same experiment in the *clock's* frame. (a) The left-hand end of the moving stick passes the clock when the clock reads $t = 0$. (b) The right-hand end of the moving stick passes the clock when the clock reads $t = (D/v)\sqrt{1 - v^2/c^2}$.

that the clock reads $t = 0$ in this picture.[1] Similarly, we also know that the clock will read $t = (D/v)\sqrt{1 - v^2/c^2}$ when the right-hand end of the stick passes, as shown in Fig. 5.3(b).

---

1. It might have been that the left-hand end of the stick punched a button on the clock as it passed, for example, that started the clock going in the first place. So if the clock reads zero when it passes the left-hand end of the stick, that will be true in all frames of reference.

How can this be? We would *expect* that in the clock's frame, the time for the stick to pass by would be simply $t = \text{distance/speed} = D/v$. Why then is the *actual* time for the stick to pass by equal instead to $t = \text{distance/speed} = (D/v)\sqrt{1 - v^2/c^2}$?

One possibility is that while the speed of the clock in the stick's frame is "v," the speed of the stick in the clock's frame is different! In fact, note that we would get the correct answer in the formula above if the speed of the stick in the clock's frame were not v but $v/\sqrt{1 - v^2/c^2}$: That would account for the square-root factor. However, this possible answer is inconsistent with the principle of relativity, which is the first postulate of Einstein. If the relative velocity between two objects (here the clock and the stick) depended upon which object was measuring it, we would have an absolute way to distinguish between the two frames of reference. We could say that one frame was "better," because the relative velocity was smaller, say, in that frame.[2] In fact, the relative velocity between two objects is just that: If I am moving at speed v from your point of view, then you are moving at speed v from my point of view.

The other possibility is that the *distance* is different in the two frames. That is, suppose that observers in the clock's frame find that the length of the moving stick is not D, but the shorter length $D\sqrt{1 - v^2/c^2}$. That would also explain why, in the clock's frame, the time is only $t = \text{distance/speed} = D\sqrt{1 - v^2/c^2}/v$ for the stick to pass by. The possibility that an object is longest in its rest frame does *not* violate the principle of relativity, because it does not specify a preferred reference frame in any fundamental sense. It is true that an object's rest frame could be taken as a preferred frame *for that object,* but a different object might have a different rest frame, so no overall preferred frame could be specified.

What we have described is the *contraction of longitudinal lengths.* If an object has length D in its rest frame, its length is only

$$d = D\sqrt{1 - v^2/c^2} \tag{5.1}$$

in a frame moving at speed v relative to the rest frame. Speeding along at $(4/5)c$ relative to the Sun, a spaceship is able to get to the nearest star, 4 light-years away, in only 3 years according to the travelers. We discovered this in Chapter 4 by calculating the required time in the Sun–star frame (5 years) and then using the time-dilation factor to discover that onboard clocks advance by only 3 years. How do the travelers explain this? From their point of view their clocks run at the normal rate, so how are they able to reach their objective in only 3 years? The answer is that the *distance* to the star is smaller to the spaceship! If the Sun and a star are a distance D apart in their rest frame

2. There is another objection to this proposed solution of the puzzle. If the speed of the stick measured in the clock's frame were $v/\sqrt{1 - v^2/c^2}$ instead of v, then if the clock were moving at speed v according to the stick, where v was in the range $c > v > c/\sqrt{2}$, the stick would be moving at a speed greater than the speed of light according to the clock! In this case, according to the time-dilation formula, a second clock that was at rest in the stick's frame would read *imaginary time* according to the observers at rest with respect to the first clock. That is nonsensical.

(which would be the length of a hypothetical string stretched between them), then the distance to the star is only

$$d = D\sqrt{1 - v^2/c^2} = 4\,c \cdot \text{yrs}\sqrt{1 - (4/5)^2} = 2.4\,c \cdot \text{yrs} \tag{5.2}$$

from the point of view of the spaceship. That is the reason why clocks on the ship read only

$$t = \frac{\text{distance}}{\text{speed}} = \frac{D\sqrt{1 - v^2/c^2}}{v} = \frac{2.4\,c \cdot \text{yrs}}{(4/5)c} = 3\text{ yrs} \tag{5.3}$$

when the ship and star meet, according to observers on the ship.

Similarly, we stand on Earth watching muons rain down upon us, passing through many kilometers of atmosphere even though they ought to be able to travel only about 660 m on average. This we interpret as a verification of Einstein's prediction that moving clocks run slow. A muon lasts longer than when it is at rest, which is why it can move so far. But what is going on in the muon's reference frame? In its own rest frame, it decays in the standard time of 2.2 microseconds, while the atmosphere rushes past it. The atmosphere can't possibly travel many kilometers from its top to its bottom, even if it is moving at nearly the speed of light, because $c(2.2\ \mu\text{s})$ is only 660 m! From the point of view of the muon, how is the ground able to travel up to meet it in such a short time? The answer is that muons and the ground can meet, in spite of the muon's short lifetime, because the height of the atmosphere is contracted by the square-root factor. The atmosphere rushing by them is severely squashed in the vertical direction.

The contraction effect is called the "Lorentz–FitzGerald contraction," or the "Lorentz contraction" for short, because the idea was proposed independently by H. A. Lorentz (in 1891) and G. F. FitzGerald (in 1889) to try to explain the Michelson–Morley experiment. The idea was not completely understood and integrated with other relativistic effects until Einstein's theory appeared in 1905. The fact is, objects moving past us with speed v are contracted in their direction of motion by the factor $\sqrt{1 - v^2/c^2}$. Equivalently, if we are moving past some object, the object is contracted by the same factor when measured in our reference frame. The *rest length* of an object is the length measured in the frame in which the object is at rest. In any frame moving parallel to the object's length, the measured length will be *shorter* than the rest length. The "object" might be Earth's atmosphere, or it might be a hypothetical string extending from the Sun to Alpha Centauri.

5.3 The Longitudinal Light Clock

In Chapter 4 we used a transverse light clock to derive time dilation. The light clock was *transverse* because in its rest frame—the spaceship frame in which the flashbulb and mirror were at rest—light traveled in directions perpendicular to the relative velocity between the spaceship frame and the Earth frame, as shown in Fig. 5.4(a). In Earth's

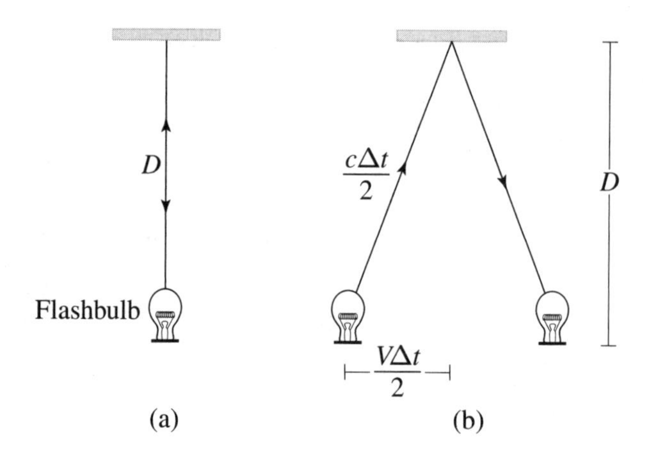

FIGURE 5.4
A transverse light clock in (a) its rest frame, (b) a frame in which it is moving at speed V.

frame the light had to travel farther, as shown in Fig. 5.4(b), so the time between ticks was longer than in the spaceship frame.

The moving transverse light clock runs slow, so logical consistency, as explained in Chapter 4, requires that *any* clock moving with the same speed must run slow by the same factor $\sqrt{1 - V^2/c^2}$.

Now suppose we rotate the transverse light clock by 90° so that it becomes a *longitudinal* light clock, with light moving back and forth in a direction parallel to the relative frame direction, as shown in Fig. 5.5(a). The distance between flashbulb and mirror is D, so light goes back and forth in the time interval $\Delta t = 2D/c$; this is the time between successive ticks according to observers in the clock's rest frame. Now if the clock is inside a spaceship moving past us at speed V, what is the tick rate of the clock as we observe it? To be consistent with other moving clocks, the time between successive ticks must be *larger* than $\Delta t = 2D/c$, so the moving clock will run slow by the usual square-root factor. That is, the time interval between successive ticks must be

$$\Delta t = \frac{2D/c}{\sqrt{1 - V^2/c^2}} \tag{5.4}$$

from our point of view. The longer time between successive ticks means the clock runs slow to us.

Now we can show that the moving longitudinal light clock runs slow by just the right amount to be consistent with all other moving clocks. In our frame the distance between the flashbulb and mirror is only $D\sqrt{1 - V^2/c^2}$, since it is Lorentz contracted. Figure 5.5(b) shows this at time $t = 0$, when the flashbulb sends a light pulse toward the mirror on the right. While the light is traveling to the right the mirror moves to the right as well, so the light reaches the mirror at time Δt_R, as shown in Fig. 5.5(c). The light has moved a distance $c\Delta t_R$ while the mirror has moved the smaller distance

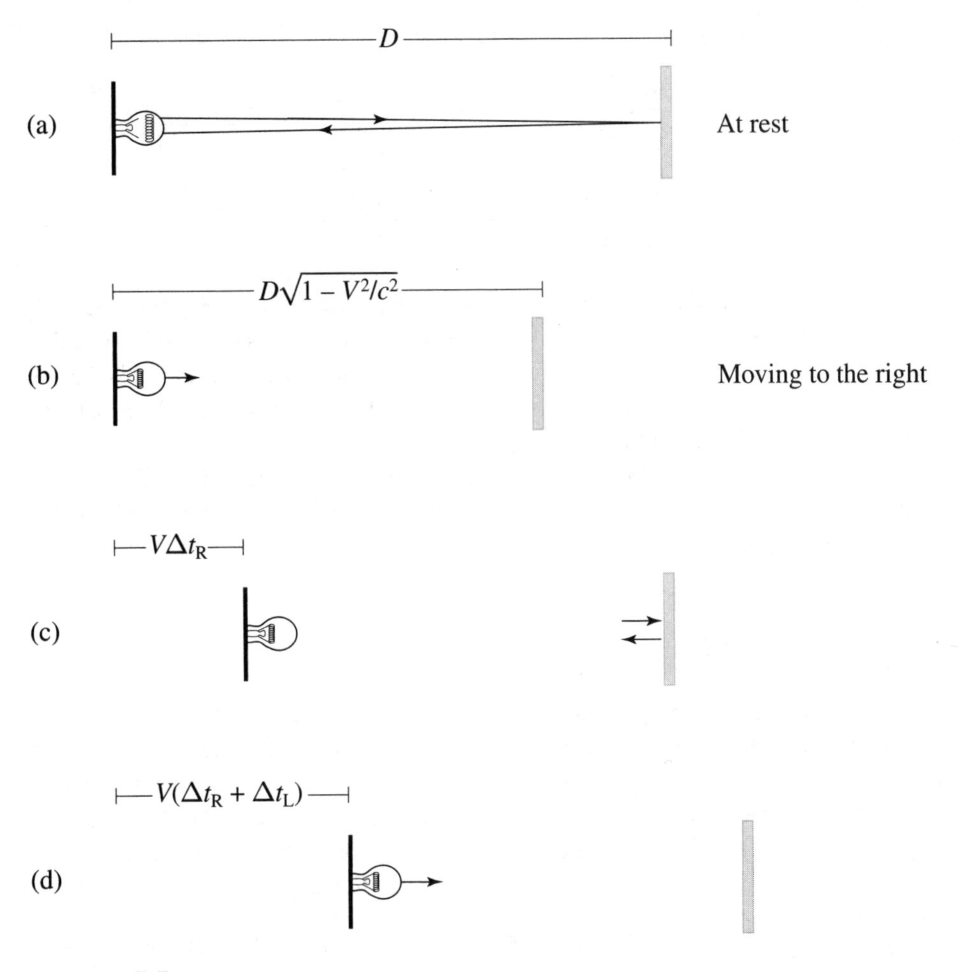

FIGURE 5.5
(a) A longitudinal light clock in its rest frame. Light travels from the flashbulb to the mirror, and reflects back again to the flashbulb. (b) The light clock moving to the right at speed V, at the moment when a light flash is emitted by the flashbulb. (c) A time Δt_R later, when the light flash reaches the mirror. (d) A time Δt_L later still, when the light flash returns to the flashbulb.

$V\Delta t_R$, where V is the speed of the ship. Note from the diagram that

$$c\Delta t_R = D\sqrt{1 - V^2/c^2} + V\Delta t_R, \tag{5.5}$$

so $\Delta t_R = D\sqrt{1 - V^2/c^2}/(c - V)$.

On the return trip, if the light requires a time Δt_L to return to the flashbulb, the flashbulb moves a distance $V\Delta t_L$ during this same time, as shown in Fig. 5.5(d). Therefore the total distance traveled by the light on the return trip is only

$$c\Delta t_L = D\sqrt{1 - V^2/c^2} - V\Delta t_L, \tag{5.6}$$

so $\Delta t_L = D\sqrt{1 - V^2/c^2}/(c + V)$. The total time between ticks is therefore

$$\Delta t = \Delta t_R + \Delta t_L = \frac{D\sqrt{1-V^2/c^2}}{c-V} + \frac{D\sqrt{1-V^2/c^2}}{c+V} = \frac{2Dc\sqrt{1-V^2/c^2}}{c^2-V^2}$$

$$= \frac{(2D/c)\sqrt{1-V^2/c^2}}{1-V^2/c^2} = \frac{(2D/c)}{\sqrt{1-V^2/c^2}}. \tag{5.7}$$

So we have found that the time between ticks of the longitudinal light clock is *exactly the same as the time between ticks of the transverse light clock,* as expected. We had to use the longitudinal contraction of lengths to get the right answer. The moral is that we cannot have time dilation without length contraction. If one is true, logical consistency requires that the other must be true as well.

So far we have derived two rules from Einstein's postulates, time dilation and length contraction. There is one more rule, perhaps the most surprising of all. That is the topic of Chapter 6.

Sample Problems

1. Barnard's star is 6.0 light-years from the Sun in the rest frame of the Sun and star. A spaceship departs the Sun at speed $(4/5)c$ bound for the star, when both the Sun clock and spaceship clock read $t = 0$. In the rest frame of the *spaceship,* (a) How far apart are the Sun and star? (b) Draw two pictures in the spaceship frame, one when the ship is beside the Sun, and one when the star has reached the ship. (c) What will the ship clock read when the star arrives at the ship?

Solution: (a) In the spaceship frame, the distance between Sun and star has been Lorentz-contracted to $D\sqrt{1 - v^2/c^2} = 6.0\, c \cdot \text{yrs}\sqrt{1 - (4/5)^2} = 3.6\, c \cdot \text{yrs}$.

(b) The two pictures are as shown.

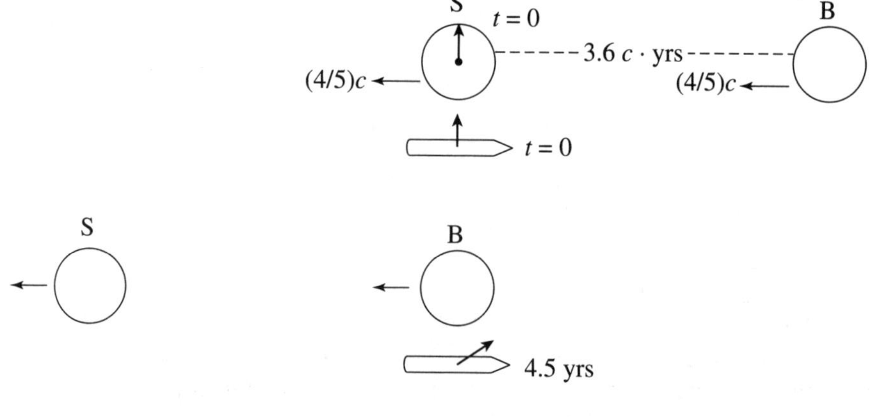

(c) The star moves a distance $d = 3.6\, c \cdot \text{yrs}$ between the first and second pictures, and its speed is $v = (4/5)c$. The time it takes to arrive at the ship is therefore $t = d/v = 3.6\,c \cdot \text{yrs}/(4/5\,c) = 4.5$ yrs. The ship clock began at $t = 0$ and stays at rest in the pictures, so it will read 4.5 yrs when Barnard's star reaches it. (Note that this is the same answer that we found in *Sample Problem* 1 of Chapter 4 for the reading of the ship clock when the ship and star get together. The reasoning there was quite different, however. In the *Sun's* rest frame, we found the 4.5 yrs using the fact that the ship clock ran slow. In the *ship's* rest frame, we found the 4.5 yrs using the fact that the Sun–star distance had been Lorentz contracted.)

2. A light clock of rest length $D = 10.0$ m is moving past us longitudinally, from left to right, at velocity $v = (3/5)c$. (a) What is the distance in our frame between the flashbulb and the mirror? (b) If the flashbulb flashes when our clocks read $t = 0$, what do our clocks read when the light flash reaches the mirror? (c) According to our clocks, how much time does it take the flash to return to the flashbulb? (d) Therefore, what is the total time it takes for the moving clock to tick once, according to our clocks? How does this compare with the time to tick once in the rest frame of the clock? (e) By what factor does the moving clock run slow?

Solution: (a) The clock is Lorentz contracted, so has the shorter length

$$10.0\,\text{m}\sqrt{1-(3/5)^2} = 10.0\,\text{m}(4/5) = 8.0\,\text{m}.$$

(b) Three pictures are shown: When the flash first leaves the flashbulb, when the flash reaches the mirror, and when the flash returns to the flashbulb.

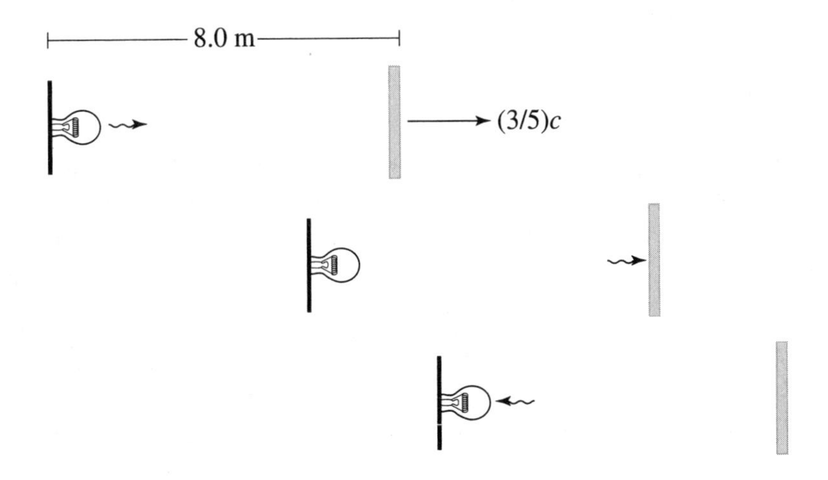

Between the first and second pictures the light moves a distance

$$d_R = 8.0\,\text{m} + (3/5\,c)\Delta t_R,$$

which is the length of the clock (in our frame) plus how far the mirror moves during a time interval Δt_R on our clocks. The distance the light moves is also the speed of light times this same time interval on our clocks, that is, $d_R = c\Delta t_R$.

Setting the two expressions equal to one another gives $(c - 3/5\,c)\Delta t_R = 8.0$ m, so

$$\Delta t_R = \frac{8.0\text{ m}}{(2/5)c} = 20.0\frac{\text{m}}{c}.$$

Our clocks started at $t = 0$ when the flashbulb flashed, so our clocks read $t = 20.0\text{ m}/c$ when the flash arrives at the mirror.

(c) Similarly, between the second and third pictures the light moves a distance $d_L = 8.0\text{ m} - (3/5\,c)\Delta t_L$, the length of the clock (in our frame) *minus* how far the flashbulb moves during a time interval Δt_L on our clocks. The minus sign comes from the fact that the flashbulb is moving *toward* the mirror, decreasing the distance the light has to travel. The light of course also moves the distance $d_L = c\Delta t_L$ in the same time interval. Setting the two distances equal to one another gives $(c + 3/5\,c)\Delta t_L = 8.0$ m, so

$$\Delta t_L = \frac{8.0\text{ m}}{(8/5)c} = 5.0\frac{\text{m}}{c}$$

is the time interval for the return trip.

(d) The total time for the moving clock to tick once as observed from our frame is therefore $\Delta t = \Delta t_R + \Delta t_L = 25.0\text{ m}/c$. In the clock's *own rest frame* it takes only time $\Delta t = (2D)/c = (2 \times 10.0\text{ m})/c = 20.0\text{ m}/c$ to tick once, the total distance the light travels, divided by its speed.

(e) The clock takes longer to tick once in our frame, so runs slow by the factor $(20.0\text{ m/c})/(25.0\text{ m/c}) = 4/5$, as expected for a clock moving at speed $(3/5)c$.

Problems

5–1. The visible disk of our Milky Way galaxy is about 10^5 light-years in diameter. A cosmic-ray proton enters the galactic plane with speed $v = 0.990c$. (a) How long does it take the proton to cross the galaxy from our point of view? (b) How long do observers in the proton's frame think it takes? (c) How wide is the galactic plane to the proton, in its direction of motion?

5–2. A spaceship of rest length 100 m passes by Earth at a speed such that only $(1/3) \times 10^{-6}$ s is required for it to pass by a given point, as measured by clocks on Earth. (a) How fast is it moving? (b) How long is the ship to observers on Earth?

5–3. A rogue planet from a distant galaxy passes by Earth. If the planet's rest frame diameter is 9,000 km and it passes by a single Earth clock in time 0.04 s, how fast is the planet moving relative to Earth?

5–4. Electrons in the 2-mile linear accelerator of SLAC (Stanford Linear Accelerator Center) reach a final velocity of about 0.999 999 999 7 c. (a) What is the length of the accelerator to such an electron? (b) Therefore how much time would it take to reach the end of the accelerator to such an electron, assuming the velocity stays constant? (c) How long would it take the electron to make the trip as seen by Stanford?

5–5. The mean lifetime of tau (τ) leptons is 2.9×10^{-13} s in their own rest frame. If a particular τ particle has this rest frame lifetime, and the shortest distance that can be resolved in a particular detector in the laboratory is about 1 mm $= 10^{-3}$ m, how fast must this τ travel for us to see it? By the time it decays, how far would the detector have moved in the rest frame of the τ?

5–6. The star Vega is 25 light-years from the Sun, in the (assumed) mutual star–Sun rest frame. According to passengers on a spaceship traveling from the Sun to Vega, the distance between these stars is only 12.5 light-years. (a) How fast is the ship moving relative to the Sun? (b) What will ship clocks read when they arrive at Vega, assuming they started at zero when they left the Sun?

5–7. The mean lifetime of a π^+ meson in its own rest frame is $\tau = 2.6 \times 10^{-8}$ s. Suppose that a pion with this lifetime is created at altitude 100 km in the atmosphere by the collision of an incoming cosmic-ray proton with an atmospheric nucleus, and that it has lifetime τ. (a) How fast would this meson have to move to reach the ground before decaying? Express the velocity in the form $v/c = 1 - \epsilon$, where $\epsilon \ll 1$. *Hint:* Use the binomial approximation of Appendix A. (b) What is the distance between the point of creation and the ground in the pion's rest frame?

5–8. A circle of radius R is cut out of a piece of cardboard and the cardboard is made to travel at a high speed along a line past a camera. At the instant the circle is at its nearpoint to the camera, a distance $r (r \gg R)$ from the camera, the plane of the circle is perpendicular to a line from the camera to the center of the circle. At the slightly later time $t = r/c$, the shutter is briefly opened. The image captured by the camera shows an elliptical hole in the cardboard, where the ratio of the major axis to the minor axis is 2.0. How fast was the piece of cardboard moving?

5–9. As we shall see in later chapters, measurement of the energy of a particle gives us the quantity $\gamma \equiv 1/\sqrt{1 - v^2/c^2}$, where v is the speed of the particle. A cosmic-ray proton with a world record energy was (indirectly) detected in 1991, having $\gamma = 6 \times 10^{14}$. (a) Find the speed of this proton, expressed in the form $v/c = 1 - \epsilon$ where ϵ is a very small quantity. That is, find the numerical value of ϵ, valid to one significant figure. (b) Suppose that instead of striking atomic nuclei (as it did) the proton had moved through 30 km of atmosphere all the way to the ground. How thick would this atmosphere have been to an observer in the proton's frame? Compare this thickness with the radius of the hydrogen atom, about 0.5×10^{-10} m.

5–10. Return to the story told in Problem 4–7. (a) How far apart are the Sun and the star Gliese 581 in the spaceship frame? (b) The Sun and spaceship clocks both read zero when the ship and the Sun separate. In the spaceship frame, what will the spaceship clock read when the ship and star meet?

5–11. Return to the story told in Problem 4–10. "The Sun and another star are 60 light-years apart and are at rest relative to one another. At time $t = 0$ (on both the Sun clock and the spaceship clock) a spaceship leaves the Sun at $v = (4/5)c$ headed for the star. Just as the ship arrives, a light signal from the Sun indicates that the Sun has exploded." (a) Draw a set of three pictures in the rest frame of the *spaceship,* illustrating the three important events. Then answer the following questions, all from the point of view of observers at rest in the spaceship frame. (b) How far apart are the Sun and star? (c) When the ship arrives, what does the ship clock read? (d) When the ship arrives, what does the Sun clock read?

CHAPTER 6

Simultaneity

THE PRECEDING CHAPTERS HAVE SHOWN, contrary to classical physics and "common sense," that moving clocks run slow and moving objects are contracted. But the job of demolition (and reconstruction) has only just begun. Everything in physics must be viewed in the light of Einstein's postulates, possibly to be modified or even scrapped entirely. As we have seen, even concepts that Newton and others thought were a priori and absolute, like space and time, have had to be brought under physical investigation and changed. The topic to be discussed now is more upsetting to most people's intuitions than any other conclusion of relativity.

6.1 The Relativity of Simultaneity

We will find in this section that simultaneity is relative. In other words, if two events are simultaneous in one frame of reference, they are generally not simultaneous in another frame of reference. Suppose for example that two supernovae are born in the universe in different galaxies. After correcting for the fact that light takes longer to reach us from the more distant supernova, we decide that supernova 1 blew up before supernova 2. Will other observers agree with our conclusion, or might they decide (again after careful correction for any difference in light travel times) that the two explosions were simultaneous, or even that supernova 2 blew up first?

To answer the question whether the order of events is absolute or relative to the observer, consider the following "experiment." We are calmly sitting in our spaceship in the midst of empty space. Suddenly two other (identical) spaceships approach from opposite directions and pass each other, as shown in Fig. 6.1. Ship "A" moves to the right and ship "B" moves to the left, with equal and opposite velocities as we watch them. Just as they pass, we set off bolts of energy at points x_1 and x_2, which explode between the two ships just as the nose of one reaches the tail of the other. To us, both explosions happen at the same time, so we say the two events are *simultaneous*. Are the explosions also simultaneous to the passengers in A and B?

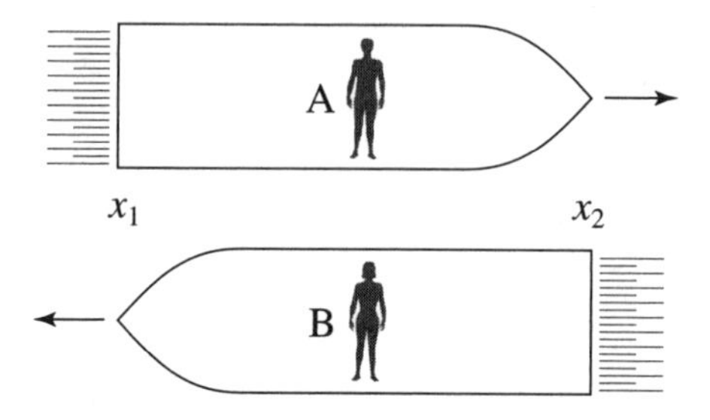

FIGURE 6.1
Two ships A and B pass by us with equal but opposite velocities. When they are in the position shown, we fire explosives at points x_1 and x_2 simultaneously from our point of view.

Suppose there are observers in the middle of each ship. They know they are in the middle of their ships, because they have carefully measured their positions using metersticks. First consider the observers in A. During the time that light from x_1 and x_2 moves toward them at speed c, they move somewhat to the *right,* so they will actually see the explosions from x_2 before that from x_1. They can then say:

> "We're halfway between x_1 and x_2, and we saw the light from x_2 first, so the explosion at x_2 must have happened earlier than the one at x_1."

On the other hand, the observers in B move to the left while the light is reaching them, so the light from x_1 gets to them before the light from x_2, allowing them to say:

> "We're halfway between x_1 and x_2 and we saw the light from x_1 first, so the explosion at x_1 must have happened earlier that the one at x_2."

The reasons for these results are easy to understand from the point of view of either set of observers. As seen by observers in B, for example, ship A is very short, so if explosion x_1 happens when the nose of B is beside the tail of A, and if explosion x_2 happens when the nose of A is beside the tail of B, then the two events can't *possibly* be simultaneous as seen by B. Figure 6.2 shows the ships in two positions as seen by B.

It is clear that observers in B claim that event x_1 happened before event x_2. In short, the question "which event *really* happened first?" will be answered differently by different observers. No absolute answer can be given. From A's point of view, explosion x_2 *really* happens before explosion x_1, B's ship is *really* shorter than A's, and B's clocks *really* run slow. The observers in A know these things, because they have found them out by careful and well-defined measurements. But they would be cautious not to ascribe their reality to everybody, and would say only that certain facts are correct from their standpoint. From the point of view of observers in B, explosion x_1 *really* happens before explosion x_2. From *our* point of view the explosions are *really simultaneous,* but we have to admit that observers in A and B have an equal right to do

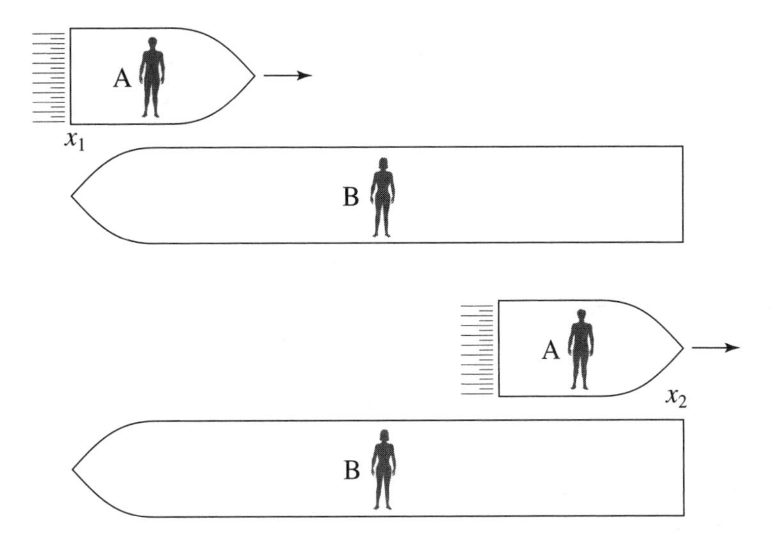

FIGURE 6.2
The events seen from the point of view of an observer in spaceship B. From B's point of view, it is clear that explosion x_1, which happens when the tail of A passes the nose of B, happens before explosion x_2, which happens when the nose of A passes the tail of B.

experiments and make conclusions from them, and that they will find the explosions are not simultaneous in their frames of reference.

6.2 Clock Synchronization in a Single Reference Frame

The outcome of the spaceship experiment indicates that simultaneity is relative, and that clocks in one frame are not synchronized with those in another. To understand more clearly how this comes about, we will search for a satisfactory method of synchronizing two or more clocks in a single frame of reference, and then show that these clocks are *not* synchronized to observers in a different frame of reference.

We are presented with two clocks, at rest with respect to us and separated by a distance D. How can we synchronize them? We try four different approaches, of which only two will turn out to be satisfactory.

1. An observer is stationed beside clock A and another observer is stationed beside clock B. A possible definition of synchronization to the observer beside A would be for both clocks A and B to always read the same, as seen by that observer. That is, if she looks over at clock B, it will read the same time as her own clock A, as illustrated in Fig. 6.3.

The trouble with this definition is that if the clocks are set so that they *look* synchronized to the observer at A, they will *not* look synchronized to the observer at B! The definition neglects the fact that the light from clock B requires a finite time interval $\Delta t = D/c$ to reach A, so that by the time a signal from B reaches A, clock B

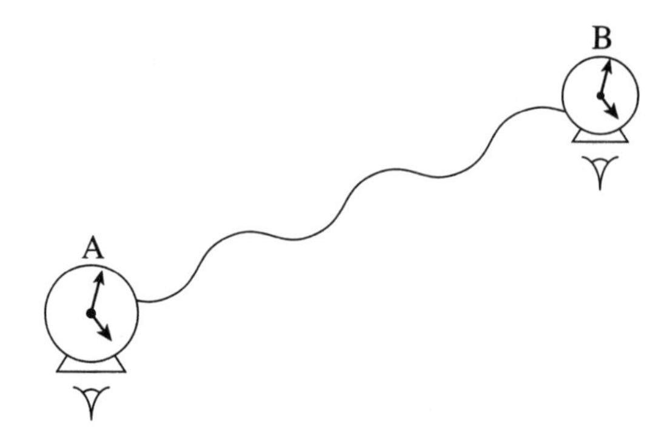

FIGURE 6.3
Judging by appearances.

reads a later time, according to an observer at B. In fact, if the observer at A uses this definition of synchronization, in which both clocks read the same to her, the observer at B will see clock A lag behind clock B by an amount $2D/c$!

2. If we want to synchronize a lot of clocks, we can begin with them all at one place, and just set them to read the same. Then we can carry them out to various locations, as illustrated in Fig. 6.4, and by definition assert that they are all synchronized.

The rather obvious problem with this definition is that the clock readings will depend upon exactly how the clocks are carried to their final locations. The time-dilation effect will ensure that all the clocks will run slow with respect to a stationary observer, but some may run slower than others, depending upon how fast they are carried and for how long a time they are carried. A second set of clocks, synchronized at a different location and dispersed to the same locations as the first group, will generally disagree with the first set.

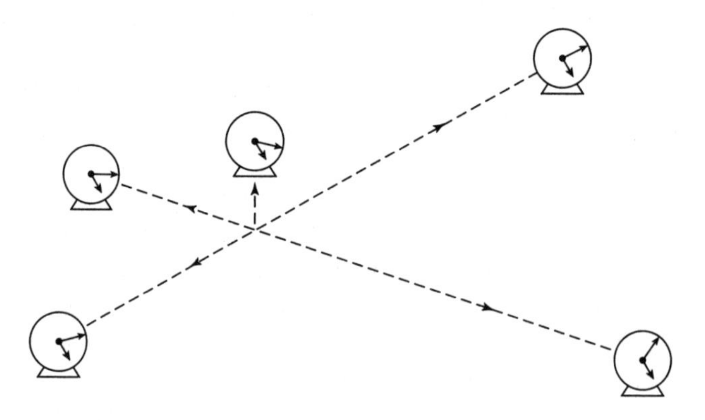

FIGURE 6.4
Transporting previously synchronized clocks.

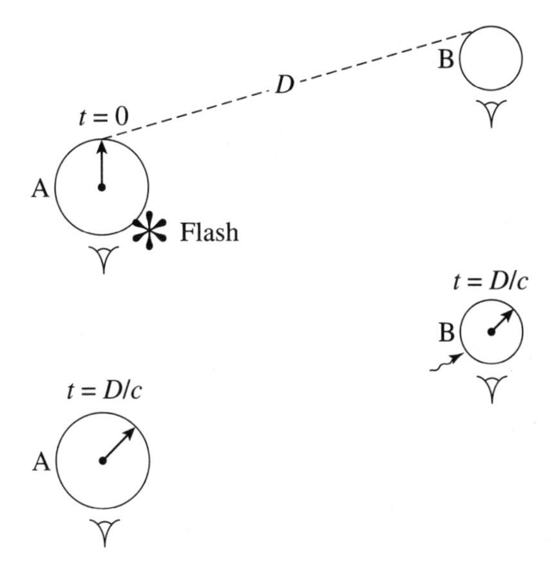

FIGURE 6.5
Accounting for light travel time.

Only if two clocks are carried away in a highly symmetrical manner will these two clocks be synchronized when they reach their destinations. So this method can be used in principle, but only with great care.

3. Our next attempt to synchronize two clocks involves taking account of the time needed for signals to pass between them. Let us carefully measure the distance D between clocks A and B. As in the first method, an observer is stationed beside each clock, and in addition each observer is equipped with a flashbulb that can be fired. They then agree on the following procedure: When clock A reads $t = 0$, observer A will set off her flashbulb. The flash will be seen by observer B, who will immediately set his clock to $t = D/c$, thus accounting for the light transmission time, as illustrated in Fig. 6.5. We claim that clocks A and B are synchronized. This is an entirely consistent definition, because at any later time t another flashbulb can be set off by B whose light will reach A at time $t + D/c$, which is what A's clock actually *will* read when A receives the light.

4. A fourth approach to clock synchronization is the "halfway between" method, which is actually equivalent to method 3. We put two observers with clocks at A and B, measure the distance between them, and put a flashbulb at the halfway point, as shown in Fig. 6.6. Both observers have previously agreed to set their clocks to $t = 0$ when the flash arrives. The bulb is then fired. The light takes equal time to reach A and B, so the observers will be justified in believing their clocks are synchronized. Given an additional flashbulb, the reader should be able to show that methods 3 and 4 give the same result, so either procedure can serve as a means of synchronizing two clocks.

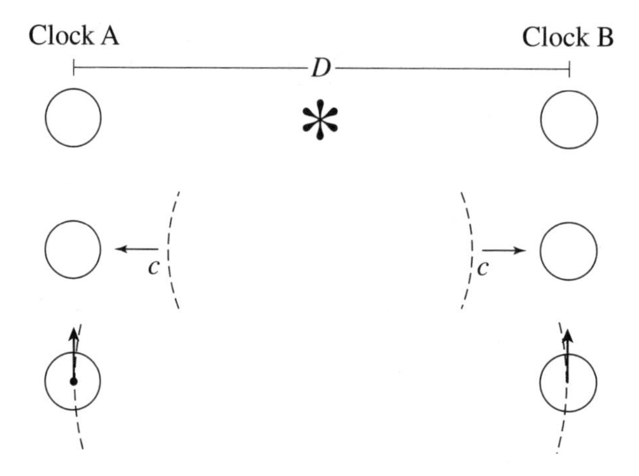

FIGURE 6.6
Clocks A and B are synchronized in their rest frame by the "halfway between" method.

We have been very careful in our definition of clock synchronization, because of the lack of simultaneity in two different frames of reference. We've found that it is possible to synchronize *two* clocks in the same reference frame by a straightforward procedure. It is also necessary to show that it is possible to synchronize three or more clocks in the same frame. Clearly, if we can synchronize clocks A and B, it is also possible by a similar procedure to synchronize clocks B and C. It is left for the reader to show that if this is done, clocks A and C will automatically be synchronized as well. Therefore a well-defined means of synchronizing clocks can be worked out, so that simultaneity is a meaningful idea in a single reference system.

Two events are simultaneous if the clocks placed beside them read the same when the events take place. Observers throughout a single spaceship (as described in the preceding section) can synchronize their clocks, and agree whether or not two onboard explosions occur simultaneously simply by comparing the readings of the two clocks on the ship that are in proximity to the explosions when they go off.

6.3 In the Very Process of Synchronizing Two Clocks, a Moving Observer Disagrees

Suppose we ourselves are in a train moving to the *left* at uniform velocity v relative to two clocks A and B that are being synchronized in the frame of the ground. To us clocks A and B move to the right, and the distance between them is contracted to $D\sqrt{1 - v^2/c^2}$. We want to picture, from our frame of reference, the process of synchronizing the clocks. Four stages in this process are shown in Fig. 6.7. In our train frame of reference, clock A intercepts the flash before clock B, because A is moving toward the light source, whereas B is moving away from it. Therefore, since the reception of the flash is the cue for each clock to be set to $t = 0$, clock A reads

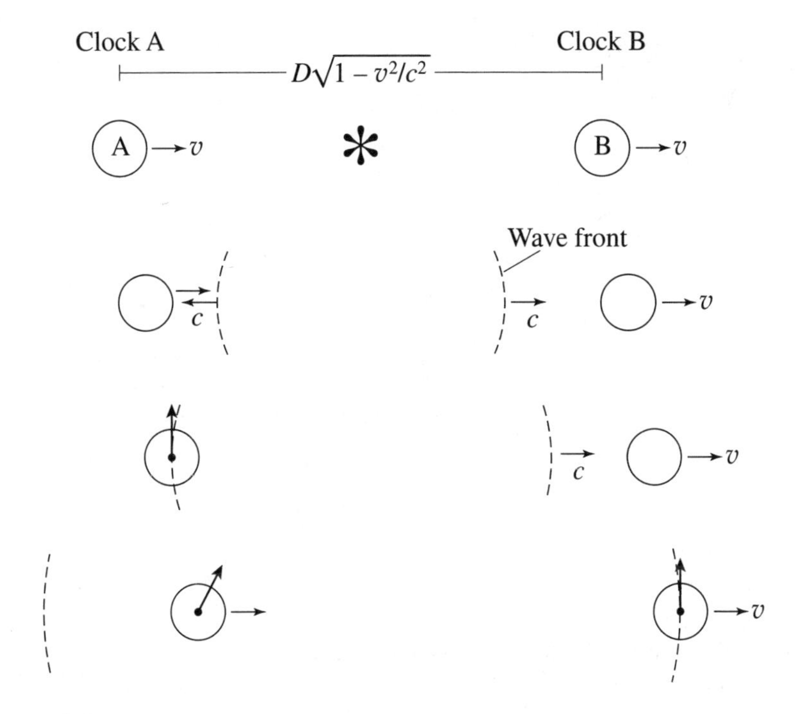

FIGURE 6.7
The process of synchronizing two clocks in their mutual rest frame by the "halfway between" method. From our point of view, clock A is set before clock B, so the two clocks are not synchronized to us.

ahead of clock B as seen from our frame of reference. Thus in the very process of synchronizing the clocks in their rest frame by the most well-defined method, they come out unsynchronized to us!

It is straightforward to calculate how *much* the two clocks will differ in our frame of reference. To begin with, suppose our clocks read $t = 0$ just as the bulb fires.[1] Between stages one and three of Fig. 6.7, both clock A and the left-hand light flash have moved, as shown in Fig. 6.8(a). Let the time when they meet be t_3 on our clock, so the light flash has moved a distance ct_3 and clock A has moved a distance vt_3, which together amount to half the distance between the two clocks, which is $(D/2)\sqrt{1 - v^2/c^2}$. That is, $ct_3 + vt_3 = (D/2)\sqrt{1 - v^2/c^2}$, giving

$$t_3 = \frac{D\sqrt{1 - v^2/c^2}}{2(c + v)}. \tag{6.1}$$

---

1. That is, our clocks have been previously synchronized with one another, and our clock that happens to be right beside the bulb when the flash occurs reads $t = 0$.

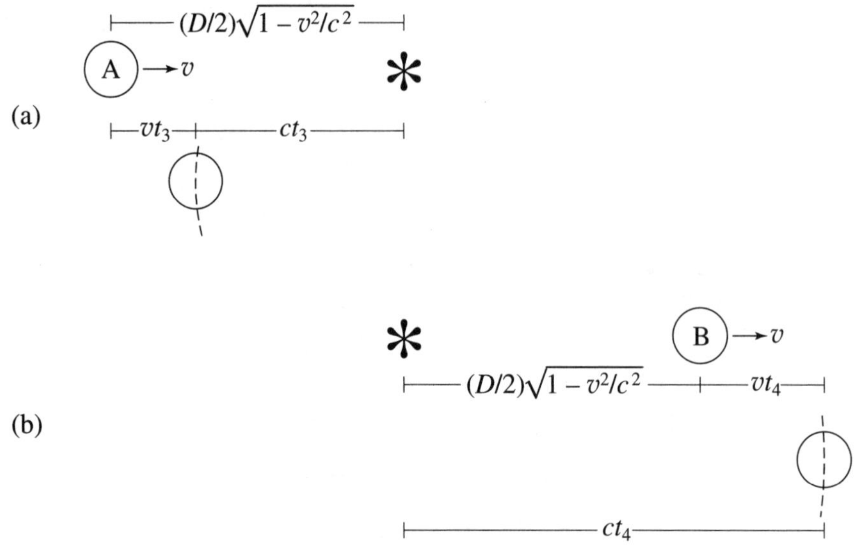

(a)

$(D/2)\sqrt{1 - v^2/c^2}$

A $\rightarrow v$

vt_3 ct_3

(b)

B $\rightarrow v$

$(D/2)\sqrt{1 - v^2/c^2}$ vt_4

ct_4

FIGURE 6.8
(a) Light traveling to clock A, and the method of finding the time t_3 when light reaches A.
(b) Light traveling to clock B, and the method of finding the time t_4 when light reaches B.

At stage four, when the light reaches clock B, suppose our clocks read $t = t_4$. Between the first and fourth pictures, the right-hand flash has traveled a distance ct_4 and clock B has traveled a distance vt_4. Clock B started a distance $(D/2)\sqrt{1 - v^2/c^2}$ to the right of the flashbulb, so it follows that $ct_4 = vt_4 + (D/2)\sqrt{1 - v^2/c^2}$, as illustrated in Fig. 6.8(b). Therefore

$$t_4 = \frac{D\sqrt{1 - v^2/c^2}}{2(c - v)}. \tag{6.2}$$

The time *difference* between stages three and four is then

$$\Delta t = t_4 - t_3 = \frac{D\sqrt{1 - v^2/c^2}}{2(c - v)} - \frac{D\sqrt{1 - v^2/c^2}}{2(c + v)} = \frac{Dv\sqrt{1 - v^2/c^2}}{c^2 - v^2} \tag{6.3}$$

as measured by our clocks. Clock A is a moving clock to us, so it runs slow by the factor $\sqrt{1 - v^2/c^2}$ during the time interval Δt, and so to us it will read time

$$t_{\mathrm{A}} = \Delta t\sqrt{1 - v^2/c^2} = \frac{Dv(1 - v^2/c^2)}{c^2 - v^2} = \frac{vD}{c^2} \tag{6.4}$$

when clock B reads $t_{\mathrm{B}} = 0$. That is, from our point of view clock A is set to zero first, and by the time clock B is set to zero, clock A has already advanced to time $t_{\mathrm{A}} = vD/c^2$. In short, the moving clocks will be out of synchronism by an amount vD/c^2 from our point of view, with the "leading clock" (clock B in Figs. 6.5 and 6.6) reading behind (that is, "lagging") in time. Note that D is the *rest distance* between the clocks, in their direction of motion. Clocks moving along side-by-side, neither leading the other, will be synchronized in our frame if they have been synchronized in their rest frame.

6.4 Overall Summary

Our results for the readings of clocks and metersticks have led to the following *three rules.*

1. Moving clocks run slow by the factor $\sqrt{1 - v^2/c^2}$.
2. Moving objects are contracted in their direction of motion by the same factor $\sqrt{1 - v^2/c^2}$.
3. Two clocks synchronized in their own rest frame are not synchronized in other frames, except for those special frames in which they are spatially separated only in a direction perpendicular to their direction of motion. The clock in front (the *leading* clock) reads an earlier time (*lags*) the chasing clock by an amount

$$\Delta t = vD/c^2, \tag{6.5}$$

where *D is the rest distance* between them along their direction of motion. The rule is "*leading* clocks *lag*."

Note a difference between rule 1 and rule 3. In rule 1, an observer with a clock finds that a moving clock runs at a different rate, by the *multiplicative* factor $\sqrt{1 - v^2/c^2}$. In rule 3, an observer finds that two moving clocks, mutually at rest, run at the same rate as one another; however, the observer has to *add* vD/c^2 to the reading of the leading clock to get the reading of the chasing clock.

6.5 No Universal Now!

Standing in the sunshine, you wonder what a friend in Frankfurt, a neighbor in Nairobi, or a brother in Beijing is doing *right now.* Or standing at night, looking at the stars, you wonder what some intelligent creature on a distant planet is thinking about *right now.* Then you start to walk, pondering the same questions. And because you are walking, the answers will be somewhat different! Simultaneity is not absolute, but depends on the frame of reference. The answers will not be very different for people nearby, because in that case vD/c^2 is tiny. However, for a creature on the other side of our galaxy, 100,000 light-years away, if you walk at 2 m/s($\sim$ 4 mi/hr), then

$$\frac{vD}{c^2} = \frac{2\text{ m/s}(10^5\, c \cdot \text{yrs})}{c^2} = \frac{2\text{ m/s}(10^5\text{ yrs})}{(3 \times 10^8\text{ m/s})} = \frac{2}{3} \times 10^{-3}\text{ yrs} \sim 6\text{ hours.}$$

The creature may be thinking something very different *now,* depending upon whether you are walking or standing still. There is no such thing as a universal *now,* independent of the frame of reference of the observer.

Sample Problems

1. A spaceship of rest length $D = 500$ m moves by us at speed $v = (4/5)c$. There are two clocks on the ship, at its nose and tail, synchronized with one another in the ship's frame. We on the ground have three clocks A, B, and C, spaced at 300 m intervals, and synchronized with one another in our frame. Just as the nose of the ship reaches our clock B, all three of our clocks as well as clock N in the nose of the ship read $t = 0$, as shown.

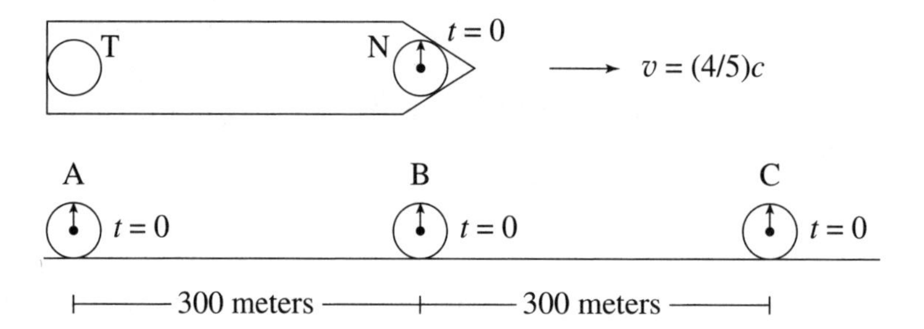

(a) At this time $t = 0$ (to us) what does the clock in the ship's tail read? (b) How long does it take the ship's tail to reach us at B? (c) At this time, when the tail of the ship has reached us, what do the clocks in the nose and tail read?

Solution: (a) The ship is Lorentz contracted in our frame, so has a length of only

$$D\sqrt{1 - v^2/c^2} = 500\text{ m} \cdot \sqrt{1 - (4/5)^2} = 500\text{ m} \cdot (3/5) = 300\text{ m}.$$

We have placed clock A next to the ship's tail and clock B next to its nose, 300 m apart. We have placed a third clock C a distance 300 m farther to the right, since the nose of the ship will reach that point when the ship's tail reaches B. The nose clock N leads the tail clock T in position, so it must lag T in time by

$$vD/c^2 = (4/5\, c)(500\text{ m})/c^2 = 400\text{ m}/c.$$

That is, since N reads zero, T must read 400 m$/c$.

The following sketch shows all clocks when N passes B. So far, we have used rule 2 of Section 6.4 to find the ship's length, and rule 3 to find the reading of T.

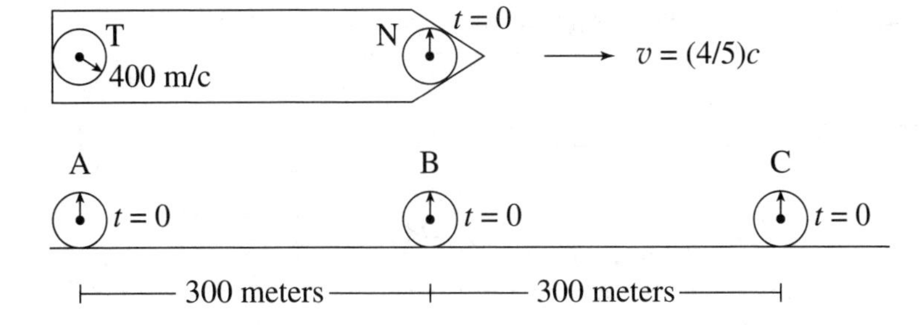

(b) The time it takes the ship's entire length to pass by clock B is

$$t = \frac{\text{distance}}{\text{speed}} = \frac{300\text{ m}}{(4/5)c} = 375\text{ m}/c.$$

(c) The clock in the nose started at *time* = *zero* in the first picture and ran slow by the factor $\sqrt{1 - v^2/c^2} = 3/5$ during the time $375\text{ m}/c$ it took to arrive in the second picture. So when the tail of the ship reaches B, the nose clock N reads

$$(3/5) \cdot 375\text{ m}/c = 225\text{ m}/c.$$

The clock in the nose lags the clock in the tail by $vD/c^2 = 400\text{ m}/c$, so the tail clock reads

$$225\text{ m}/c + 400\text{ m}/c = 625\text{ m}/c.$$

The ship's position and the readings of all clocks when T passes B are shown below.

T
t = 625 m/c
N
t = 225 m/c
A
B
C
t = 375 m/c
t = 375 m/c
t = 375 m/c

2. Sketch the spaceship of Sample Problem 1 in the *ship's* frame (a) when clock B passes clock N, and (b) when clock B passes clock T. In each sketch label the readings of all five clocks, A, B, C, N, and T.

Solution: (a) When B passes N, the situation in the ship's frame is as shown.

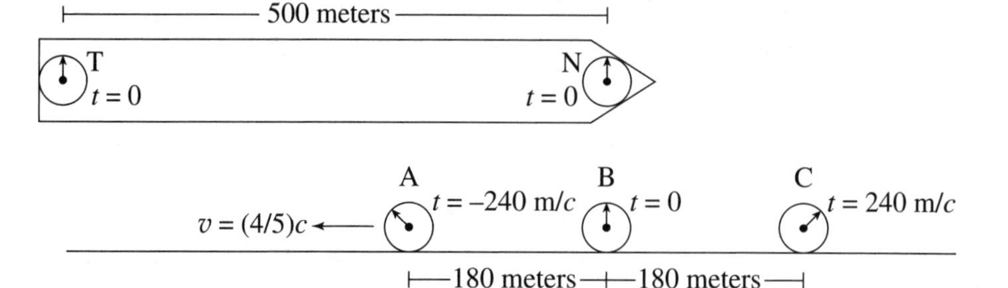

Note that the ship has rest length 500 m, and N and B both read $t = 0$, as given. Clocks N and T are synchronized in their rest frame, so T must also read zero. Clocks A, B, and C are separated by 300 m in their rest frame, so in the ship's frame they are only a distance

$$300\text{ m}\sqrt{1 - v^2/c^2} = 300\text{ m} \cdot 3/5 = 180\text{ m}$$

apart by rule 2, as shown. Using rule 3 we can find the readings of clocks A and C: That is, since A, B, and C are moving to the left in the picture, A leads B, so lags B in time by $vD/c^2 = (4/5\,c)(300\text{ m})/c^2 = 240\text{ m}/c$, and B leads C, so lags C in time by $240\text{ m}/c$. Therefore since B reads zero, A reads $-240\text{ m}/c$ and C reads $+240\text{ m}/c$ in the picture above.

(b) When B passes T, the situation in the ship's frame is as shown below. Note that since B must have traveled a distance 500 m at speed $(4/5)c$, the time in the ship's frame has advanced by time

$$t = \frac{\text{distance}}{\text{speed}} = \frac{500\text{ m}}{(4/5)c} = 625\text{ m}/c.$$

T
t = 625 m/c
N
t = 625 m/c
A
t = 135 m/c
B
t = 375 m/c
C
t = 615 m/c

Therefore that is the time read by N and T, as shown. According to rule 1, clocks A, B, and C have run slow by the factor 3/5, so have advanced by only $(3/5) \cdot 625\text{ m}/c = 375\text{ m}/c$. So B reads $375\text{ m}/c$, A reads $-240\text{ m}/c + 375\text{ m}/c = 135\text{ m}/c$, and C reads $+240\text{ m}/c + 375\text{ m}/c = 615\text{ m}/c$, as shown.

The fact that B reads $375\text{ m}/c$ when T reads $625\text{ m}/c$ is no surprise! That is exactly what those two clocks *did* read when they passed, when calculated in the frame of A, B, and C in Sample Problem 1. The moral is: *When two clocks pass right by one another, whatever times they read when calculated in one frame must be the same times they read when calculated in any other frame.*

To convince yourself of this, suppose the passing clocks nick one another slightly when passing, so that each clock stops ticking. Each clock then reads what it read then forevermore, which can be observed by anyone in any frame at any time.

3. Explorers board a spaceship and proceed away from the Sun at $(3/5)c$. Their clocks read $t = 0$, in agreement with the clocks of earthly observers, at the start of the journey. When the explorers' clocks read 40 years, they receive a light-message from Earth indicating that the government has fallen. (a) Draw a set of three pictures in the Sun's frame, one for each of the three important events in the story. (b) Draw a set of three pictures in the ship's frame, one for each of the three important events. (c) In the explorers' frame, how far are they from the Sun when they receive the signal? (d) According to observers in the explorers' frame, what time was the message sent? (e) At what time do the stay-at-homes say the message was sent? (f) According to the explorers, how far from the Sun were they when the message was sent? (g) According to the stay-at-homes, how far away was the spaceship when the message was sent?

Solution:

(a) Sun-frame pictures

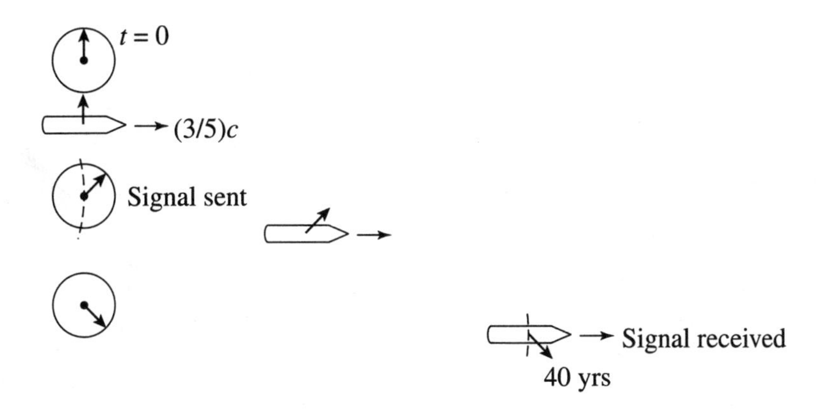

(b) Ship-frame pictures

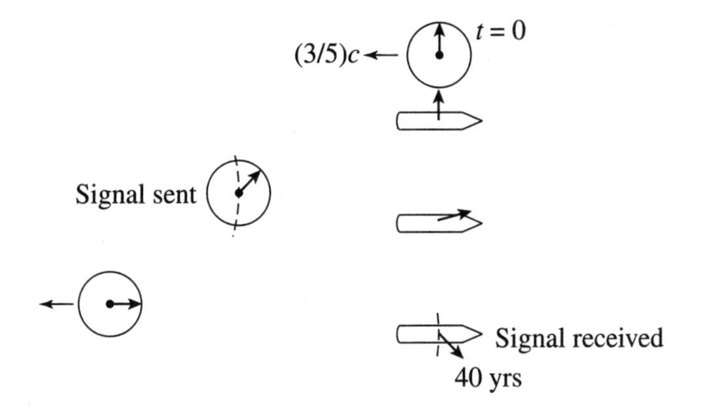

(c) The Sun has been traveling for 40 yrs at $(3/5)c$, so its distance is $24\ c \cdot$ yrs.

(d) Call this time t_0, in the second picture of (b). Note that the distance the Sun has traveled in time t_0 is the distance light has traveled in time 40 yrs – t_0, so $(3/5)c \cdot t_0 = c(40 \text{ yrs} - t_0)$. Therefore $(8/5)t_0 = 40$ yrs, giving $t_0 = 25$ yrs.

(e) In the final picture of part (a), the Sun clock reads $40 \text{ yrs}/(4/5) = 50$ yrs, so the distance between Sun and ship is then $(3/5)c \cdot 50 \text{ yrs} = 30\ c \cdot \text{yrs}$. Let the Sun time be t_1 when the signal is sent. The light signal moves the distance $30\ c \cdot$ yrs in time $(50 \text{ yrs} - t_1)$, so $30\ c \cdot \text{yrs} = c(50 \text{ yrs} - t_1)$, giving $t_1 = 20$ yrs.

(f) The distance in the second picture of (b) is

$$(3/5\ c)t_0 = (3/5\ c)25 \text{ yrs} = 15\ c \cdot \text{yrs}.$$

(g) The distance in the second picture of (a) is

$$(3/5\ c)t_1 = (3/5\ c)20 \text{ yrs} = 12\ c \cdot \text{yrs}.$$

Problems

6–1. Two clocks have been previously synchronized in our frame of reference. We stand beside one and look at the other, which is 30 m away. (a) What will the other clock appear to read when the clock beside us reads $t = 0$? (b) Now the distant clock is carried to us at the constant speed $v = 30$ m/s. By how much will the two clocks differ when they are side-by-side?

6–2. Believing a rumor that the Sun is about to become a supernova, we blast off for the star Sirius. Just as the journey is half over, we see explosions from the Sun and also from Sirius at the same instant! Are we justified in concluding that in our (that is, the spaceship's) frame of reference the two explosions were simultaneous? Explain why or why not, with the help of pictures.

6–3. Synchronized clocks A and B are at rest in our frame a distance 1 light-hour apart. Clock C passes A at speed $(12/13)c$ bound for B, when both A and C read $t = 0$ in our frame. (a) What time does C read when it reaches B? (b) How far apart are A and B in C's frame? (c) In C's frame, when A passes C, what time does B read?

6–4. Synchronized clocks A and B are at rest in our frame of reference a distance 2 light-minutes apart. Clock C passes A at speed $(4/5)c$ bound for B, when both A and C read $t = 0$ in our frame. (a) What time does C read when it reaches B? How far apart are A and B in C's frame? (c) In C's frame, when A passes C, what time does B read?

6–5. Clocks A, B, and C are all mutually at rest. Show by outlining a conceivable experiment that if clocks A and B are synchronized and if clocks B and C are synchronized, then clocks A and C will be synchronized as well.

6–6. The S' frame moves to the right at speed V relative to the S frame. In S' a straight rod is kept parallel to the x' axis, and made to move in the y' direction with constant velocity v'. (a) Explain why, according to observers at rest in the S frame, the rod is *tilted* relative to the x axis. (*Hint:* Attach clocks to the two ends of the rod.) (b) Find the tilt angle in S relative to the x axis, in terms of V and v'.

6–7. A spaceship of rest length $L = 1000$ m moves by us at speed $v = (3/5)c$. There are two clocks on the ship, at its nose and tail, which have been synchronized with one another in the ship's frame. We on the ground have a number of clocks, all synchronized with one another in our frame. Just as the nose of the ship reaches us, both our clock and the clock in the ship's nose read $t = 0$. (a) At this time $t = 0$ (to us) what does the clock in the ship's tail read? (b) How long does it take the tail of the ship to reach us? (c) At this time, when the tail of the ship has reached us, what does the clock in the tail read? (d) At this time, what does the nose clock read?

6–8. A very long stick ruled with meter markings is placed in empty space. A spaceship of rest length $L = 100$ m runs lengthwise alongside the stick. Two space travelers equipped with knives and synchronized watches station themselves fore and aft. At a prearranged time, each reaches through a porthole and slices through the stick. If the relative velocity of the stick and ship is $v = (4/5)c$, how many meter marks are on the cut-off portion of the stick? Do the calculation first in the frame of the ship, and then do the calculation over in the frame of the stick. In each case, draw careful pictures.

6–9. Two atoms are at rest 1 meter apart in the laboratory. A photon is emitted from one atom, travels at the speed of light, and is absorbed by the other. Show that there is *no* frame of reference (whose velocity relative to us is V, with $V < c$) in which the emission and absorption are simultaneous.

6–10. Return to the story told in Problem 4–10. "The Sun and another star are 60 light-years apart and are at rest relative to one another. At time $t = 0$ (on both the Sun clock and the spaceship clock) a spaceship leaves the Sun at speed $v = (4/5)c$ headed for the star. Just as the ship arrives, a light signal from the Sun indicates that the Sun has exploded." (a) Draw a set of three pictures in the rest frame of the *spaceship,* illustrating the three important events. Then answer the following questions, all from the point of view of observers at rest in the spaceship frame. (b) How far apart are the Sun and star? (c) When the star arrives at the ship, what does the ship clock read? (d) When the star arrives at the ship, what does the Sun clock read? (e) When the star arrives at the ship, what does the star clock read? (f) When the Sun explodes, what does the ship clock read? (g) When the Sun explodes, what does the Sun clock read? (h) When the Sun explodes, what does the star clock read?

6–11. Suspecting that the Sun is about to blow up, scientists board a spaceship and proceed away from the Sun at $(4/5)c$. Their clocks read $t = 0$, in agreement with the clocks of earthly observers, at the start of the journey. When the travelers' clocks read 2 years, their scintillator, which detects gamma radiation, indicates that the Sun has exploded. (a) Draw a set of three pictures in the Sun's frame, one for each of the three important events in the story. (b) Draw a set of three pictures in the ship's frame, one for each of the three important events. (c) In the travelers' frame, how far are they from the Sun when they receive the signal? (d) According to observers in the travelers' frame, what time did the Sun blow up? (e) At what time do the stay-at-homes, assuming some survive, say the disaster happened? (f) According to the travelers, how far from the Sun were they when the Sun exploded? (g) According to the stay-at-homes, how far away was the spaceship when the blow-up occurred?

6–12. Mysterious signals are received from a planetary system several light-years from Earth, so an astronaut traveling at $(12/13)c$ is sent to investigate. Three years and nine months later, her husband back on Earth receives a message that she just passed the signal source and that, to complicate things, she has just given birth to a son. (a) How far from Earth is the signal source, as measured in Earth's frame? (b) When she gives birth, how long has she been gone according to (*i*) her husband's clock and (*ii*) her own clock?

6–13. Two spaceships A and B are moving in opposite directions. Each measures the speed of the other to be $(3/5)c$. They pass alongside each other and each sets its clocks to zero at that instant. Ship A receives a light-message at time $t = 320$ hours on its clocks that there has been an explosion in B that destroyed B's oxygen generating plants and also left them unable to alter their velocity. The occupants of B know that they have only enough oxygen to survive 1000 hours (on B's clocks). Ship A, immediately upon receiving the message, sends rescue ship C to B. The velocity of C, as measured by A, is $(4/5)c$. Assume that B sent the message simultaneously with the explosion and that C was dispatched simultaneously with the receipt of the message by A. (a) How long must the crew of B wait for rescue, according to A, from the time of the explosion to the time that C arrives? (b) How long is this time interval according to the clocks of B? Were they rescued or not? *Hint:* Start by drawing a sequence of four pictures in A's rest frame, each picture representing an important event in the story.

6–14. Starship *Endeavor* leaves Earth at $v = (3/5)c$; its clock and Earth's clock both read $t = 0$ at departure. When the ship clock reads $t = 1$ week, Lt. Cmdr. Jones on *Endeavor* sends a light-speed message back to Starfleet Command on Earth, requesting that they send a St. Valentine's greeting to Dr. Fong on *Endeavor*. St. Valentine's day on the ship is 3 weeks after Jones sends his request, according to *Endeavor* clocks, and Starfleet Command sends its greeting at light speed immediately upon receiving Jones's request. (a) Sketch a sequence of four pictures, illustrating the important events in the story. Draw pictures in the rest frame of Earth. (b) What does the Earth clock read when Jones sends his request? (c) What does the Earth clock read when Earth receives Jones's request? (d) What does the Earth clock read when the greeting reaches Endeavor? (e) What does the ship clock read when the greeting is received? (f) Does the greeting arrive in time?

6–15. Federation Starship *Challenger* is drifting in space without power. At midnight *Challenger* detects a radar pulse; First Officer Rosa realizes that an alien ship sent the pulse, and that the alien ship was only 2 light-hours away (in *Challenger's* frame) when the pulse was sent, and that the aliens were moving at speed $v = c/2$ toward *Challenger.* Rosa reasons that the aliens do not yet know that *Challenger* is there, but when the radar pulse returns to them they are likely to immediately fire a "photon torpedo" (at $v = c$) toward *Challenger.* (a) What is the time interval, measured by the aliens, between the emission and return of their radar pulse? (b) By what time (as measured on its own clocks) must *Challenger* restore its power, in order to immediately create a shield against the photon torpedo?

6–16. Spaceship A is accompanied by escort ships B and C. All three are moving at velocity $(3/5)c$ in our frame of reference. In our frame, ship B is above ship A a distance $d = 16\,c \cdot$ hrs, while ship C is to the *right* of ship A this same distance $d = 16\,c \cdot$ hrs in *our* frame, as shown below. The picture is drawn when the clock in A reads $t = 0$, and when all of the clocks at rest in our frame also read $t = 0$. The clocks in A, B, and C have been synchronized in their mutual rest frame. (a) What is the rest distance between A and B, in $c \cdot$ hrs? (b) What is the rest distance between A and C? (c) In the picture shown, where A reads $t = 0$, what does the clock in B read? (d) In the picture, what does the clock in C read? (e) When the clock

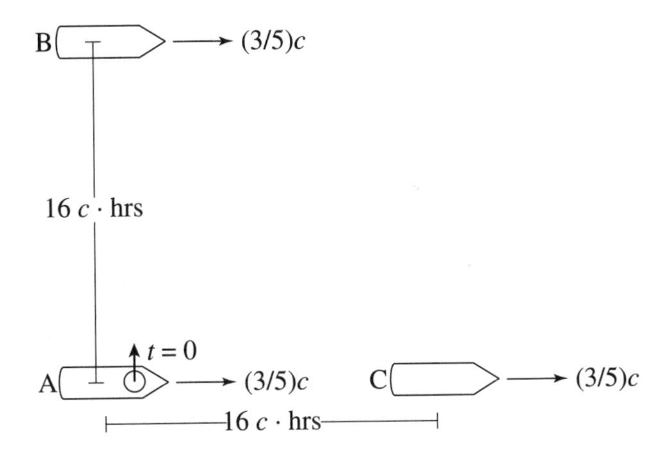

in A reads $t = 0$, the crew in A finds it has run out of potato chips, so sends distress light signals to B and C, to see if they have extra. Consider in particular the light signal sent from A to B. What is the horizontal component of the signal's velocity in our frame? (f) What is the vertical component of the light's velocity in our frame? (g) What do our clocks read when B receives the signal? (h) What does B's clock read when B receives the signal? (Upon receiving the request, the crew on B quickly finishes its remaining potato chips and reports back that it has none left.) (i) Now consider the light signal sent from A to C when A's clock reads $t = 0$. Draw a sequence of two pictures of A and C in our frame of reference, illustrating their positions when A sends the signal and when C receives the signal. (j) What do our clocks read when C receives the signal? (k) Using the pictures in our frame of reference, what does A's clock read when C receives the signal? (l) What does C's clock read when C receives the signal? Upon receiving the message from A, the more generous crew of C immediately dispatches a space probe bearing potato chips back to A.

CHAPTER 7

Paradoxes

RELATIVITY CAN BE VERY CONFUSING. The fact is, most people who understand it have had to work through one confusion after another until enlightenment eventually dawns. Some of the confusions can be related in the form of "paradoxes." Working one's way through these paradoxes, coming to understand that they do not present genuine logical contradictions, is an excellent way to understand special relativity.[1]

7.1 If Your Clock Runs Slow to Me, How Can My Clock Run Slow to You?

At first this seems impossible. Two inertial observers A and B, moving at speed V relative to one another, each have clocks in their rest frames. From A's point of view B's clock runs slow. However, since there is no preferred inertial frame, A's clock must also run slow from B's point of view. How can both be true?

The paradox is resolved when we realize that it takes two synchronized clocks in A's frame, A_1 and A_2, to measure the tick rate of a moving clock. Yet even though A_1 and A_2 are synchronized in A's frame, they are not synchronized in B's frame. And it is this lack of synchronization that allows A_1 and A_2 to run slow in B's frame, as we shall see. The details in A's frame are shown in Fig. 7.1.

Clock B passes clock A_1 when both clocks read $t = 0$. Later, clock B reaches A_2. This requires a period of time $t = D/V$, according to the rest clocks A_1 and A_2. The moving clock runs slow by the square-root factor, however, so B reads only $(D/V)\sqrt{1 - V^2/c^2}$ in this picture, as shown in Fig. 7.1(b). This is what we mean by saying that the moving clock runs slow. Note that it takes *two* observers in A's frame to make the measurements, since there needs to be a clock *right beside* B at each of

1. Definitions of "paradox" in Webster's New World Dictionary include "a statement that seems contradictory, unbelievable, or absurd, but that may actually be true in fact," and "a statement that is self-contradictory in fact, and, hence, false." The paradoxes given here are of the first type.

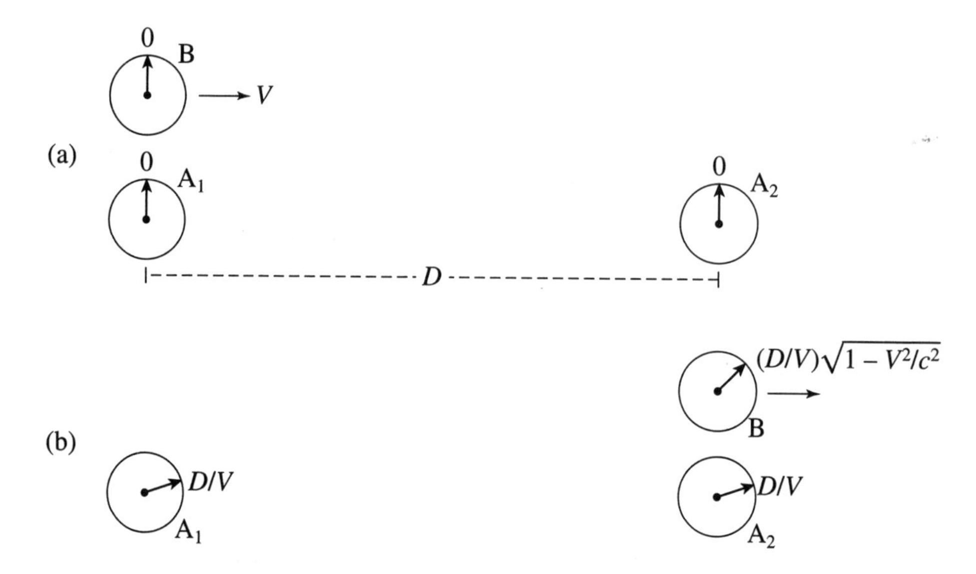

FIGURE 7.1
Clock B moves at speed V relative to A's frame. (a) Clock B passes clock A_1 when B and A_1 both read $t = 0$. (b) Clock B later reaches A_2, when A_2 reads D/V and B reads $(D/V)\sqrt{1 - V^2/c^2}$. Clearly B's clock has run slow compared with A's clocks.

the important times in the experiment. We would not want to look from afar, because the experiment would then be complicated by light travel times.

Now look at the same experiment from B's point of view. In Fig. 7.2(a), clock A_1 passes clock B when both read $t = 0$. Note that according to B, the distance between A_1 and A_2 is only $D\sqrt{1 - V^2/c^2}$ due to Lorentz contraction. Later on, as shown in Fig. 7.2(b), clock A_2 arrives at B. This requires a time interval

$$\Delta t = \frac{\text{distance}}{\text{speed}} = \frac{D\sqrt{1 - V^2/c^2}}{V}, \tag{7.1}$$

which will be the time clock B (the rest clock in these pictures) reads when A_2 arrives. This agrees with the time that clock B reads upon arrival, according to the pictures (Fig. 7.1(b)) in A's frame, as it must.

In A's frame we found that clock A reads time $t = D/V$ when B and A_2 meet. This must also be true in B's frame, because these two clocks are right beside one another when they meet. They must both agree that A_2 reads D/V and B reads $(D/V)\sqrt{1 - V^2/c^2}$. Why? Because they might, for example, nick each other slightly as they pass, stopping the ticking in both clocks so that anyone in any reference frame can verify that clock B reads $(D/V)\sqrt{1 - V^2/c^2}$ forevermore, while clock A_2 reads D/V. So far so good.

Now the critical question: In B's frame we found that B's clock reads a *lesser* time at the end than clock A_2, even though B's clock has been at rest and clock A_2 has been

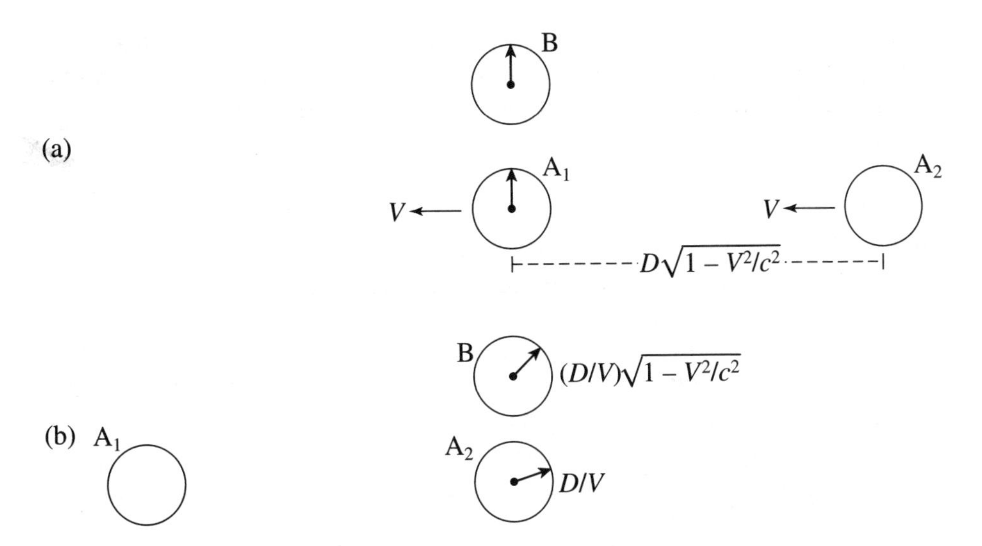

FIGURE 7.2
The same experiment according to observers in B's frame. (a) Clock A_1 passes B when both these clocks read $t = 0$. (b) Clock A_2 reaches B, when B's clock reads $t =$ distance/speed $= \left(D\sqrt{1 - V^2/c^2}\right)/V$.

moving. Shouldn't it be the *moving* clock that runs slow, and so shouldn't clock A_2 read a lesser time?

There is a way out of this apparent contradiction. Figure 7.2(a) shows that when clock A_1 passes clock B, both read zero. But we didn't say what A_2 reads in this picture! By the third "rule" of special relativity, the *leading* clock A_1 *lags* the following clock A_2 in time, as shown in Fig. 7.2(a). Therefore A_2 should read the later time VD/c^2 in that picture. That is, clock A_2 has a *head start*! That is why it can read a later time at the end, even though it has been running slow between pictures 7.2(a) and 7.2(b).

Let's see how this idea really works. What should clock A_2 read in Fig. 7.2(b)? It should read its starting value VD/c^2, as shown in Fig. 7.2(a), *plus* how much it advances between the two pictures. The time between the two pictures is $(D/V)\sqrt{1 - V^2/c^2}$ according to B's rest-clock, so A_2 should advance by

$$\Delta t_A = \left(\frac{D}{V}\sqrt{1 - V^2/c^2}\right)\sqrt{1 - V^2/c^2}, \tag{7.2}$$

with an extra $\sqrt{1 - V^2/c^2}$ since it is a moving clock. Therefore in Fig. 7.2(b) clock A_2 should read the time

$$\Delta t_{A_2} = \frac{VD}{c^2} + \Delta t_A = \frac{VD}{c^2} + \frac{D}{V}(1 - V^2/c^2) = \frac{D}{V}, \tag{7.3}$$

which is exactly what it *does* read, as we learned by analyzing the experiment in A's frame. When they pass, clock B and clock A_2 read $t_B = (D/V)\sqrt{1 - V^2/c^2}$ and $t_{A_2} = D/V$, respectively, independent of which frame we use to calculate.

The apparent contradiction has been resolved. It *is* possible for A_2 to read a later time than B, even though it has been running slow according to B, because the reading of A_2 had a *head start.* And so it is indeed true that your clock runs slow to me, even though my clock runs slow to you.

7.2 If Your Meterstick Is Short to Me, How Can My Meterstick Be Short to You?

Two inertial observers A and B, moving at speed $V = (4/5)c$ relative to one another, each has a meterstick. From A's point of view B's stick is Lorentz contracted. However, since there is no preferred inertial frame, A's meterstick must be Lorentz contracted from B's point of view. How can both be true?

Again, the secret is to look carefully at how observers at rest in A's frame can actually *measure* the length of B's stick. Let there be multiple observers at rest in A's frame, lined up side-by-side, all with synchronized clocks. Suppose that just as their clocks read $t = 0$, the left-hand side of B's stick whizzes by the left-hand side of A's stick. Observers A_1 and A_2 are positioned to make simultaneous measurements of the locations of the left and right ends of B's stick at $t = 0$. Suppose also that clock B_1, at the left-hand end of B's stick, reads $t = 0$ at this time, as shown in Fig. 7.3.

Observers on B see that A_1 and A_2 make measurements of the two ends of B's stick, but they do not agree that the measurements were made simultaneously. In fact, clock B_2 at the front of B's stick reads only $-VD/c^2 = -(4/5) \cdot (1\,\text{m}/c)$ in Fig. 7.3, according to the leading-clocks-lag rule. Therefore from B's point of view, the measurement by A_2 was made *before* the measurement by A_1, so no *wonder* the observers on A got the "wrong answer" for the length of B's stick.

Now why does B think that A's stick is short? Observers on B need to make simultaneous measurements from their point of view of the two ends of A's stick, let's say when their clocks read zero. Drawing the pictures again from A's point of view, B_1 measures the left end of A's stick in Fig. 7.3; only later does clock B_2 read zero. How

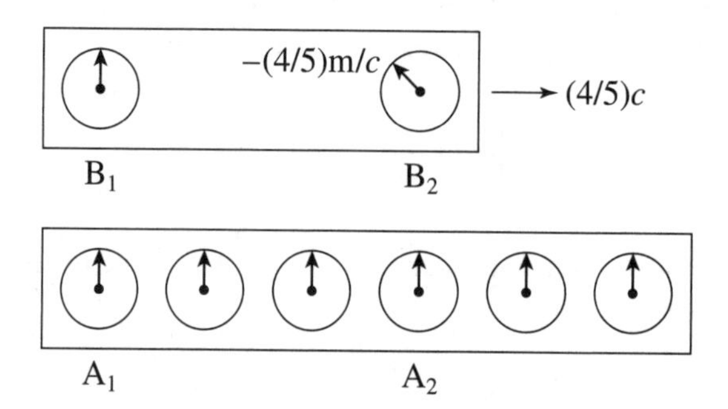

FIGURE 7.3
Stick B moves at speed $V = (4/5)c$ relative to A's frame, so has a length only $1\,\text{m} \cdot \sqrt{1 - (4/5)^2} = 0.60\,\text{m}$. Clock B_1 passes clock A_1 when both clocks read $t = 0$.

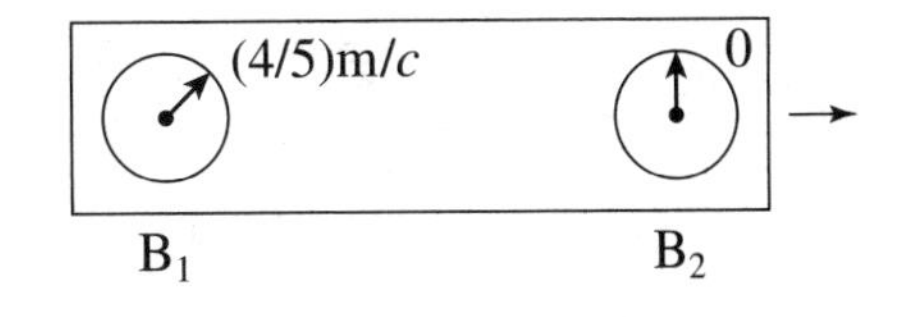

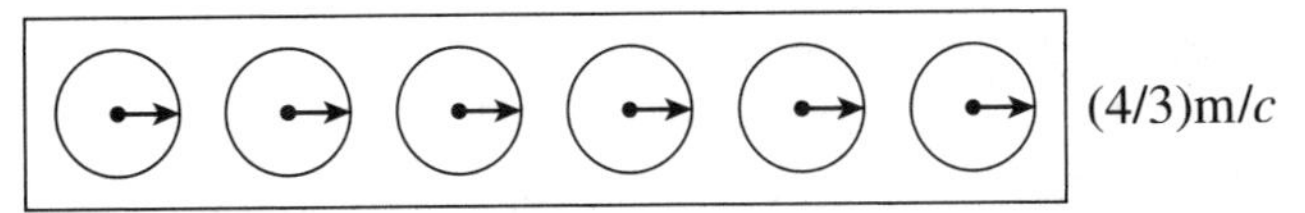

FIGURE 7.4
From A's point of view, B's stick moves to the right by the time clock B_2 reads zero.

long does it take for clock B_2 to advance to read zero? It must advance by $(4/5)\ \mathrm{m}/c$, as shown in the preceding paragraph. However, since B's clocks run slow to A, it will take a time

$$\frac{4/5\ \mathrm{m}/c}{\sqrt{1-(4/5)^2}} = \frac{4}{3}\frac{\mathrm{m}}{c} \tag{7.4}$$

in A's frame for clock B_2 to advance to zero. During this time B's stick moves to the right a distance $d = Vt = (4/5)c \cdot (4/3)\ \mathrm{m}/c = 16/15\ \mathrm{m}$, as shown in Fig. 7.4. The right-hand side of B's stick began at $x = 3/5\ \mathrm{m}$ in Fig. 7.3, so when clock B_2 reads zero the right-hand end of B's stick is located at

$$x = \frac{3}{5}\ \mathrm{m} + \frac{16}{15}\ \mathrm{m} = \frac{25}{15}\ \mathrm{m} = \frac{5}{3}\ \mathrm{m}, \tag{7.5}$$

so that the right-hand end of B's stick is way past the right-hand end of A's stick when B_2 makes the measurement. Therefore A's stick is shorter than B's stick in B's frame; in fact, A's stick is only 3/5 as long as B's own stick, consistent with the Lorentz contraction.

7.3 The Magician's Assistant

The magician's assistant is 5 feet tall. She lies horizontally on a table, perfectly straight. Towering above her are two masked medieval executioners, each wielding a sharp-bladed axe above his head (Fig. 7.1). They stand 3 feet apart, and have agreed to swing their blades down upon the table when their previously synchronized watches read $t = 0$. Can she come through the experience in one piece, while continuing to lie perfectly straight?

The magician knows relativity, and so plans to send her and the table past the executioners at velocity $v = (4/5)c$. She should then be Lorentz contracted to a length of only $5.0\ \mathrm{ft} \cdot \sqrt{1-(4/5)^2} = 5.0\ \mathrm{ft}\,(3/5) = 3.0\ \mathrm{ft}$ in the frame of the room, barely

FIGURE 7.5
The magician's assistant and two executioners.

fitting between the descending blades, so that only stray hairs and toes would be at risk.

The assistant is not happy with this plan. She reasons that in *her* reference frame, she will have her full length of 5.0 ft, while the distance between the executioners' blades will be Lorentz contracted to only $3.0\text{ ft}\sqrt{1-(4/5)^2} = 3.0\text{ ft }(3/5) = 1.8\text{ ft}$. So when they swing down their blades she will likely emerge in two or possibly three pieces, an uncomfortable outcome for her.

It would appear that in one frame she emerges whole and in the other she gets chopped up. This constitutes a relativistic "paradox," because both cannot be true. Either she comes out in one piece or she does not; anyone in any reference frame can verify the result. Does she come through all right or not? If she comes out whole, something is wrong about *her* argument. If she comes out in pieces, something is wrong with the *magician's* argument. Who is right, and what was the fault in the other's reasoning?

As with many "paradoxes" in relativity, we have to look carefully at simultaneity. In her frame her length is 5.0 ft, and the distance between the blades is only 1.8 ft. However, in her frame the executioner's watches are not synchronized. In fact, the watch of the left-hand executioner is the leading clock, so it lags the other by $vD/c^2 = (4/5)(3.0\text{ ft})/c = (12/5)\text{ ft}/c$.

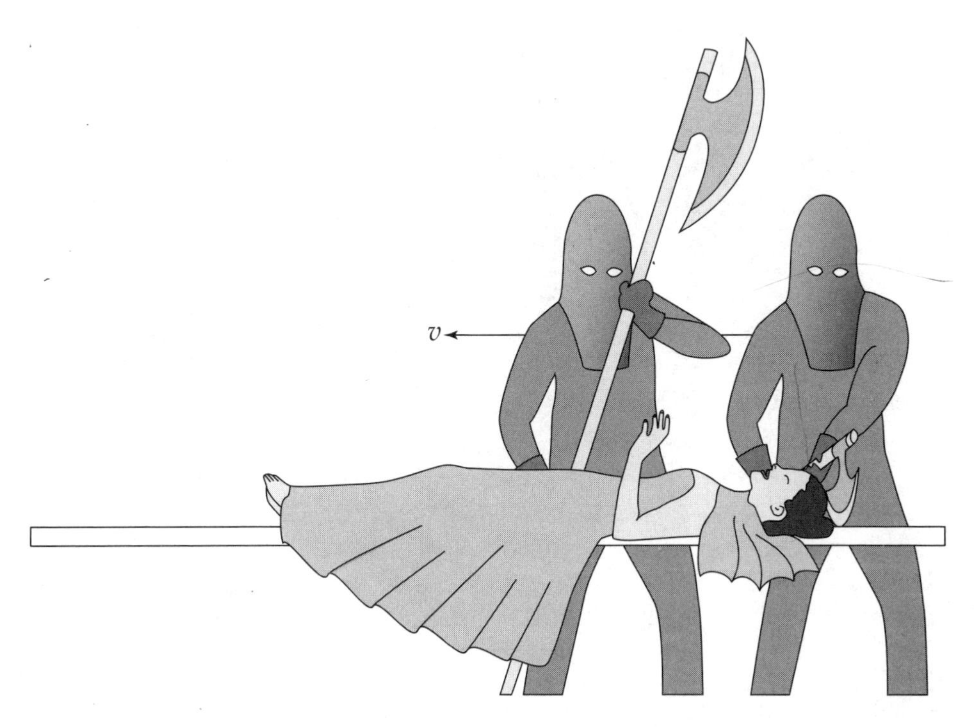

FIGURE 7.6
The situation from the assistant's point of view.

The leading left-hand watch, lagging in time, will not read $t = 0$ until *after* the right-hand watch reads zero, so the right-hand blade descends first, just missing her head. At this instant to her, the left-hand blade is 1.8 ft from her head, and the left-hand watch reads $-(12/5)$ ft$/c$. The left-hand executioner has not yet swung his blade, as shown in Fig. 7.6. How long must she wait for him to do so?

If the left-hand executioner's (moving) clock advances by (12/5) ft$/c$, so that it reaches zero, clocks in the assistant's rest frame must advance by the *larger* amount,

$$\Delta t = \frac{(12/5)\ \text{ft}/c}{\sqrt{1 - v^2/c^2}} = \frac{(12/5)\ \text{ft}/c}{3/5} = 4\ \text{ft}/c. \tag{7.6}$$

During that time interval, the executioners (moving at velocity $(4/5)c$ from her point of view) will move a distance $d = v\Delta t = (4/5\ c) \cdot 4\ \text{ft}/c = 16/5\ \text{ft} = 3.2$ ft.

So now it is clear what happens from her point of view. The right-hand executioner brings down his blade, just missing her head. At that instant the left-hand executioner's blade is 1.8 ft from her head. By the time the left-hand watch reaches $t = 0$, the executioners have moved a distance 3.2 ft. So when the left-hand blade descends, the left-hand blade is $1.8\ \text{ft} + 3.2\ \text{ft} = 5.0$ ft from her head, so it will slice down just to

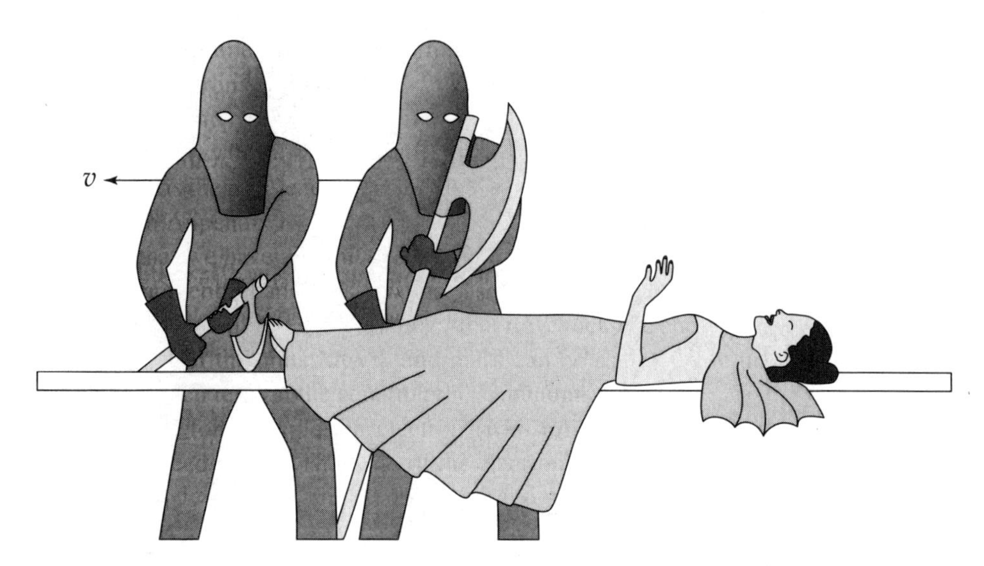

FIGURE 7.7
The left-hand blade descends, just missing her toes.

the left of her toes, and she emerges unscathed, as shown in Fig. 7.7. There is no contradiction; when the calculation is made right, the blades miss her in both frames.

7.4 Rigid Bodies, a Pole Vaulter, and a Barn

One of the most convenient notions in classical mechanics is that of a rigid body. A rigid body is simply an object where the distance between any two particles in the object always stays the same: It doesn't bend, twist, or stretch, but moves as a unit. For many purposes we can treat a rigid body as though it were a single particle localized at the center of mass.

Now it is evident that there is really no such thing in nature as a rigid body, because nothing is infinitely "stiff." Every real object can bend and twist to some extent, so that for example when you spin it, it will stretch slightly. Nevertheless in many cases these effects are small, so that the rigid body approximation is pretty good, and of course very useful.

Special relativity now enters the picture, telling us that not even in principle can a perfectly rigid body exist! Because all signals travel with only finite speed ($v < c$), no object can be completely stiff. For example, if one end of a steel bar is struck with a hammer, the other end of the bar cannot possibly feel the blow immediately. Thus one end of the bar may begin to move while the other end remains at rest, because the signal hasn't yet had time to propagate from one end to the other. So this object is not a rigid body at all. It cannot be described as a single particle.

This inherent lack of rigidity is a help in understanding some of the relativistic "paradoxes." An example is the paradox of the pole vaulter and the barn. It is worth

FIGURE 7.8
A pole vaulter and a barn.

thinking about not only because it casts light on the "rigid body" notion, but also because it graphically demonstrates the fact that the simultaneity of two events is a relative concept, depending on the observer.

The paradox begins on a farm. On the farm is a smallish barn of length 50 ft that has a front door that can be opened and closed, and a back wall that is unusually sturdy. A friend of ours, a pole vaulter, owns a pole that is 80 ft long (unusually long, actually). The pole cannot fit inside the barn. The pole vaulter has the virtue of being extremely speedy, so we request that our friend walk quite a distance from the barn, and then (with the pole in a horizontal position) run toward the barn at speed $(4/5)c$, as shown in Fig. 7.8. The front door of the barn opens inward, so we open the door but stand behind it inside. Our friend then runs toward the barn at $(4/5)c$, so that from our point of view the pole's length is only $(3/5) \cdot 80 \text{ ft} = 48 \text{ ft}$. It easily fits inside the barn. Just as the far end of the pole strikes the back of the barn, we slam the front door shut, thereby capturing the pole and pole vaulter inside the barn with a pole that isn't supposed to fit!

The back wall of the barn is very sturdy, so the pole is brought to a halt, at which time it must expand back to its rest length of 80 ft. Assuming the pole doesn't shatter, it must shove the pole vaulter back through the closed front door, knocking it off its hinges and down onto the ground.

The puzzle arises if we consider this series of events from the standpoint of the pole vaulter. To our friend the pole retains its full length of 80 ft, while the barn, moving at speed $(4/5)c$, has a length of only $(3/5)(50 \text{ ft}) = 30 \text{ ft}$. So it seems impossible for the pole and pole vaulter to fit inside the barn. Yet we saw our friend in the barn, and we saw the front door knocked flat on the ground. The fact that the front door subsequently lies on the ground cannot depend on the reference frame. It either ends up on the ground or it doesn't. So what really happened from the pole vaulter's point of view? Does the pole vaulter get inside the barn or not?[2]

2. See Problem 7–8.

7.5 The Twin "Paradox"

The most famous relativistic "paradox" is the *paradox of twins*. Consider the twins Al and Bertha. Al stays at home while Bertha travels off in a straight line at velocity $(4/5)c$ for 5 years according to Al's clock, thereby reaching the star Alpha Centauri, a distance $(4/5)c \cdot 5\text{ yrs} = 4\,c \cdot$ yrs away in Al's frame. During this time she ages only $5\text{ yrs} \cdot 3/5 = 3$ yrs using time dilation. She then quickly turns around and returns at the same speed, covering the $4\,c \cdot$ yrs in 5 years according to Al's clock, but only 3 years according to her own clock, as shown in Fig. 7.9. The net result is that the twins are reunited, and while Al is 10 years older, Bertha is only 6 years older. All calculations have been made in Al's frame of reference.

Now the "paradox" is this: Since motion is relative, why can't Bertha argue as follows? "I (Bertha) did not move; it was Al who moved away from me and returned. Therefore since Al's clocks are moving from my point of view, he will come back younger than me."

The problem of course is that they can compare clocks when they have reunited, and both will have to agree which clock reads ahead of the other. It would seem that from Al's point of view his clock should be ahead, but from Bertha's point of view her clock should be ahead. Both cannot be true! They are standing right next to one another and can plainly see which clock reads ahead of the other. If they are young enough at departure, they can simply look at one another and see who is younger at the end. If the trip starts when they are 6 years old, one will be 16 and the other only 12 when they reunite. It would be obvious who had aged the most.

How is the "paradox" resolved? The resolution points out an important aspect of special relativity. The usual simple rules of special relativity are valid only for observers at rest in some *inertial frame,* that is, in a *nonaccelerating* frame. In particular, an accelerating observer cannot use time dilation alone to understand the rate of a nonaccelerating clock. Clearly at least one of the twins must have accelerated, if they move apart and then reunite. And acceleration is not relative! They can both agree who is accelerating and who is not accelerating.[3] They can carry simple accelerometers, for example, each consisting of a mass attached to one end of a spring, as in Fig. 7.10. A twin holds the other end of the spring and watches what happens. If the spring never stretches or compresses at any time during the story, that twin has remained inertial throughout. On the other hand, if one of the twins finds that her/his spring stretches or compresses, that means that she/he is accelerating.[4]

During the period of acceleration, the accelerating twin cannot understand the rate of the other twin's clock using only the usual time-dilation formula. It is *Al* in the story

3. Subtleties can occur if either experiences a gravitational force. So we assume here that neither experiences any real gravity during the journey.

4. Strictly speaking, each will want to carry *three* spring–mass systems oriented in orthogonal directions, to detect acceleration in any direction.

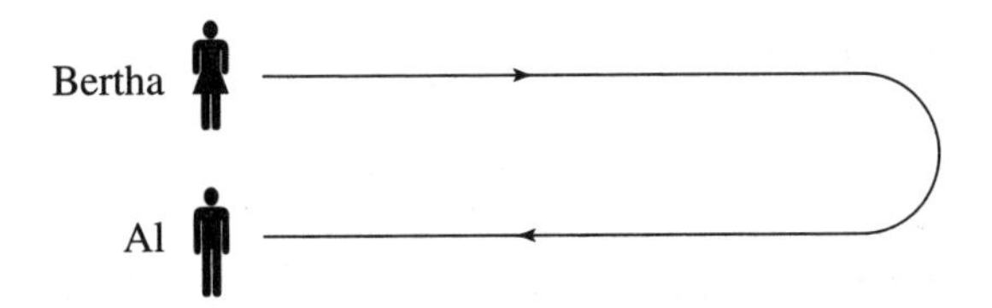

FIGURE 7.9
Bertha travels away and returns younger than her twin Al who stayed at home.

FIGURE 7.10
An accelerometer.

who did not accelerate, so he can use the formula throughout Bertha's trip. He reasons (correctly) that Bertha will be younger at the end, and so will she be.

Bertha cannot use the usual rules alone, so calculations in her frame are harder to do. Appendix D includes much more on the twin paradox—for example, it shows how the story plays out in Bertha's frame, and why from *her* point of view she is younger than Al when they get back together.

Want to destroy relativity? Then construct a genuine logical contradiction in which careful application of the rules fails to resolve the paradox. If you succeed, you will become famous! If you fail, you will succeed anyway, because you will either understand relativity more deeply or you will discover an interesting property of nature. See Appendix E for an example of this kind.

Sample Problem

We are standing on the African savanna when we observe a rhinoceros running at relativistic speed v with clocks attached to its horn and tail, as shown below. We also observe a tickbird, which lives on the rhino's back, flying from the rhino's horn toward its tail at $-v$ relative to the rhino, with a tiny clock around its neck. The velocities of rhino and tickbird are in opposite directions, so the tickbird is at rest relative to us. *We know that the rhino's clocks run slow from our point of view, and that the tickbird clock runs slow from the rhino's point of view, so we reason that the bird's clock should run even slower from our point of view.* However, the bird is at *rest* relative to us, so its clock should run at the same rate as our clock! Which is right? Does the tickbird's clock run slow compared with our clocks, or does it run at the same rate?

Solution: The answer has to be that the bird's clock runs at the same rate as our own, because is it at rest in our frame of reference. The interesting puzzle is to find the flaw in the argument that "the rhino's clocks run slow from our point of view, and the bird's clock runs slow from the rhino's point of view, so the bird's clock should run even slower from our point of view." Let us draw two pictures in our reference frame, one when the bird starts flying from the rhino's horn, and the other when it reaches the tail. Pictured are four clocks: ours, the bird's, and two of the rhino's, one at the horn and the other at the tail. The first three of these clocks read $t = 0$ in the first picture, when they are all essentially at the same location. The clock on the rhino's tail reads vD/c^2 by the "leading-clocks-lag rule," where D is the rest length of the rhino.

The second picture shows when the rhino's *tail* reaches our clock. The tickbird is there also, because it has not moved relative to us. By the time of the second picture, our clock has advanced by $\Delta t = \text{distance/speed} = D\sqrt{1 - v^2/c^2}/v$, since the distance the rhino has moved in our frame is its Lorentz contracted length.

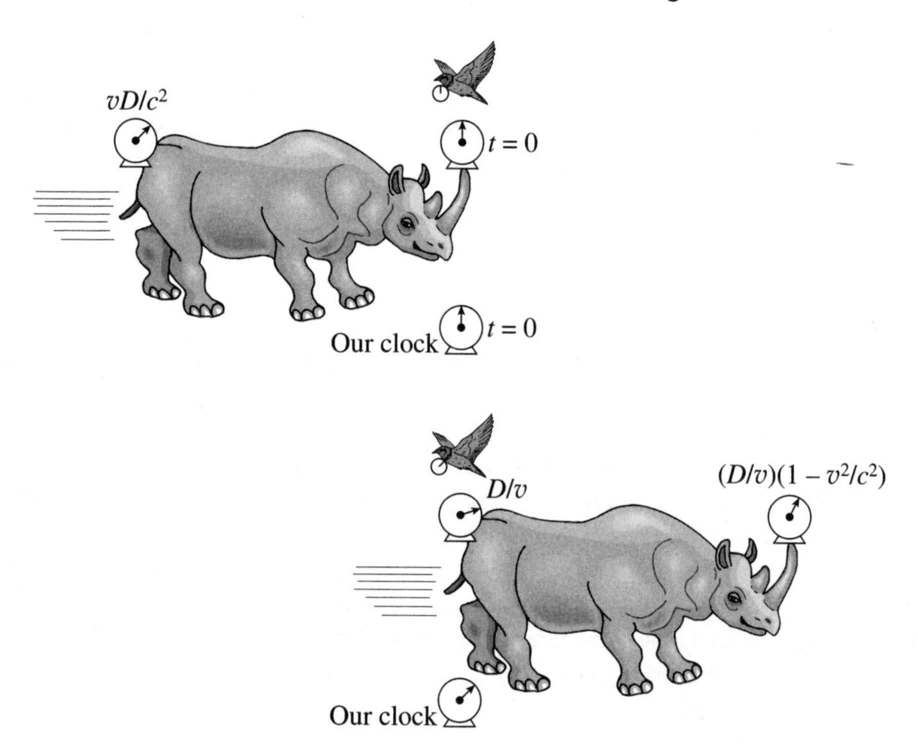

The tickbird's clock has advanced by this same amount, since it is at rest in our frame. The rhino's clocks have run slow, however, so each of them advances by the lesser time

$$\left(D\sqrt{1 - v^2/c^2}/v\right)\sqrt{1 - v^2/c^2} = (D/v)(1 - v^2/c^2).$$

The horn clock therefore reads $(D/v)(1 - v^2/c^2)$ in the second picture, while the tail clock reads the later time $vD/c^2 + (D/v)(1 - v^2/c^2) = D/v$, since it had a head start of vD/c^2 in the first picture.

In our frame, the clock on the rhino's tail had a head start, so (when it meets the tickbird) it will read a larger time than the clock on the bird, even though the rhino's clock ran slow compared with the bird's clock. We can understand why the rhino thinks it was the bird clock that ran slow, since the rhino's tail-clock reads D/v while the bird's clock reads only $(D/v)\sqrt{1 - v^2/c^2}$ when the clocks pass one another. So now we can see the fallacy in the argument presented, according to observers in our reference frame. It is *not* a contradiction for (*i*) the rhino's clocks to run slow from our point of view, and (*ii*) the bird's clock to run slow from the rhino's point of view, while (*iii*) the bird's clock runs *at the same rate* as our clock. For the rhino to know that the bird's clock is running slow in its frame, the rhino requires *two* clocks, and these clocks are not synchronized in our frame. (We can also survey the situation in the rhino's reference frame: See Problem 7–9.)

Problems

7–1. Identical twins A and B are separated at birth, A staying home and B traveling at constant speed $(12/13)c$ toward a distant star. Upon arrival, B quickly turns around and travels back home at the same speed. When B arrives home, she finds that she is 32 years younger than A. (a) How old is each twin when B arrives back home? (b) How far away is the star in A's frame? (c) Since A is moving from B's point of view, why can't B argue that A should be younger than B when they get back together?

7–2. Al and Bertha are identical twins! When she is 18 yrs old, Bertha travels to a distant star at constant speed $(24/25)c$, turns quickly around, and returns at the same speed. When she arrives home she is 25 yrs old. (a) How old is Al when she returns? (b) How far away was the star in Al's frame?

7–3. A radical rhino named Robespierre (R) is being sent to a special double-guillotine, which has two blades a horizontal distance 8 ft apart. The two blades descend simultaneously in the frame of the ground. R is 10 ft long, including an 8 ft body and a 2 ft head, counting his horn, so if R is placed in the guillotine so that one blade just misses his tail, the other will neatly behead him. Meanwhile R's compatriots conspire to place him on a cart and push him at speed $(3/5)c$, reasoning that in the ground frame R's length will then be only $4/5(10 \text{ ft}) = 8$ ft, so if one blade just misses his tail, the other will just miss his horn. R himself is worried, however, because he reasons that from his point of view he is 10 ft long and the distance between the two blades will be only $4/5(8\text{ ft}) = 6.4$ ft, so that he will not escape execution. Who is right, R or his compatriots? Then be sure your decision can be explained satisfactorily in both R's frame and the ground frame.

7–4. Two pieces of wood are cut from a board; the first piece has a projection of length D, and the second piece has a hole of depth $D + \epsilon$, where ϵ is small, so the two pieces nearly fit together. A bug of a size less than ϵ is able to live between the two pieces. Now the first piece is brought at a relativistic velocity toward the second piece. The projection is Lorentz contracted, so it will not fill the hole, leaving plenty of space for the bug, as seen in the frame of the second piece. In the frame of the first piece, however, the projection has its full length D while the hole is Lorentz contracted. Therefore when the pieces come together, as seen by the first piece there will not be enough room for the small bug, and it gets squashed. Which is right? Is the bug squashed or not?

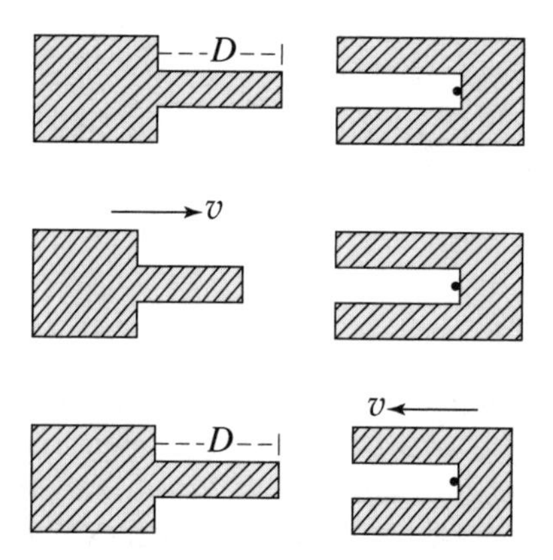

7–5. In our frame of reference, a meterstick is moving in a straight horizontal line parallel to the orientation of the stick itself. A metal plate with a 1-meter-diameter circular hole in it is rising vertically, perpendicular to the stick, as shown below. Suppose that both the meterstick and plate have negligible thickness, and that at some instant the center of the meterstick is projected to coincide with the center of the hole. The meterstick is moving so fast in our frame that it has the Lorentz-contracted length of 10 cm, and so should easily fit through the 1-meter hole in the rising plate, and so should later be found beneath the plate. In the rest frame of the stick, however, the stick is 1 meter long, while the plate is moving so rapidly in the opposite horizontal direction that the hole is Lorentz contracted to 10 cm. Therefore the one-meter stick should not be able to fit through the 10 cm hole, so cannot later be found beneath the plate. Does the stick get through the hole or not? Explain in both reference frames. (See Martin Gardner's "April Fools" jokes in *Scientific American,* April 1, 1975.)

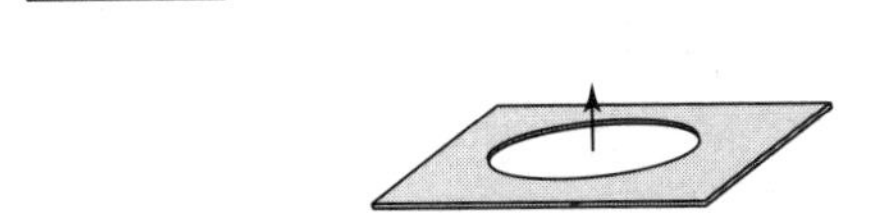

7–6. We invent a new way to synchronize clocks. Two clocks, each with a button on top, are placed on a table some distance apart. Each starts at zero when its button is pressed. A long stick, oriented horizontally, is held at rest above the two clocks and then released. The stick falls and hits both buttons, starting both clocks. Since the falling stick remained horizontal, the buttons are pressed simultaneously, so the clocks are synchronized in their rest frame. Now view the procedure in a frame moving from left to right at speed V. Wouldn't the buttons also be pressed simultaneously in this frame as the stick falls, so wouldn't the clocks be synchronized in the new frame as well as the frame of the table? So doesn't this violate the leading-clocks-lag rule, since the clock on the right is leading the clock on the left in this frame? Are the clocks synchronized or not in this frame?

7–7. A 1-meter hole is cut out of a very thin tabletop, as shown. A very thin meterstick moves at a relativistic velocity along the table toward the hole. In the frame of the table the meterstick is Lorentz contracted, so will easily fall through the hole. In the frame of the stick it is the hole that is Lorentz contracted, so the hole is not wide enough to accept the stick. Does the meterstick fall through or does it not fall through, and how is this understood in both frames? (Note that this problem has similarities to Problem 7–5, but its resolution is quite different!)

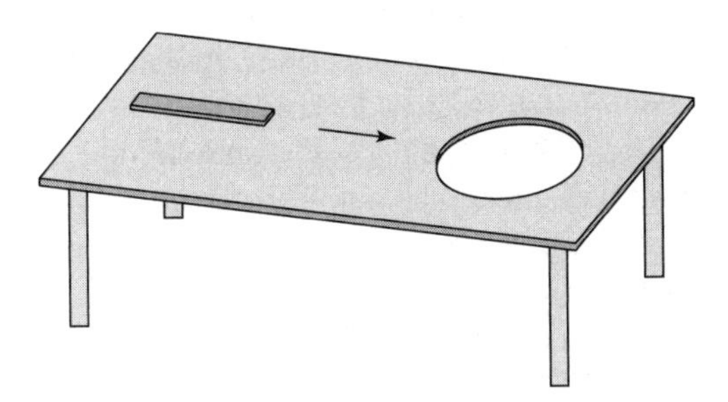

7–8. Resolve the "pole vaulter and barn" paradox. That is, show that either the barn door ends up on the ground, and explain it in both frames, or else it does not end up on the ground, and explain that in both frames. *Hint:* The rear of the pole cannot react instantaneously when the front of the pole strikes the rear of the barn. As you resolve the paradox, it is also helpful to picture clocks attached to the front and rear of the pole, which have been synchronized with one another in the original rest frame of the pole. Then construct the sequence of events in each frame.

7–9. Returning to the scenario of the Sample Problem, draw two pictures of the rhino and tickbird, their clocks and our clock, from the perspective of observers at rest in the rhino's frame of reference. Show that our clock and the tickbird's clock run at the same rate, even though the bird's clock runs slow to the rhino, and the rhino's clocks run slow to us.

7–10. The Train in the Tunnel: A Paradox (Thanks to D. C. Petersen)
Cast of characters:

a relativistic commuter train, rest length 300 m, traveling at $(4/5)c$;
a dark tunnel, rest length 400 m;
the train's diligent crew;
some nefarious saboteurs

The situation: The train must pass through the tunnel. The saboteurs have decided to blow up the train in the tunnel. They put photodetectors on the top of the train at the front and the back. When either photodetector is in darkness, it sends a laser beam to a third detector on the top of the train, located exactly midway between the front and back of the train. If this detector sees both laser beams simultaneously it sends a signal to detonate a bomb on the train.

The paradox: As the train is hurtling down the track at $(4/5)c$ toward the tunnel, the engineer is informed of the plan in detail. Not knowing any relativity, he foresees both ends of the train are soon to be in the dark at the same time and panics. "Stop the train!"

His fireman, who has been reading a little relativity during breaks between shoveling coal into the firebox, is even more worried. "It's even worse than you think," he tells the engineer. "We're moving at $(4/5)c$, so our train is only $(300\text{ m})\sqrt{1-(4/5)^2} = 180\text{ m}$ long. Clearly, we'll perish! Put on the brakes!"

The brakeman, being somewhat lazy and not wanting to slow the train, has also been reading his relativity. "Don't worry," he says. "When we go through the tunnel we'll see that the tunnel, which is, after all, moving relative to us, is only $(400\text{ m})\sqrt{1-(4/5)^2} = 240\text{ m}$ long. By the time the tail end of the train gets to the tunnel, the front will already be out. The detector won't see both laser beams simultaneously and won't detonate the bomb. In fact, the faster we go, the safer we'll be. More coal!"

The question: Does the train blow up or not?

CHAPTER 8

The Lorentz Transformation

LET US NOW explore systematically how the coordinates of an event in one reference frame are related to the coordinates of the same event in a different reference frame. In particular, we set up two systems of coordinates that have x axes in common, and are moving with relative velocity V in the x direction. Using the same conventions introduced in Chapter 1, we call these two sets of coordinates S and S', with S' moving to the right, as shown in Fig. 8.1.

By now it is obvious that the Galilean transformation (1.3) of Chapter 1,

$$\begin{aligned} x' &= x - Vt \\ y' &= y \\ z' &= z \\ t' &= t \end{aligned} \tag{8.1}$$

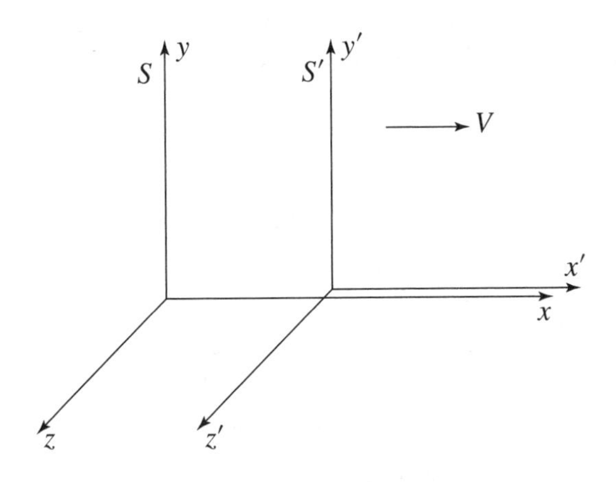

FIGURE 8.1
Frame S' moves to the right at speed V relative to frame S.

cannot be correct. The equations predict that lengths and time intervals do not depend on the observer's frame of reference, and that two events that are simultaneous in one frame are simultaneous in all frames. Also the Galilean velocity transformation $v'_x = v_x - V$ predicts that light travels at different speeds in different frames, because we can take the special case of a light wave traveling with $v_x = c$. Can we find an alternative transformation that *is* compatible with length contraction, time dilation, and all the other results obtained so far?

8.1 Derivation of the Lorentz Transformation

The correct transformation is called the Lorentz transformation, named for the Dutch physicist H. A. Lorentz mentioned in Sample Problem 3 of Chapter 2. We can derive it from our knowledge of time dilation, length contraction, and the leading-clocks-lag rule. As in the Galilean transformation, we set the clocks at the origins of the S and S' frames both equal to zero just as the origins coincide. Sometime later, an event takes place (say, a flashbulb goes off) at a point with coordinates (x, y, z, t) in S and (x', y', z', t') in S', as shown in Fig. 8.2.

We will first try to find x in terms of x'. From the diagram, we see that the origins of the two frames are separated by a distance Vt according to observers in S. Also to observers in S, the distance x' will be contracted by the factor $\sqrt{1 - V^2/c^2}$. That is, observers in S' say the x distance from their origin to the event is x', but observers in S measure it to be only $x'\sqrt{1 - V^2/c^2}$. So we then have $x = Vt + x'\sqrt{1 - V^2/c^2}$. Solving for x',

$$x' = \frac{x - Vt}{\sqrt{1 - V^2/c^2}}. \tag{8.2}$$

This is the same as the Galilean transformation for x', except for the denominator $\sqrt{1 - V^2/c^2}$, which is almost unity for frame velocities $V \ll c$.

There is no contraction of the y and z components, because they are perpendicular to the direction of motion, so the next two equations of the Lorentz transformation are $y' = y$ and $z' = z$, just as in the Galilean transformation.

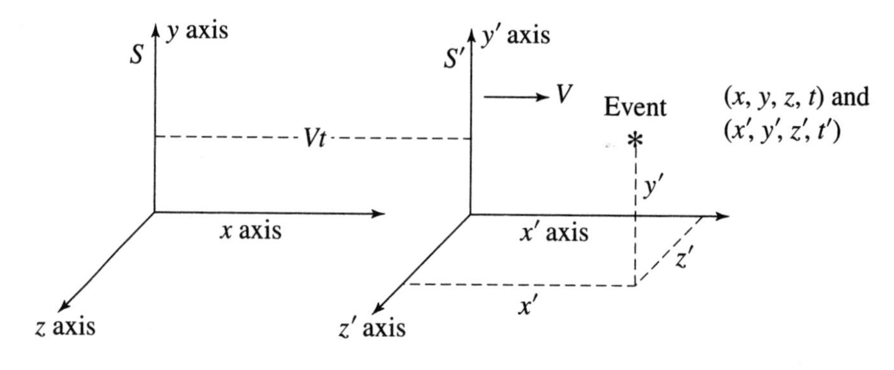

FIGURE 8.2
An event with coordinates (x, y, z, t) in S and (x', y', z', t) in S'.

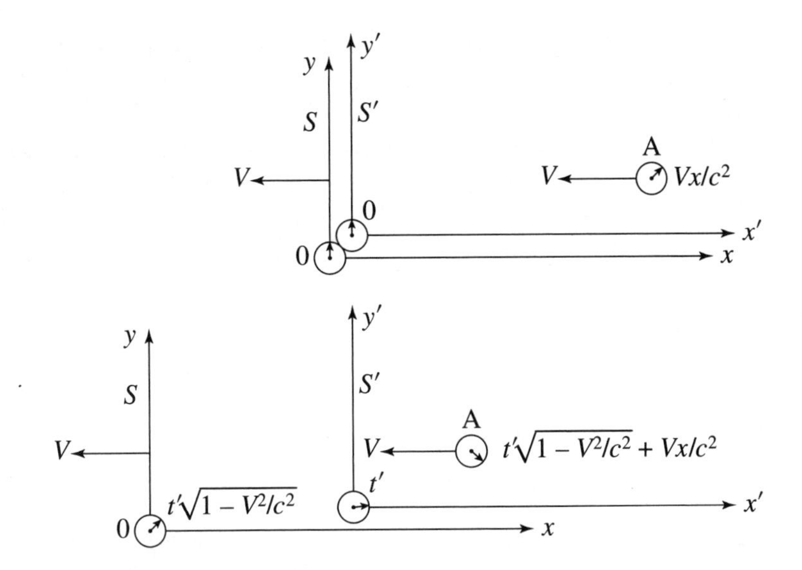

FIGURE 8.3
Clock times from the S' frame, (a) when the origins of S and S' coincide, (b) at some arbitrary later time t'.

Finally we come to the time transformation, which we expect to be more interesting than the old $t' = t$ equation. We can get this equation by placing a clock A at rest at some arbitrary location (x, y, z) in the S frame, and observing it from the S' frame, in which the clock and entire S frame are moving to the left, as shown in Fig. 8.3(a). As usual, when the origins of S and S' coincide, clocks placed at the two origins are set to $t' = t = 0$, as shown. Since the clock at the origin (which reads zero and is *leading* clock A) *lags* clock A by Vx/c^2, clock A reads $t = Vx/c^2$ in this picture. Later on, when clocks at rest in S' all read some time t', frame S has moved to the left, as shown in Fig. 8.3(b). The unprimed clock at $x = 0$ has run slow, so reads only $t = t'\sqrt{1 - V^2/c^2}$, while clock A has been running slow as well, yet had a "head start" by Vx/c^2, so reads $t = t'\sqrt{1 - V^2/c^2} + Vx/c^2$ in this picture. That is, for clock A, placed at an arbitrary position (x, y, z), we can solve for t' to find

$$t' = \frac{t - Vx/c^2}{\sqrt{1 - V^2/c^2}}. \tag{8.3}$$

As expected, the time transformation *is* more interesting than the nonrelativistic Galilean result $t' = t$. Collecting all our results, we have the complete Lorentz transformation

$$\begin{aligned} x' &= \gamma(x - Vt) \\ y' &= y \\ z' &= z \\ t' &= \gamma(t - Vx/c^2), \end{aligned} \tag{8.4}$$

where $\gamma \equiv 1/\sqrt{1 - V^2/c^2}$. Note that these equations reduce approximately to the Galilean transformation for small relative frame velocities; that is, when V/c approaches zero.

The Lorentz transformation of Eqs. 8.4 is useful if we know the coordinates x, y, z, and t in the unprimed frame, and want to find the corresponding coordinates x', y', z', and t' in the primed frame. We can invert the equations to solve algebraically for x, y, z, and t in terms of x', y', z', and t', which gives the *inverse* Lorentz transformation

$$\begin{aligned} x &= \gamma(x' + Vt') \\ y &= y' \\ z &= z' \\ t &= \gamma(t' + Vx'/c^2). \end{aligned} \tag{8.5}$$

Note that the inverse equations can be found by simply switching the primed and unprimed coordinates, and replacing V with $-V$. The minus sign is not surprising: The primed frame moves to the *right* as seen from the unprimed frame, while the unprimed frame moves to the *left* as seen from the primed frame.

8.2 Time Dilation, Length Contraction, and Leading Clocks

Now we will unwind the Lorentz transformation to make sure it agrees with what we know. First, do moving clocks run slow? Set a clock at the origin of the S' frame, which reads $t' = 0$ as it passes the origin of the S frame. We will stand in the S frame and see what the moving clock reads some time t later, as illustrated in Fig. 8.4.

The clock has moved a distance $x = Vt$, and (since it stays at rest at the S' origin) x' remains zero. So from the final Lorentz transformation Eq. 8.4, we have

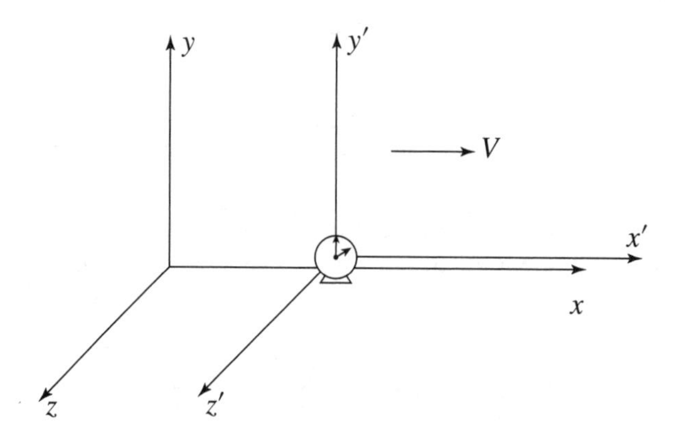

FIGURE 8.4
A clock that remains at rest at the origin of the primed frame.

$$
\begin{aligned}
t' &= \gamma(t - Vx/c^2) = \gamma(t - (V/c^2)Vt) \\
&= \gamma t(1 - V^2/c^2) = t\sqrt{1 - V^2/c^2}
\end{aligned}
\tag{8.6}
$$

since $\gamma = (1 - V^2/c^2)^{-1/2}$. The moving clock does indeed run slow when measured in our frame, by the expected factor.

Now let's see what happens to the length of a typical passing object. A rhinoceros is at rest in frame S', with its head at x'_1 and its tail at x'_2, so its rest length is $D = x'_1 - x'_2$. It moves at speed V with respect to the unprimed S frame, as shown in Fig. 8.5. With synchronized watches two of us stand at rest in the S' frame, one at x_1 and the other at x_2, which are the positions where the rhino's head and tail will be when our watches both read the same time $t_1 = t_2$.

From the first of the Lorentz-transformation equations, we find that

$$
D = x'_1 - x'_2 = \frac{x_1 - Vt_1}{\sqrt{1 - V^2/c^2}} - \frac{x_2 - Vt_2}{\sqrt{1 - V^2/c^2}} = \frac{x_1 - x_2}{\sqrt{1 - V^2/c^2}}, \tag{8.7}
$$

using the fact that $t_1 = t_2$, since we measure the position of its head and tail at the same time in our frame. Thus the rhino's length in our frame is $x_1 - x_2 = D\sqrt{1 - V^2/c^2}$, contracted in our frame of reference by the usual factor. This is no surprise, because we used length contraction in deriving the Lorentz transformation in the first place.

Finally, we need to see if the Lorentz transformation contains the leading-clocks-lag rule. Set two clocks at rest in the S' frame, one at x'_1 and the other at x'_2. The rest distance between the clocks is $D = x'_1 - x'_2$. Both move at speed V relative to us in frame S, as shown in Fig. 8.6. Using the final Lorentz transformation equation twice, once for each moving clock, we have

$$
t'_1 - t'_2 = \frac{(t_1 - t_2) - (V/c^2)(x_1 - x_2)}{\sqrt{1 - V^2/c^2}}. \tag{8.8}
$$

If we, at rest in the unprimed frame, make *simultaneous* measurements of the times read by the moving clocks (that is, $t_1 = t_2$), then the difference in the times read by the two moving clocks is

$$
t'_1 - t'_2 = -\left(\frac{V}{c^2}\right)\frac{(x_1 - x_2)}{\sqrt{1 - V^2/c^2}}. \tag{8.9}
$$

FIGURE 8.5
The length of a moving object is contracted in our frame.

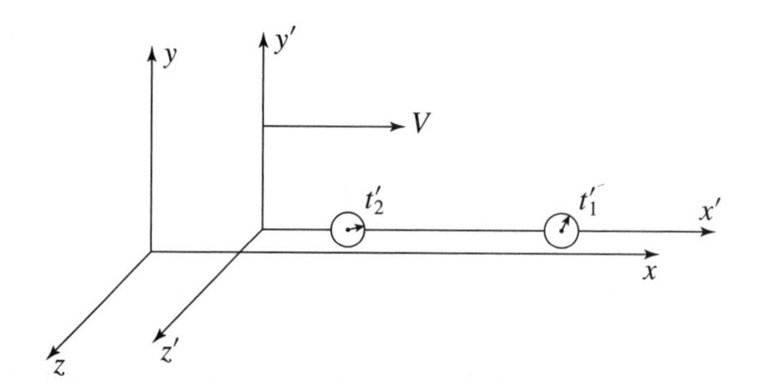

FIGURE 8.6
Two clocks at rest in the primed frame, moving to our right at speed V.

The distance between the two clocks from our point of view is $x_1 - x_2 = D\sqrt{1 - V^2/c^2}$, where D is the rest distance between the clocks; therefore

$$t_1' - t_2' = -\left(\frac{V}{c^2}\right) D, \tag{8.10}$$

which is just the leading-clocks-lag rule. The moving clock in front reads an earlier time than the clock in the rear, by the amount VD/c^2. *The Lorentz transformation therefore contains all three of the rules we have discovered in the preceding chapters.*

We now have *two* methods to translate times and lengths between one frame and another. On the one hand, we can use the three physical rules governing the behavior of clocks and metersticks. On the other hand, we can use the Lorentz transformation. Sometimes one approach is easier, and sometimes the other. It is sometimes useful to use *both* methods! The physical rules usually give a more intuitive, physical insight into why things happen as they do. The Lorentz transformation is often easier for calculations as long as we really understand what all the symbols mean, and it has a formal power that we will find especially useful in upcoming chapters.

8.3 The Velocity Transformation

Given the coordinates of an event in one reference frame, the Lorentz transformation tells us what the coordinates of the same event are in a different reference frame. As a particle moves, it traces out a *sequence* of events; this sequence of events is related to the velocity of the particle. We can use the Lorentz transformation to relate the velocity of the particle in one frame to its velocity in another frame.

The velocity components of a particle are $(dx/dt, dy/dt, dz/dt)$ in the unprimed frame and $(dx'/dt', dy'/dt', dz'/dt')$ in the primed frame. We can relate these components by differentiating the Lorentz transformation equations. For example, using the chain rule of differential calculus,

$$\frac{dx'}{dt'} = \frac{dx'}{dt}\frac{dt}{dt'} = \frac{dx'/dt}{dt'/dt}. \tag{8.11a}$$

We can then evaluate the derivatives in the numerator and denominator of Eq. 8.11a using the Lorentz transformation equations $x' = \gamma(x - Vt)$ and $t' = \gamma(t - Vx/c^2)$. This gives

$$\frac{dx'}{dt'} = \frac{\gamma\,(dx/dt - V)}{\gamma\,\big(1 - (V/c^2)\,dx/dt\big)}, \tag{8.11b}$$

so the velocity transformation for the x component of the velocity is

$$v'_x = \frac{v_x - V}{1 - Vv_x/c^2}. \tag{8.12}$$

Similarly,

$$\frac{dy'}{dt'} = \frac{dy'/dt}{dt'/dt} = \frac{dy/dt}{\gamma\,\big(1 - (V/c^2)\,dx/dt\big)}, \tag{8.13a}$$

so

$$v'_y = \frac{v_y\sqrt{1 - V^2/c^2}}{1 - Vv_x/c^2}, \tag{8.13b}$$

with a similar formula for v'_z. Collecting the results, we have found that

$$v'_x = \frac{v_x - V}{1 - Vv_x/c^2}, \quad v'_y = \frac{v_y\sqrt{1 - V^2/c^2}}{1 - Vv_x/c^2}, \quad v'_z = \frac{v_z\sqrt{1 - V^2/c^2}}{1 - Vv_x/c^2}. \tag{8.14}$$

Note two interesting features of these equations: (i) It is the x *component* of velocity v_x that appears in the denominator of all three expressions, including the transformations for v_y and v_z. (ii) If the velocity of light were infinite (that is, if $c \to \infty$), these expressions would reduce to $v'_x = v_x - V$, $v'_y = v_y$, $v'_z = v_z$, the nonrelativistic (Galilean) velocity transformation equations. This illustrates the fact that Einstein's relativity would be unnecessary if the speed of light were infinite.

We can find the *inverse* velocity transformation by inverting Eqs. 8.14,

$$v_x = \frac{v'_x + V}{1 + Vv'_x/c^2}, \quad v_y = \frac{v'_y\sqrt{1 - V^2/c^2}}{1 + Vv'_x/c^2}, \quad v_z = \frac{v'_z\sqrt{1 - V^2/c^2}}{1 + Vv'_x/c^2}, \tag{8.15}$$

which simply interchange primed with unprimed velocity components and change the sign of V. If we know the velocity of a particle in the primed frame, these equations tell us the velocity of the particle in the unprimed frame.

Let us explore the velocity transformation in some special cases. We had first better make sure that *light* has the *same* speed in every frame. Take a light beam moving in the x direction, for example, with $v_x = c$, $v_y = v_z = 0$. Then

$$v'_x = \frac{c - V}{1 - Vc/c^2} = c, \quad v'_y = 0, \quad v'_z = 0. \tag{8.16}$$

So far so good. The beam travels in the x direction in the primed frame also, with the same velocity c. Next, take a beam moving in the y direction, with $v_y = c$, $v_x = v_z = 0$. Then

$$\begin{aligned} v'_x &= \frac{-V}{1-0} = V, \\ v'_y &= \frac{c\sqrt{1 - V^2/c^2}}{1-0} = c\sqrt{1 - V^2/c^2}, \quad v'_z = 0. \end{aligned} \tag{8.17}$$

The beam does *not* travel purely in the y' direction, which is not surprising, since the primed and unprimed frames are moving in the x direction relative to one another. This is just the aberration of light effect discussed in Chapter 2, similar to watching snowflakes fall toward the ground from inside a moving car. The velocity of the beam is still c, however, because

$$v' = \sqrt{(v'_x)^2 + (v'_y)^2 + (v'_z)^2} = \sqrt{V^2 + c^2(1 - V^2/c^2)} = c, \tag{8.18}$$

as illustrated in Fig. 8.7.

As another example, suppose a supersonic aircraft moves in the x direction at velocity 2000 m/s with respect to the ground. Another moves in the same direction at velocity 500 m/s. In nonrelativistic physics, you could confidently predict that the first aircraft moves at velocity 2000 m/s – 500 m/s = 1500 m/s from the second's point of view, a direct result of the Galilean transformation of velocities. In reality, if they could

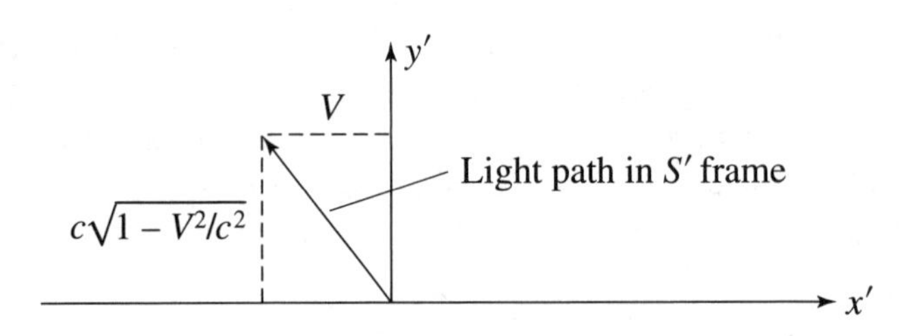

FIGURE 8.7
Light moving along the y axis with speed c has a tilted path in the primed frame. The speed is still c, however.

measure with arbitrarily high precision, people in the second aircraft would find that the first aircraft's speed is

$$v'_x = \frac{v_x - V}{1 - Vv_x/c^2} = \frac{2000 - 500}{1 - (500)(2000)/(3 \times 10^8)^2} \text{ m/s} = 1500.000000017 \text{ m/s}. \quad (8.19)$$

The *numerator* of this expression is the nonrelativistic prediction; realistic measurements made in the second aircraft are obviously unlikely to detect the effect of the *denominator* in this case.

The most interesting application of the velocity transformation is to show that no object can travel faster than the speed of light. Actually we should qualify this, and say that if a particle is traveling with some speed $v < c$ in a particular reference frame, it will travel with $v' < c$ in any other reference frame as well. (Of course, our second reference frame must *not* travel with speed $V > c$ relative to the first, because then the famous factor $\sqrt{1 - V^2/c^2}$ would become imaginary, which makes no sense. Lengths and times, instead of being contracted and dilated, would become imaginary.)

As a particular example, picture two rhinoceroses destined for a head-on collision at unusually high velocities, as shown in Fig. 8.8. One rhino (A) approaches from the left at $v = 0.99\,c$ and another (B) approaches from the right with $v = -0.99\,c$, the minus sign referring to the fact that B is running to the left. We ask the question: "What is the velocity *difference* between A and B?" The answer is simple in the old Galilean relativity. The difference in their velocity would be $0.99\,c + 0.99\,c = 1.98\,c$, regardless of the reference frame in which the events are viewed. But in Einsteinean relativity, we have to be careful to specify the observer's frame of reference. If we stand on the ground (as in the top picture of Fig. 8.8) with the beasts approaching one another, each at $0.99\,c$, the difference in velocities is $1.98\,c$, *by definition.* Of course, there is no *object* moving at this velocity. But now take the point of view of one of

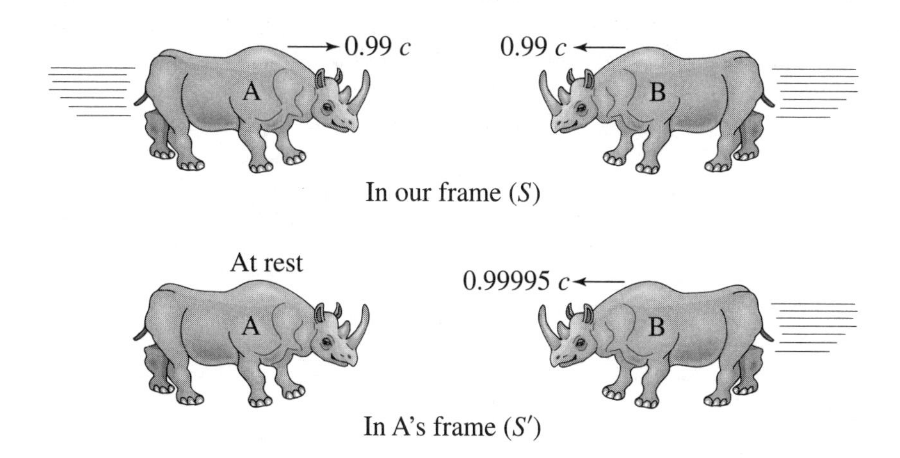

FIGURE 8.8
Rhinos in the ground frame and in the frame of the left-hand rhino. Lorentz contractions are not shown!

the two rhinos. In its frame of reference, the difference in their velocities is just the velocity of the other rhino.

Let the ground frame be the S frame, in which rhino A runs to the right with $v_x = 0.99\,c$, and B runs to the left with $v_x = -0.99\,c$. Choose the S' frame to be moving with rhinoceros A to the right: that is, A is at rest in S'. Therefore the velocity of B with respect to A will be simply B's velocity in the S' frame, which is v_x'. So, since the relative velocity between frames S and S' is $V = 0.99\,c$, we have

$$v_x' = \frac{v_x - V}{1 - Vv_x/c^2} = \frac{-0.99\,c - 0.99\,c}{1 - (0.99\,c)(-0.99\,c)/c^2} = -(0.999949\ldots)c. \quad (8.20)$$

In A's frame, rhino B is running with velocity $v_x' = -(0.999949\ldots)c$, negative because B is moving toward the left. The result is a difference in velocity that is *less* that c. This is illustrated in Fig. 8.8, showing again that velocities do not simply add up in relativity. To translate the velocity of an object from one frame to another it is necessary to use the Lorentz velocity transformation.

The Lorentz transformation summarizes all the results of special relativity arrived at so far. It contains within it the rules for how positions, times, velocities, and accelerations change from one inertial frame to another. In addition, an interesting property of the transformation led one of Einstein's former professors to suggest an entirely new way to look at things. That professor was Hermann Minkowski, and the new way to look at things is the subject of the next chapter.

Sample Problems

1. A primed frame moves at velocity $V = (3/5)c$ relative to our unprimed frame along their mutual x axes, and the frame origins coincide when both our clock and the primed clock at the origin read time = *zero*. A laser emits a pulse at location $(x, y, z) = (40.0\text{ m}, 0, 0)$ in our frame, at time $t = 2.0 \times 10^{-7}$ s. When and where is the pulse emitted according to observers in the primed frame?

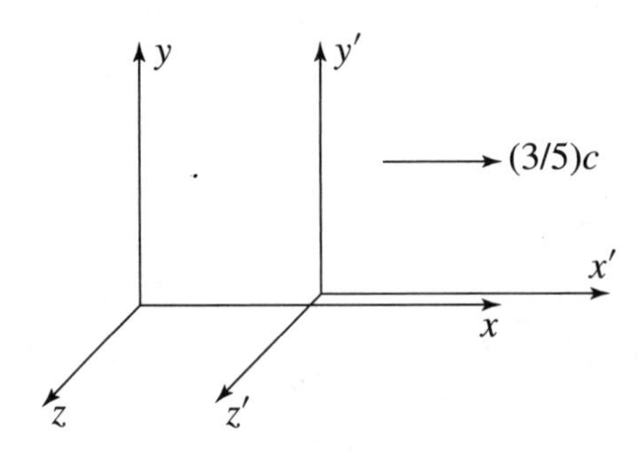

Solution: We can use the Lorentz transformation Eqs. 8.4, since the relative motion is along mutual x axes, and the origins of the two frames coincide when the origin clocks read $t = t' = 0$. The locations are on the x' axis, since $y' = y = 0$

and $z' = z = 0$, and the γ factor in the equations is $\gamma = 1/\sqrt{1 - V^2/c^2} = 5/4$. Therefore the location in the primed frame is

$$x' = \gamma(x - Vt) = \frac{5}{4}\left(40.0\,\text{m} - \frac{3}{5}(3.0 \times 10^8\,\text{m/s})(2.0 \times 10^{-7}\,\text{s})\right)$$
$$= \frac{5}{4}(40.0\,\text{m} - 36.0\,\text{m}) = 5\,\text{m}.$$

The time in the primed frame is

$$t' = \gamma(t - Vx/c^2) = \frac{5}{4}\left(2.0 \times 10^{-7}\,\text{s} - \frac{3}{5c}(40.0\,\text{m})\right)$$
$$= \frac{5}{4}(2.0 \times 10^{-7}\,\text{s} - 0.8 \times 10^{-7}\,\text{s}) = 1.5 \times 10^{-7}\,\text{s}.$$

In summary, the location and time of the pulse as measured by observers in the primed frame are $(x', y', z') = (5\,\text{m}, 0, 0)$ and $t' = 1.5 \times 10^{-7}$ s.

2. A bus of rest length 15.0 m is barreling along the interstate at $V = (4/5)c$. The driver in the front and a passenger in the rear have synchronized their watches. Parked along the road are several Highway Patrol cars, with synchronized clocks in their mutual rest frame. Just as the rear of the bus passes one of the patrol cars, both the clock in the patrol car and the passenger's watch read exactly 12:00 noon. (a) When its clock reads 12:00 noon, a second patrol car happens to be adjacent to the *front* of the bus. How far is the second patrol car from the first? (b) What does the bus driver's watch read according to the second patrol car?

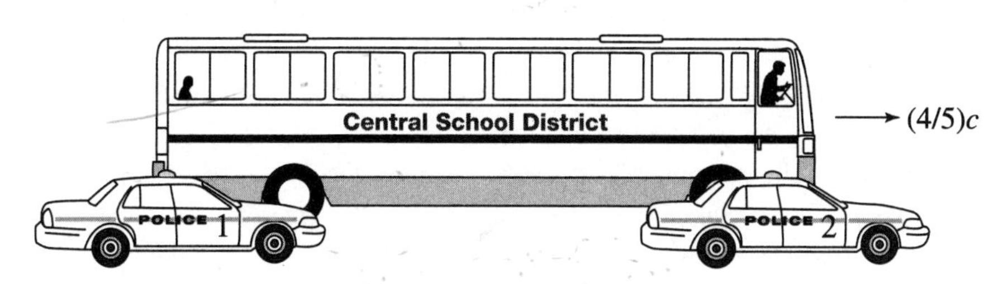

Solution: We will first answer each part using one or more of the three "rules": time dilation, length contraction, and "leading clocks lag." Then we will answer the same question using the Lorentz transformation instead. Let the origin of the unprimed (patrol car) frame be located at the rear patrol car, and the origin of the primed (bus) frame be located at the rear of the bus—this is appropriate, since the passenger's watch and the rear patrol car's clock both read 12:00 when they pass one another. (We will identify 12:00 noon with time $t = 0$.)

(a) Using the length-contraction rule, the bus will have the shorter length $15.0\,\text{m}\sqrt{1 - V^2/c^2} = 15.0\,\text{m}(3/5) = 9.0\,\text{m}$ in the patrol-car frame. Using the Lorentz transformation instead, we need to find the position x of the front of the bus (that is, in the unprimed patrol-car frame) when $t = 0$. We know that the front of the bus is located at $x' = 15.0$ m in the bus's frame. Knowing t and x' suggests that we use the Lorentz transformation equation $x' = \gamma(x - Vt)$ to solve for x. That is, $x' = 15.0\text{ m} = \gamma(x - V \cdot 0)$, so $x = 15.0\text{ m}/\gamma = 15.0\text{ m}\sqrt{1 - V^2/c^2} = 9.0$ m. The length contraction rule got us there quicker, but the Lorentz transformation also worked.

(b) Using the leading-clocks-lag rule in the patrol-car frame, the bus driver's watch lags the passenger's watch by $VD/c^2 = (4/5\,c)(15\text{ m})/c^2 = 12\text{ m}/c = 12\text{ m}/(3.0 \times 10^8\text{ m/s}) = 4.0 \times 10^{-8}$ s. The driver's watch therefore reads -4.0×10^{-8} s, again taking 12:00 noon to be $t = 0$. Using the Lorentz transformation instead, we want to find t', the driver's watch time, when $t = 0$ and $x' = 15.0$ m. An equation involving all three of these quantities is the inverse equation $t = \gamma(t' + Vx'/c^2)$, which gives $0 = \gamma(t' + (4/5c)(15.0\text{ m})/c^2)$. Therefore the driver's watch reads $t' = -(4/5c)(15.0\text{ m})/c^2 = -12\text{ m}/c = -4.0 \times 10^{-8}$ s, consistent with the leading-clocks-lag rule. Alternatively, since we know that $x = 9.0$ m from part (a), we can use $t' = \gamma(t - Vx/c^2) = \gamma(0 - (4/5c)(9.0\text{ m})/c^2)$, which gives the same result.

3. The passenger who had been sitting at the back of the speeding bus in Sample Problem 2 starts to run forward at speed $(3/5)c$ relative to the bus. How fast is the passenger moving as measured by the parked patrol cars?

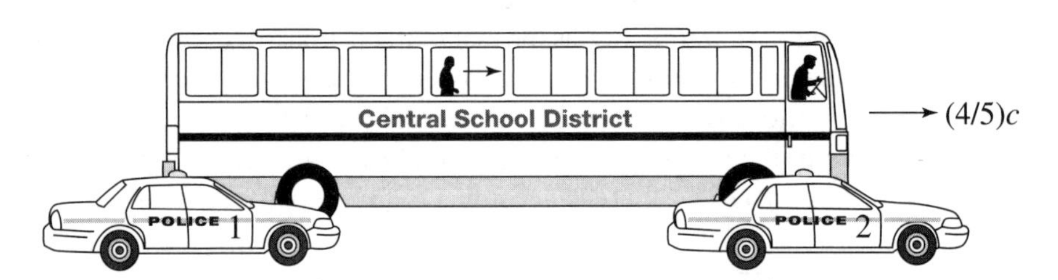

Solution: Here we use the relativistic velocity transformation. As in Sample Problem 2, the primed frame is the frame of the bus, because it is moving to the right. The unprimed frame is the frame of the patrol cars. We know that the passenger's velocity in the bus frame is $v'_x = (3/5)\,c$, and the relative velocity between the two frames is $V = (4/5)c$. The velocity of the passenger in the patrol-car frame is therefore

$$v_x = \frac{v'_x + V}{1 + Vv'_x/c^2} = \frac{(3/5)c + (4/5)c}{1 + (4/5)(3/5)} = \frac{(7/5)c}{1 + 12/25} = \frac{35}{37}c.$$

Problems

8–1. A clock moving past us at speed $v = c/2$ reads $t' = 0$ when our clocks read $t = 0$. Using the Lorentz transformation, find out what our clocks will read when the moving clock reads $t' = 10$ s.

8–2. Two clocks are at rest in the primed frame; they are both attached to the x' axis, one at $x' = 0$ and the other at $x' = 1$ m. The primed frame moves past us to the right at speed $v = (3/5)c$. The clock at $x' = 0$ reads $t' = 0$ just as it passes our origin $x = 0$ at $t = 0$. (a) Find the x coordinate (in our frame) of the other clock at $t = 0$. (b) Find the reading of the other clock at $t = 0$.

8–3. Two clocks are at rest in the primed frame; they are both attached to the x' axis, one at $x' = 0$ and the other at $x' = -2$ m. The primed frame moves past us to the right at speed $V = (4/5)c$. The clock at $x' = 0$ reads $t' = 0$ just as it passes our origin $x = 0$ at $t = 0$. (a) Find the x coordinate (in our frame) of the other clock at $t = 0$. (b) Find the reading of the other clock at $t = 0$.

8–4. A primed frame moves at $V = (4/5)c$ to the right relative to an unprimed frame. Primed-frame clocks have been synchronized with one another, as have unprimed-frame clocks. Just as the origins pass, clocks at the origins of both frames read zero, and a flashbulb explodes at that point. Later, the flash is seen by observer A at rest in the primed frame, whose position is $(x', y', z') = (1\,\text{m}, 0, 0)$. (a) What does A's clock read when A sees the flash? (b) When A sees the flash, where is A located according to unprimed observers? (c) To unprimed observers, what do their clocks read when A sees the flash?

8–5. A spaceship of rest length 100 m moving at speed $v = (3/5)c$ relative to us contains a passenger in the ship's tail. The passenger fires a bullet toward the nose of the ship, at speed $(3/5)c$ relative to the ship. (a) How fast is the bullet traveling relative to us? (b) How long is the ship in (i) our frame, (ii) the ship's frame, (iii) the bullet's frame? (c) How much time does it take the bullet to reach the nose of the ship, as measured by (i) passengers in the ship, (ii) us?

8–6. Derive Eqs. 8.5 algebraically from the Lorentz transformation Eqs. 8.4. That is, solve Eqs. 8.4 for x in terms of x' and t' and for t in terms of x' and t'.

8–7. A K^0 meson moves past us to the right at speed $v = c/2$. In the K^0 rest frame it decays into two π^0's, one going straight "up," and the other straight "down," with equal and opposite velocities v_0. (a) In terms of v_0, how fast do the π^0's move in the laboratory? (b) At what angle will they move with respect to the vertical?

8–8. Two rhinos charge one another. According to observers on the ground, the left-hand rhino moves to the right at $(3/5)c$, and the right-hand rhino moves to the left at $(4/5)c$. How fast is the right-hand rhino moving in the frame of the left-hand rhino?

8–9. Two rhinos charge one another. According to observers on the ground, the left-hand rhino moves to the right at speed $(1 - \epsilon)c$ and the right-hand rhino moves to the left at $(1-\epsilon)c$, where ϵ is in the range $0 < \epsilon < 1$. How fast is the right-hand rhino moving in the frame of the left-hand rhino? Show that this is less than c, no matter how small ϵ is.

8–10. A peculiar dart game is in progress on a spaceship of rest length 800 m. Players located in the tail and nose of the ship launch darts (with speeds $(4/5)c$ relative to the ship), and these darts land simultaneously on opposite sides of a thin target located in the middle of the ship. The ship has a velocity $(3/5)c$ according to "stationary" observers, and just as the tail of the ship passes their origin of coordinates, they note that the player in the tail of the spaceship launches his dart.

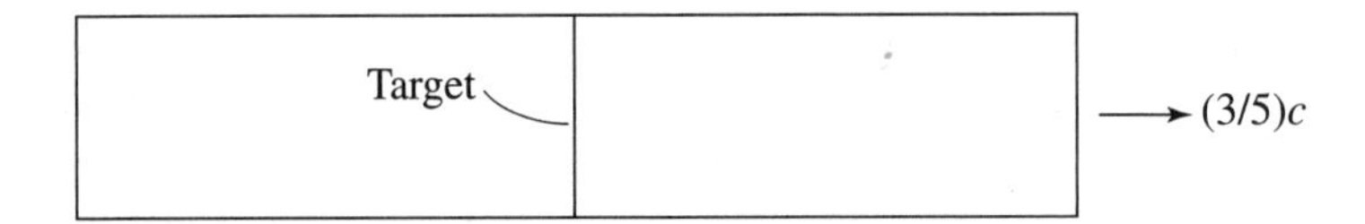

At this time, all clocks in the "stationary" system read $t = 0$. All questions below refer to the "stationary" reference system. (a) What is the velocity of the dart from the tail? (b) When is the dart from the nose thrown? (c) What is the velocity of the dart from the nose? (d) When do the darts get to the target? (e) How far does the nose dart travel in getting to the target?

8–11. A beam of light moves in the x, y plane, at angle θ to the x axis, according to observers in the unprimed frame. If a primed frame moves to the right at velocity V along the x axis, at what angle θ' to the x' axis does the beam move, according to observers in the primed frame? (This is the relativistic aberration of light effect, discussed further in Chapter 13.)

8–12. An ideal superball clock features a ball bouncing back and forth between two walls in gravity-free space. The walls are a distance D apart, and the ball moves with constant speed v_0, so the round-trip time is $T = 2D/v_0$. An identical clock in a spaceship moves past us to the right at speed V. The walls are separated by a distance D in the ship frame, and the ball moves with speed v_0 relative to the ship. From our point of view the ball moves diagonally, as shown below. (a) What is the horizontal component of the ship-ball's velocity, as measured in our frame? (b) What is the vertical component of the ship-ball's velocity, in our frame? (c) How long does it take the ship-clock to tick once (that is, for the ship-ball to bounce once back and forth?) (d) Using these results, does the ship superball clock run slow, fast, or at the same rate as our own clock, as measured in our frame? Explain. Does the result agree with time-dilation formulas or not?

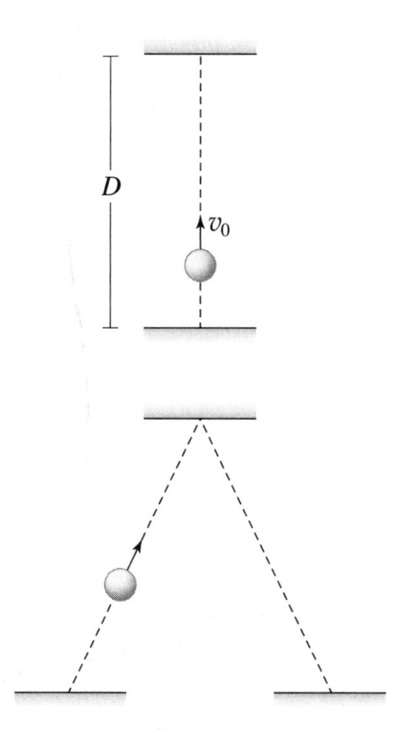

8–13. Two spaceships suddenly blast off from Earth with equal relativistic speeds v, moving at right angles to one another. What is the speed of each ship in the frame of the other?

8–14. Derive Eqs. 8.15 algebraically from Eqs. 8.14. That is, solve Eqs. 8.14 for v_x, v_y, and v_z in terms of v'_x, v'_y, and v'_z.

8–15. By differentiating the velocity transformation equations we can obtain transformation laws for acceleration. (a) Show in this way that the x components of the acceleration of an object as measured in two frames are related by

$$a'_x = a_x \frac{(1 - V^2/c^2)^{3/2}}{(1 - v_x V/c^2)^3}.$$

(b) What is the acceleration transformation in nonrelativistic physics (that is, in the limit $V << c$)? (c) What is the inverse transformation, for a_x in terms of a'_x and v'_x?

8–16. A spaceship transporting passengers bound for a planet in the Vega system leaves Earth at speed $(3/5)c$, when both its clock and the clock on Earth read $t = 0$. Twenty weeks later, according to their own clock, Earthlings realize that a saboteur may be aboard the ship, so they immediately transmit a radio message at $v = c$ toward the ship, informing the ship of their suspicion. Upon receiving the message, the ship's computer immediately discovers the saboteur lurking in one of the ship's lifeboats, so instantly fires the lifeboat at speed $(35/37)c$ *relative to the ship,* back toward Earth, returning the saboteur to stand trial. (a) Draw a series of four pictures in Earth's reference frame, illustrating the important events in the story. Then, calculating in Earth's frame, answer the following. (b) What does the ship-clock read (in weeks) when Earth sends the original message? (c) How far is the ship from Earth (in light-weeks) when the message is sent? (d) What does the Earth's clock read (in weeks) when the ship receives the message? (e) What does the ship's clock read when it receives the message? (f) How far is the ship from Earth when it receives the message? (g) How fast does the lifeboat travel in Earth's frame of reference? (h) What does Earth's clock read when the lifeboat arrives? (i) Suppose the lifeboat clock reads zero when it is launched from the ship. What time does it read when it reaches Earth? (j) How much younger is the saboteur than she would have been had she stayed on Earth instead of going off on the ship and returning in the lifeboat?

8–17. Pythagorean triples (PTs) are sets of integers (m, n, p), with $m < n < p$, such that $m^2 + n^2 = p^2$. Examples include $(m, n, p) = (3, 4, 5)$ and $(12, 35, 37)$. The particular "quantized" relativistic velocities $v/c = m/p$ or n/p are often convenient to use, because then $\sqrt{1 - v^2/c^2} = n/p$ or m/p, respectively, which are simple fractions: in particular, $\sqrt{1 - (4/5)^2} = 3/5$ and $\sqrt{1 - (12/37)^2} = 35/37$. Now transform such a PT velocity v'_x in the primed frame into an unprimed-frame velocity v_x, where the relative velocity V between the two frames is also formed from a PT. For example, if $v'_x = (4/5)c$ and $V = (3/5)c$, then

$$v_x = \frac{v'_x + V}{1 + v'_x V/c^2} = \frac{(4/5)c + (3/5)c}{1 + (4/5)(3/5)} = \frac{(7/5)c}{37/25} = \frac{35}{37}c,$$

which is also formed from a PT! Is this coincidental? (a) Combine three additional sets of PTs in this way, to see if the combination of two PTs gives another PT in those cases as well. (b) State a general theorem, and prove it. (c) Suppose the integers m, n, p can be negative as well as positive. Do positive *or negative* PT velocities necessarily transform into positive or negative PT velocities?

CHAPTER 9

Spacetime

IN CLASSICAL PHYSICS, space and time never get mixed up. The two concepts are separate and distinct—there isn't even any particular reason to mention them both in the same sentence. The Galilean transformation of classical mechanics can be written with the time equation off to one side, or even left out entirely because it appears so self-evident:

$$\begin{aligned} x' &= x - Vt \qquad\qquad t' = t \\ y' &= y \\ z' &= z. \end{aligned} \tag{9.1}$$

The transformation guarantees that the time at which any event happens is the same in all inertial frames, and that the length of an object is likewise an absolute quantity.

In relativistic physics, however, space and time become intertwined through the Lorentz transformation

$$\begin{aligned} x' &= \gamma(x - Vt) \\ y' &= y \\ z' &= z \\ t' &= \gamma(t - Vx/c^2). \end{aligned} \tag{9.2}$$

Here the time equation is as important as the others and cannot be retired to one side. Time intervals and space intervals are not the same to all observers, but instead become mixed with one another. What is purely a distance to one observer may correspond to both a distance and a time interval to an observer in a different frame of reference.

This situation sounds much like viewing an object from two different angles—how much of an object is in its "height," "breadth," or "depth" depends on the point of view. A meterstick oriented in the x direction to one observer might be partially along the y direction to an observer using a rotated coordinate system. If two sets of

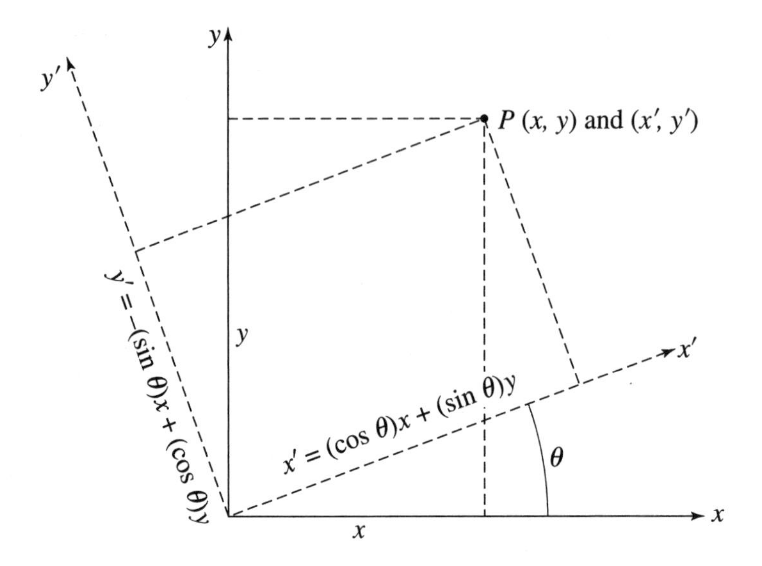

FIGURE 9.1
The coordinates (x', y') of a point P in a primed coordinate system compared with the coordinates (x, y) of the same point in an unprimed coordinate system.

(x, y) coordinates have the same origin but are relatively rotated by an angle θ, the coordinates of a point P in the primed system are given in terms of the coordinates of the same point in the unprimed system by the equations

$$x' = (\cos\theta)x + (\sin\theta)y$$
$$y' = -(\sin\theta)x + (\cos\theta)y, \tag{9.3}$$

as shown in Fig. 9.1. Note that the primed coordinates are linear combinations of the unprimed coordinates.

Neglecting for a moment the y and z directions, we may write the Lorentz transformation (9.2) in a somewhat similar linear form,

$$x' = (\gamma)x - (\gamma V)t$$
$$t' = -(\gamma V/c^2)x + (\gamma)t, \tag{9.4}$$

which looks as though it might be a kind of rotation in an "x–t plane." This analogy between the (x, t) Lorentz transformation and an (x, y) rotation in space is far from perfect, however. First of all, any of the coefficients γ, γV, or $\gamma V/c$ can exceed unity, so cannot be identified as sines or cosines. (The latter two coefficients are not even dimensionless, as sines and cosines must be, which can be traced to the fact that x is a length while t is a time.) Also, the sign of the second term on the right-hand side of the x' equation is different from the sign in the rotation equations. In spite of these difficulties, since x and t are so closely coupled, it might be useful to think of them as coordinates in some kind of space having unusual rules for rotation.[1] If y and z are thrown in as well, we would have a four-dimensional "spacetime," in which changing from one inertial frame to another plays the role of a rotation between coordinate systems. Time and space would become a matter of perspective, the amount of each depending on the "angle" from which it is viewed.

9.1 Minkowski Spacetime

It was the Russian-born German mathematician Hermann Minkowski who invented "spacetime" in 1908 and thereby made one of the most important contributions to relativity theory after Einstein's original work. He showed that Einstein's theory becomes simpler and more elegant when looked at in this way. The classical complete separation of space and time is unnatural and somewhat artificial in relativity. To quote Minkowski's manifesto,

> *Space by itself, and time by itself, are doomed to fade away into mere shadows, and only a kind of union of the two will survive.*

The idea of spacetime, or "Minkowski space," has not only been important as a useful arena for special relativity, but it was *essential* in Einstein's later construction of general relativity—his theory of gravity—which also allows physics to be described from the point of view of observers at rest in *noninertial* reference frames. There Einstein carried it beyond the "flat" or "Euclidean-like" spacetime we are discussing here and investigated the consequence of allowing this space to be curved. The results showed that the phenomenon of gravitation can be ascribed to such a spacetime curvature.

The introduction of four-dimensional spacetime is clearly not very mystical. It provides a point of view that makes a lot of sense, because relativistic physics becomes simpler and more elegant, and also because it unifies concepts that were previously thought to be distinct. Nobody can fully picture four-dimensional spacetime in the same way that we grasp three-dimensional space. The chief methods one uses to understand it are to draw simple pictures, use mathematics, and become well acquainted with situations described by the mathematics. For example, since we cannot draw a complete picture of four-dimensional spacetime, we can at least draw pictures with one time and one space dimension, "$(1 + 1)$-dimensional" spacetime. Let the time axis be vertical and the space axis be horizontal.[2] To give the time and space directions the same dimensions, we use "ct" as the time axis, because we can measure ct in meters.

In this space a light ray might look as shown in Fig. 9.2. If in particular the space axis corresponds to the x axis, a line with a positive slope of 45° represents a light ray moving in the $+x$ direction. A line with the opposite slope represents a ray moving

1. See Sample Problem 1 and Problems 9–12 and 9–13.

2. This is just a convention. One could just as well make the time direction horizontal and a space direction vertical. High-energy physicists frequently draw them that way.

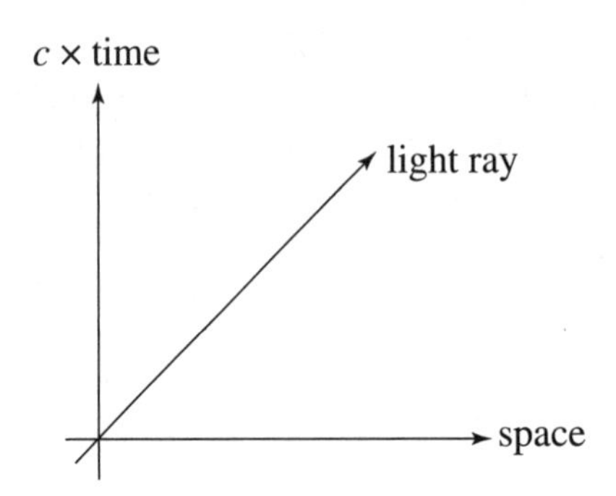

FIGURE 9.2
A light ray in (1 + 1)-dimensional spacetime.

in the $-x$ direction. If we add one more space axis (say, the y axis) projecting out of the page and take the locus of all light rays passing through some given point, they form a *cone* in a $(1+2)$-dimensional spacetime, as shown in Fig. 9.3. In three spatial dimensions, adding the z axis as well, the locus of points is still called the "light cone" but can be thought of as a sphere of light that collapses to a point and then re-expands as time evolves.

What about the motion of massive particles? As we will show later on, any such particle must move with speed $v < c$, so the path for any particle that passes through the vertex of the cone (which we'll call the origin) must be "trapped" inside the light

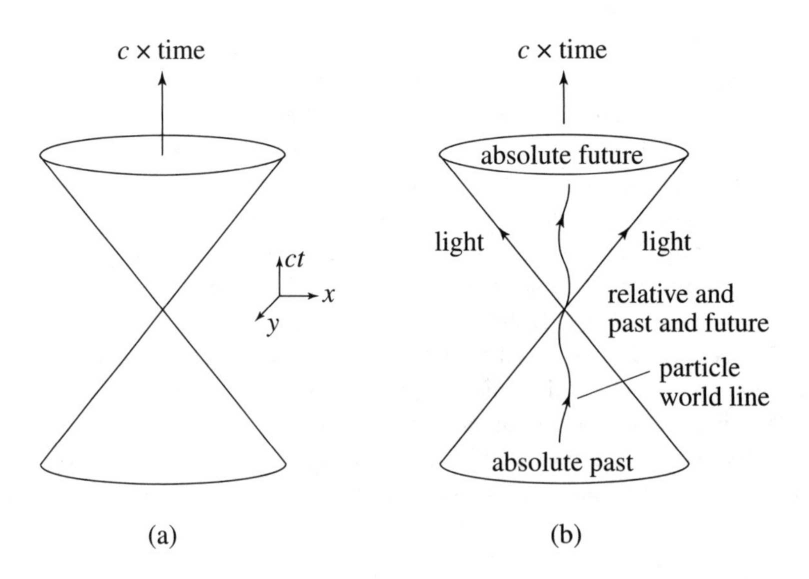

FIGURE 9.3
(a) A $(1+2)$-dimensional spacetime. (b) Absolute past, absolute future, and the relative past and future in a $(1+2)$-dimensional spacetime.

cone. That is, the path of a single particle in spacetime (called its "world line") might look as shown in Fig. 9.3(b), where at all times the magnitude of its vertical slope is greater than that of a light ray.

Consider the particle when it is located at the origin in the spacetime diagram. The *absolute past* of the particle consists of all events on or within the *lower* light cone. No event outside the cone could have influenced the particle in any way, since the influence would have had to propagate faster than light in order to reach the particle at this instant. Similarly, the *absolute future* of the particle consists of all events on or within the *upper* light cone. The particle cannot escape from this cone, nor can it influence any event outside the cone. So the region outside either cone is some kind of limbo that we'll call the *relative past and future.* Any event in this region takes place at a positive time in some frames and at negative time in other frames. The relativity of simultaneity cannot allow us to say that any particular event outside the light cone is definitely in the past or definitely in the future.

For example, suppose the Sun blew up 10 minutes ago in our frame of reference. That event is definitely in our absolute past, since the Sun is only about 8 light-minutes away, where 1 light-minute is the distance light moves in 1 minute. All observers, no matter how fast they are moving, will agree that the explosion was in our past, either by checking their clocks or just by looking at the explosion's deleterious effect upon us.

Now suppose that at time $t = 0$ to us on Earth, we sit down to eat lunch. Suppose also that the Sun blew up 4 minutes before in our reference frame. The event is therefore in our past, but not in our *absolute* past, since it has not been able to influence us when we are sitting down to eat. Figure 9.4(a) is a spacetime diagram of the events. Shown are the world lines of Earth and the Sun, and the past and future light cones centered on Earth at $t = 0$. The Sun is to the left of Earth, at a distance of $8\,c \cdot$ minutes.

Now look at the same events in a primed frame moving to the right at speed $V = (3/5)c$, as shown in Fig. 9.4(b). The Lorentz time transformation

$$t' = \gamma(t - Vx/c^2)$$

shows that for Earth, at $x = 0$, the times in the primed and unprimed frames are related by $t'_E = \gamma t_E = (5/4)t_E$, so at the instant $t_E = 0$, when we sit down to eat, it is also true that $t'_E = 0$; the clock at rest in the primed frame that happens to whiz by Earth at that moment also reads zero. Clocks at the position of the Sun obey

$$\begin{aligned} t'_{\text{Sun}} &= \gamma(t_{\text{Sun}} - Vx/c^2), \\ &= (5/4)[t_{\text{Sun}} - (3/5c)(-8c \cdot \text{min})] \\ &= (5/4)t_{\text{Sun}} + 6 \text{ minutes.} \end{aligned}$$

At the instant the Sun explodes, at $t_{\text{Sun}} = -4$ minutes,

$$t'_{\text{Sun}} = +1 \text{ minute,}$$

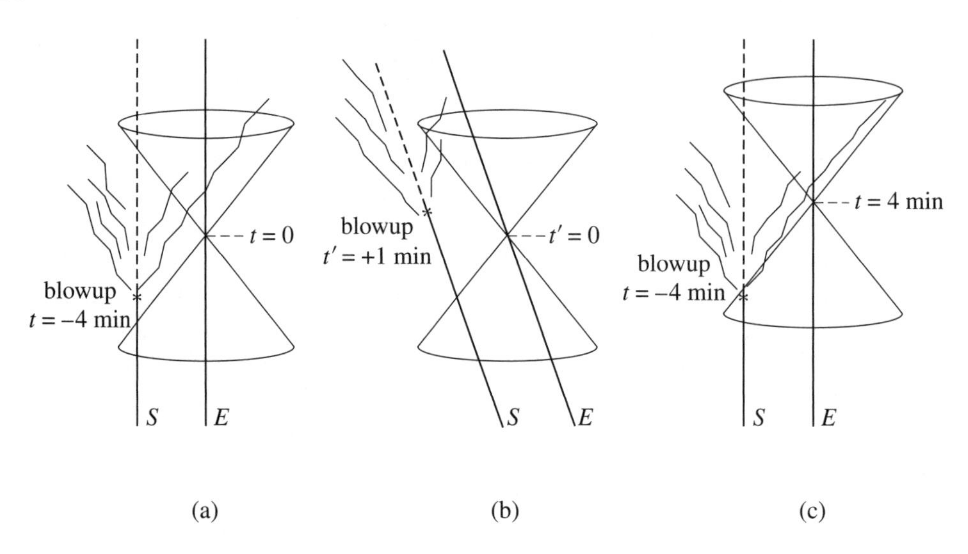

FIGURE 9.4
Spacetime diagram in (a) the frame of Earth and the Sun; (b) a frame moving to the right at $V = (3/5)c$. (c) Spacetime diagram in the frame of Earth and Sun, with the light cone drawn at time $t = 4$ minutes. Now the explosion is in our *absolute* past, because the explosion can affect us at this time.

so in the primed frame the Sun blows up 1 minute *after* we sit down to eat lunch. The sequence of events is different in the two frames. In the rest frame of Earth and the Sun, the Sun blows up *before* we sat down; in a primed frame the Sun blows up *after* we sat down. The fact that the sequence depends on the frame comes about because neither event is in the absolute past or future of the other.

As mentioned before, the region of spacetime outside the light cones is called the *relative* past and future. It is also sometimes called *elsewhere,* being neither the absolute past nor the absolute future.

If we draw a *different* light cone in the frame of Earth and Sun, centered this time when we are at $t = 4$ min, as shown in Fig. 9.4(c), the Sun's explosion is in the *absolute* past of us on Earth in this case, because it can affect us at that time.

9.2 Timelike, Null, and Spacelike Intervals

In ordinary three-dimensional space, the *displacement vector* between two particular points 1 and 2 is $\mathbf{r}_{12} = \mathbf{r}_2 - \mathbf{r}_1$, with components

$$(x_2 - x_1, y_2 - y_1, z_2 - z_1), \tag{9.5}$$

the difference between the position vectors of the two points. The vector $\mathbf{r}_{12}$ tells us how far and in what direction we have to go from point 1 to get to point 2, as shown in Fig. 9.5.

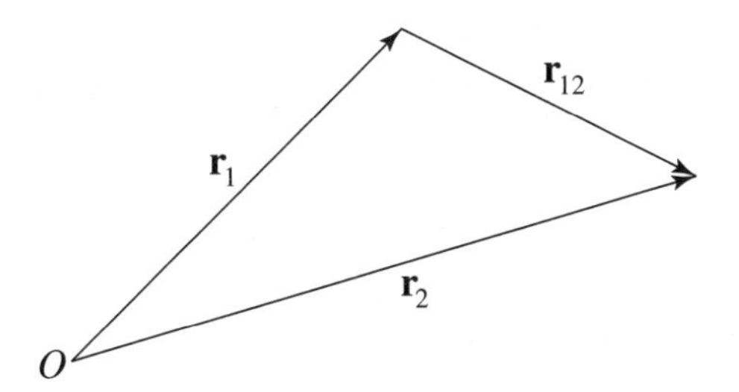

FIGURE 9.5
The displacement vector $\mathbf{r}_{12}$.

Clearly the square of the length of $\mathbf{r}_{12}$ is positive if the two points are distinct. That is, if points 1 and 2 are different,

$$|\mathbf{r}_{12}|^2 = (x_2 - x_1)^2 + (y_2 - y_1)^2 + (z_2 - z_1)^2 > 0. \tag{9.6}$$

Another feature of $|\mathbf{r}_{12}|^2$ is that it keeps the same value even if we rotate coordinates.

The square of the distance between two points does not depend upon whether we use the original coordinates or a rotated set of coordinates.

For example, if we rotate about the z axis, the x and y components of $\mathbf{r}_{12}$ change according to the rule given in Eq. 9.3, but it is easy to show that the sum of the squares of the components stays the same. That is, for any primed and unprimed frames,

$$(x_2' - x_1')^2 + (y_2' - y_1')^2 + (z_2' - z_1')^2$$
$$= (x_2 - x_1)^2 + (y_2 - y_1)^2 + (z_2 - z_1)^2. \tag{9.7}$$

We say that the square of $\mathbf{r}_{12}$ is an *invariant*; it does not change under coordinate transformations.

What is the corresponding displacement vector in four-dimensional spacetime? First, consider the position vector of a single point in spacetime, given by the components

$$(x_\mu) = (ct, x, y, z) \tag{9.8}$$

where the index μ takes on the values 0, 1, 2, 3, for the four components $x_0 = ct$, $x_1 = x, x_2 = y, x_3 = z$. We call this position vector a *four-vector*, because it is a vector in four-dimensional spacetime. The components of x_μ change when we change reference frames, as given by the Lorentz transformation.

We would like four-vectors to have similar properties to ordinary vectors in three-dimensional space. For example, the absolute length of the vector (or, alternatively, the sum of squares of the components) should be the same in all frames. Now, is it in fact true that $c^2t'^2 + x'^2 + y'^2 + z'^2 = c^2t^2 + x^2 + y^2 + z^2$? If so, the four-dimensional spacetime of special relativity would be a very simple extension of ordinary three-dimensional space.

However, it doesn't work! Using the Lorentz transformation of Eq. 9.2 to substitute in for the primed quantities on the left, it is easy to show that the hoped-for equality of the above equation is *false.* The left-hand side is *not* equal to the right-hand side. Interestingly, a *different* combination, with a minus sign on the square of each time coordinate, *is* invariant: That is, we can show that

$$-c^2t'^2 + x'^2 + y'^2 + z'^2 = -c^2t^2 + x^2 + y^2 + z^2. \tag{9.9}$$

This particular combination of the squares of the components is invariant under Lorentz transformations. (See Problem 9–4.) So although spacetime is a four-dimensional space, where Lorentz transformations can transform space partially into time or time partially into space, it is fundamentally different than a normal Euclidean space. Why? Because in spacetime we need a minus sign in front of the square of the time component. For that reason, four-dimensional spacetime is sometimes said to be "*pseudo*-Euclidean."

We can see why Eq. 9.9 makes sense in a simple special case. Suppose there is a sudden light flash at the origin, $x = 0$, $y = 0$, $z = 0$, at time $t = 0$. At any later time t the wave front of the flash has spread into a sphere of radius $R = ct$, where

$$R^2 = x^2 + y^2 + z^2, \tag{9.10}$$

so at any time t the wave front satisfies $-c^2t^2 + x^2 + y^2 + z^2 = 0$. The speed of light is the same in any primed frame as well, so the wave front also satisfies $-c^2t'^2 + x'^2 + y'^2 + z'^2 = 0$. Therefore the wave front of the flash satisfies Eq. 9.9, showing that it is valid in that special case.

Now we can write the components of a displacement vector between points 1 and 2 in four-dimensional spacetime. In spacetime this vector is called the *spacetime interval,* which has the four components ($c\Delta t$, Δx, Δy, and Δz), the differences between the position-vector components, where $\Delta t = t_2 - t_1$, and so on. The sum of the squares is not invariant under Lorentz transformations, but the peculiar combination

$$(\Delta s)^2 \equiv -(c\Delta t)^2 + (\Delta x)^2 + (\Delta y)^2 + (\Delta z)^2 \tag{9.11}$$

is invariant, and is the closest we can come to a Pythagorean theorem in spacetime. If we calculate $(\Delta s)^2$ between two points (that is, two "events") in spacetime using *one* frame of reference, it will take on the same value in any other frame of reference as well.

The most peculiar thing about $(\Delta s)^2$ is that it can be negative as well as positive or zero. And it can be zero even if the two points are not the same!

First suppose that $(\Delta s)^2 > 0$. According to Eq. 9.11, this means that the spatial separation $\sqrt{(\Delta x)^2 + (\Delta y)^2 + (\Delta z)^2}$ between the two events is larger than $c\Delta t$. This interval is said to be *spacelike,* because the spatial separation dominates the time separation. This must be true in any inertial frame, since $(\Delta s)^2$ is invariant. For such a spacelike interval, we can show that it is always possible to find a frame of reference in which $\Delta t = 0$, so that in that particular frame the two events are simultaneous, like two clocks in the same time zone each striking twelve noon. The interval between the two events can then be measured by metersticks alone, and is called the *proper length* $\Delta \ell$ between them. In this frame,

$$(\Delta s)^2 = (\Delta x)^2 + (\Delta y)^2 + (\Delta z)^2 \equiv (\Delta \ell)^2 \equiv (\text{proper length})^2. \quad (9.12)$$

That is, *the proper length is the distance between two events in a frame in which the events are simultaneous.* This can only happen if the interval is spacelike, with $(\Delta s)^2 > 0$.

Now suppose that $(\Delta s)^2 < 0$, in which case we say that the interval is *timelike.* The time term $-(c\Delta t)^2$ in Eq. 9.11 now dominates the sum of the spatial terms $(\Delta x)^2 + (\Delta y)^2 + (\Delta z)^2$. In this case it is always possible to find a frame in which the two events (say two sneezes by the same person) take place at the same location, so that $\Delta x = \Delta y = \Delta z = 0$. In that frame the interval can be measured by a clock alone, and the clock reads what is called the *proper time* $\Delta \tau$ between the events. In this frame,

$$(\Delta s)^2 = -c^2(\Delta t)^2 \equiv -c^2(\Delta \tau)^2 \equiv -c^2(\text{proper time})^2. \quad (9.13)$$

That is, *the proper time between the two events is the time difference in a frame in which the two events take place at the same location.* This can only happen if the interval is timelike, with $(\Delta s)^2 < 0$.

The seemingly surprising fact that $(\Delta s)^2$ can be negative reflects the definition (9.11), with the minus sign in front of the first term. A negative value of $(\Delta s)^2$ simply means that the interval can be measured by a clock. Suppose that a particular interval is indeed timelike, but we as observers are not at rest in the frame of the proper-time clock. To us, that clock moves through the coordinate distances $\Delta x = v_x \Delta t$, $\Delta y = v_y \Delta t$, and $\Delta z = v_z \Delta t$, where v_x, v_y, v_z are the velocity components of the moving clock in our frame and Δt is read by our clocks. Then

$$\begin{aligned}(\Delta s)^2 &= -(c\Delta t)^2 + (\Delta x)^2 + (\Delta y)^2 + (\Delta z)^2 \\ &= -(c\Delta t)^2 \left[1 - \left(v_x^2 + v_y^2 + v_z^2\right)/c^2\right] \\ &\equiv -(c\Delta t)^2(1 - v^2/c^2) = -c^2(\Delta \tau)^2, \end{aligned} \quad (9.14)$$

using Eq. 9.13, where $v = \sqrt{v_x^2 + v_y^2 + v_z^2}$ is the speed of the clock in our frame. Solving for $\Delta \tau$, we find that

$$\Delta \tau = \Delta t \sqrt{1 - v^2/c^2}, \quad (9.15)$$

which is no surprise; the proper-time clock is running slow from our point of view by the usual time-dilation factor!

Finally, suppose that $(\Delta s)^2 = 0$, in which case we say that the interval is *null*. The only way the square of the displacement vector in ordinary *three*-dimensional space can be zero is if the two points are identical. But that is not true in spacetime, because of the minus sign in Eq. 9.11. If two events are separated by a spatial distance Δx (with $\Delta y = 0$ and $\Delta z = 0$), and they take place at different times Δt, then it is obviously possible to have $(\Delta s)^2 = -(c\Delta t)^2 + (\Delta x)^2 = 0$, just by taking $\Delta x = \pm c\Delta t$. That is, null intervals correspond to the interval between two events along some light ray. Any two spacetime events on a light ray are separated by a null interval, and conversely, given that the interval between two particular spacetime points is null, a light ray could connect them. If the interval between two distinct spacetime points is null, it cannot be measured by either a meterstick alone or by a clock alone. There is neither a proper length nor a proper time for null intervals.

Suppose again that the Sun blows up, and that in our rest frame we sit down to eat lunch 4 minutes later. Those two events are separated by a spacelike interval. The Sun's blowup cannot possibly have influenced us at the time we sat down, because it takes 8 minutes for the Sun's light to reach Earth. Furthermore, since the interval is spacelike, there must be a reference frame in which the two events are *simultaneous* (see Problem 9–8). If instead we sit down to eat 8 minutes after the explosion, the two events are separated by a null interval. We "experience" the explosion at the instant we sit down. Finally, if instead we sit down to eat 10 minutes later (extremely unlikely under the circumstances), the two events are separated by a timelike interval. The Sun's blowup can certainly have influenced us, and there is even a reference frame in which the two events happen sequentially at the same place! (See Problem 9–9.)

WARNING: In representing a spacetime interval, we have used the equation

$$(\Delta s)^2 \equiv -(c\Delta t)^2 + (\Delta x)^2 + (\Delta y)^2 + (\Delta z)^2$$

for which timelike intervals are negative and spacelike intervals are positive. Some people use the opposite sign convention

$$(\Delta s)^2 \equiv (c\Delta t)^2 - (\Delta x)^2 - (\Delta y)^2 - (\Delta z)^2$$

for which timelike intervals are positive and spacelike intervals are negative. Unfortunately, there is no universal choice of signs! However, as long as one uses a particular sign convention throughout a calculation, it really doesn't matter which convention is used.

Sample Problems

1. Figure 9.1 (reproduced below) graphs how two different coordinate systems (x, y and x', y'), one rotated relative to the other, represent the position of a point in ordinary two-dimensional space. The figure helps us visualize the transformation Eqs. 9.3. Find an analogous graph for the Lorentz transformation between coordinates t, x and t' x', as given by Eqs. 9.4.

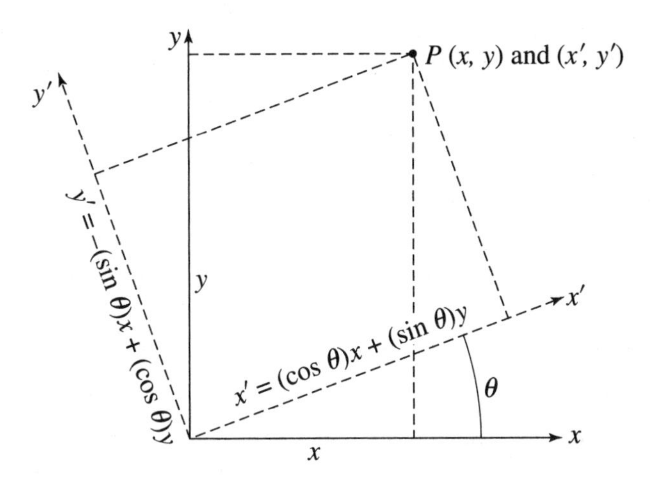

Solution: First, we will let the time axis be ct instead of simply t, so we can use the same units (meters) for both coordinates. Then the unprimed coordinates can be represented as shown, where $ct = 0$ everywhere on the x axis, and $x = 0$ everywhere on the ct axis. A light ray starting from the origin follows the dashed-line path at 45° to both axes, since $x = ct$ along that path. To find the *primed* axes we start by rewriting Eqs. 9.4 in the form

$$(1)\ x' = \gamma[x - (V/c)ct] \quad \text{and} \quad (2)\ ct' = \gamma[ct - (V/c)x].$$

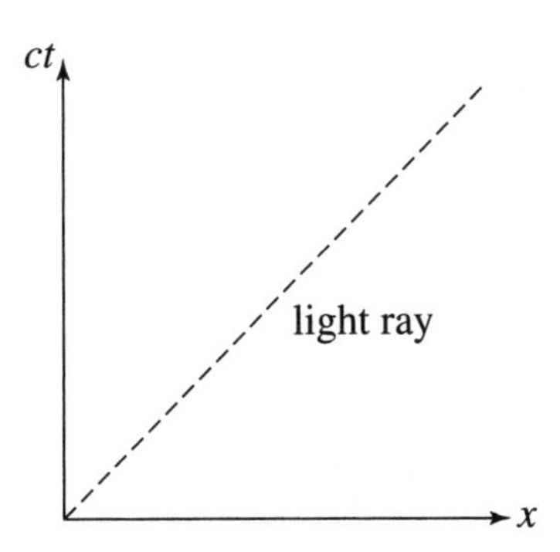

Then to draw the x' axis on the ct, x graph, we set $ct' = 0$, which gives (using Eq. 2) the result $ct = (V/c)x$. This is a straight line with slope V/c (which is < 1) so the x' axis is at an angle less than 45° from the x axis. Similarly, to draw the ct' axis, we set $x' = 0$ in Eq. (1) above, giving $x = (V/c)ct$. This line is also straight, with slope *V/c relative to the ct axis.* That is, the angle between the ct and ct' axes is the same as the angle between the x and x' axes, as shown on the next page. Note how different this graph is from the graph for rotations above; the ct' and x' axes are not perpendicular to one another! Note also that the “speed”

of the 45° dashed-line light ray is (consistently) the same in both frames, since $\Delta x/c\Delta t = \Delta x'/c\Delta t' = 1$.

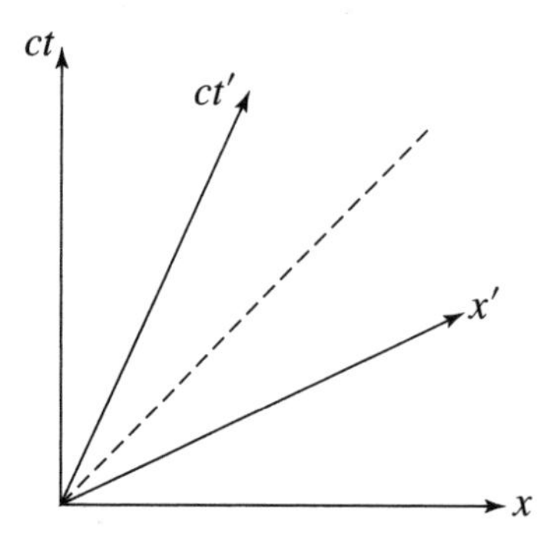

2. A *spacetime shoot-out.* Spaceships B and C, starting at the same location when each of their clocks reads zero, depart from one another with relative velocity $(3/5)c$. One week later according to B's clocks, B's captain goes berserk and fires a photon torpedo at C. Similarly, when clocks on C read one week, C's captain goes crazy and fires a photon torpedo at B. Draw two-dimensional spacetime diagrams of events in (a) B's frame (b) C's frame. Which ship gets hit first?

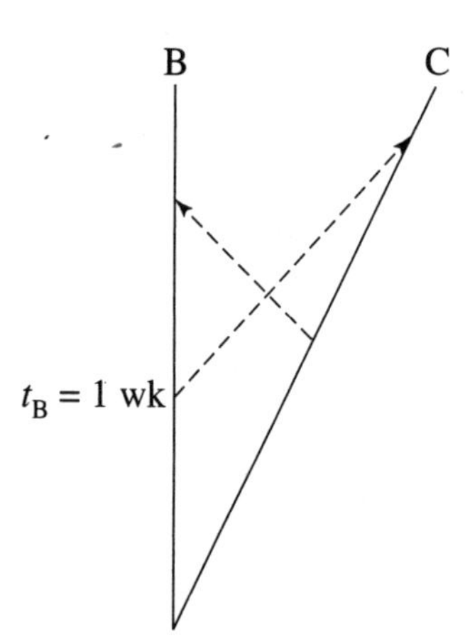

Solution: (a) The spacetime diagram in B's frame is as shown above. In B's frame, C's clock runs slow by the factor $\sqrt{1-(3/5)^2} = 4/5$, so in B's frame, C fires its photon torpedo when C's clock reads 1 wk and B's clock reads 5/4 wk. The photon torpedoes both move at an angle of 45° with respect to the horizontal, B's torpedo to the right and C's torpedo to the left. To find the exact times of arrival, consider the dashed triangles shown on the next page. The small triangle at the left (whose hypotenuse is the path of the leftward-moving torpedo) has a horizontal leg of length equal to how far C has moved in time 5/4 wk, that is, $d = vt = (3/5\,c)(5/4\,\text{wk}) = 3/4\,c \cdot \text{wk}$, where $1\,c \cdot \text{wk}$ is the distance light travels in 1 week. The triangle is a 45° right triangle, so the vertical leg is also $3/4\,c \cdot \text{wk}$. The total distance up along the ct axis from zero to arrival of the torpedo is therefore $(5/4 + 3/4)c \cdot \text{wk} = 2\,c \cdot \text{wk}$, and so B's clock reads 2 weeks when the photon torpedo arrives at B. The larger triangle on the right (whose hypotenuse is the path of the rightward-moving torpedo) has a horizontal leg of length equal to how far C has moved in time t_{arrival}, where t_{arrival} is measured in B's frame. This distance is $d = vt = (3/5\,c)t_{\text{arrival}}$. This distance is also equal to $c(t_{\text{arrival}} - 1\,\text{wk})$, which is how far light moves since it started out at 1 wk. Therefore $d = vt = (3/5\,c)t_{\text{arrival}} = c(t_{\text{arrival}} - 1\,\text{wk})$, which gives $t_{\text{arrival}} = 2\frac{1}{2}$ wks. Therefore in B's frame, C's torpedo strikes B a half-week *before* B's torpedo strikes C.

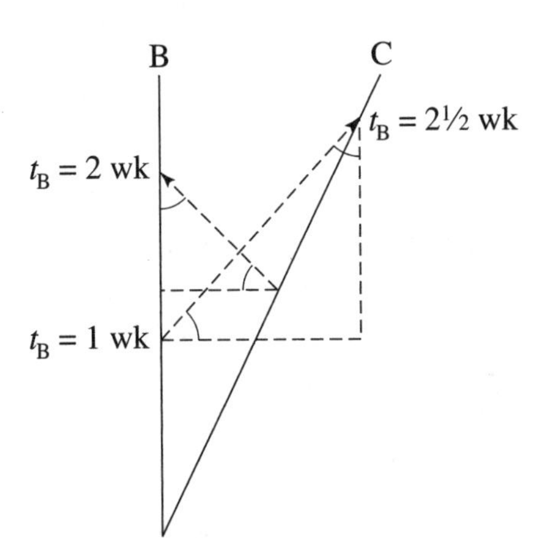

(b) The spacetime diagram of events in C's frame is shown below. All of the calculations are similar, so in C's frame B's torpedo arrives at C when C's clock reads 2 wks, and C's torpedo arrives at B when C's clock reads $2\frac{1}{2}$ weeks. So in C's frame, B's torpedo strikes C a half-week before C's torpedo strikes B. This is an example of the relativity of time ordering. In B's frame it is C's torpedo that strikes first, while in C's frame it is B's torpedo that strikes first. There is no paradox; both are true! Both ships get knocked out sooner or later; there is no advantage to either.

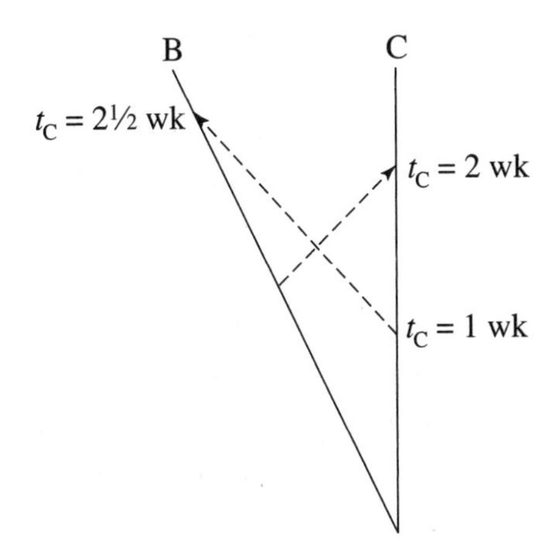

3. Suppose that the star α-Centauri, 4 $c \cdot$ yrs away, is at rest relative to the Sun, and that it suddenly emits a flare. Sometime later, in the rest frame of α-C and the Sun, you fall over a chair. Are the two events (the flare and your falling) separated by a *timelike, null,* or *spacelike* interval, if "sometime later" means (a) 3 yrs, (b) 4 yrs, (c) 5 yrs?

Solution: (a) If you fall over the chair 3 years after the flare, the two events are separated by a *spacelike* interval, because there has not been time for either light or a massive particle to travel between the two events. There is no causal interaction between the two events—the flare cannot have caused you to fall. Also, there exists a reference frame in which the two events are simultaneous!

(b) If you fall over the chair 4 years after the flare, the two events are separated by a *null* interval, because a light ray can connect the two events. The event of your falling is on the future light cone of the flare (maybe you fell over the chair *because* you were surprised when you saw the flare through a telescope).

(c) If you fall over the chair 5 years after the flare, the two events are separated by a *timelike* interval. A particle could move at less than light speeds from the flare to you (in fact, it could move at a constant $(4/5)c$), arriving at the instant you fall—as in the null case, the flare might have *caused* you to fall. Furthermore, there exists a reference frame in which the two events happen at the same *place,* namely, the frame that moves at $(4/5)c$ along with the particle mentioned above. In that frame the particle stays at rest and therefore, since it connects the two events, the events occur at the same location in that frame but at different times.

Problems

9–1. (a) A chicken crosses the road. Is the spacetime interval between the beginning and end of its journey spacelike, timelike, or null? (b) Close your eyes and then open them briefly. What set of points in spacetime do you see during that brief time interval?

9–2. Consider two spacetime events with coordinate differences $(\Delta t, \Delta x, \Delta y, \Delta z)$ as given. In each case, is the spacetime interval spacelike, timelike, or null?
(a) (1 s, 1 m, 1 m, 1 m) (b) (1 s, 1$c \cdot$ s, 0, 0) (c) (10^{-8} s, 1 m, 2 m, 3 m)

9–3. A tachyon is a hypothetical particle that travels faster than light (no such particle has actually been observed!) Suppose tachyons exist, and that a tachyon travels from the star Proxima Centauri (4 light-years distant) to us in 2 years flat. (a) Does the tachyon follow a *spacelike, timelike,* or *null* trajectory in spacetime? (b) On a spacetime diagram drawn in a frame in which we are at rest, sketch the path of this tachyon, assuming that it arrives at time $t = 0$ in our frame.

9–4. Using the Lorentz transformation, show the following:

$$\text{(a) } c^2t'^2 + x'^2 + y'^2 + z'^2 \neq c^2t^2 + x^2 + y^2 + z^2$$

$$\text{(b) } -c^2t'^2 + x'^2 + y'^2 + z'^2 = -c^2t^2 + x^2 + y^2 + z^2$$

9–5. Two events in spacetime are separated in time and also in the x direction in some particular (unprimed) frame of reference, but they have the same y and z coordinates. Prove from the Lorentz transformation (to a new primed frame propagating in the $+x$ direction) that if in the unprimed frame the interval Δs^2 between the events is (a) null, (b) timelike, (c) and spacelike, the interval *remains* (a) null, (b) timelike, (c) and spacelike in the primed frame as well.

9–6. Prove that the time order of two events is the same in all inertial frames if and only if they can be connected by a signal traveling at or below speed c.

9–7. Spaceship S leaves Earth E, moving to the right relative to E at velocity $(4/5)c$. (a) Draw a spacetime diagram in E's frame, showing both world lines S and E, and the past and future light cones centered on the point of departure. (b) Draw a spacetime diagram in S's frame of the same world lines and light cones. (c) Now suppose that both the E and S clocks read zero when S departs, and that when E's clock reads 15 days, E sends a tachyonic message to S at infinite speed in E's frame (tachyons are hypothetical particles that travel faster than light.) Add the tachyon path to the spacetime diagram in E's frame, and find what S's clock reads when the tachyon arrives. (d) Add the tachyon path to the spacetime diagram in S's frame as well. Do you notice anything strange about this path?

9–8. If the Sun blows up at some instant and four minutes later we on Earth sit down to eat lunch, those two events are separated by a spacelike interval. The Sun's blowup cannot possibly have influenced us at the time we sat down, because it takes 8 minutes for light to reach us from the Sun. And since the interval is spacelike, there must be a reference frame in which the two events are *simultaneous*. Find the speed of this frame relative to the Sun–Earth rest frame.

9–9. If we on Earth sit down to eat lunch 10 minutes after the Sun explodes (extremely unlikely under the circumstances), the two events are separated by a timelike interval, because the Sun is only 8 light-minutes from Earth. The Sun's blowup can certainly have influenced us, and there is even a reference frame in which the two events happen sequentially at the same place! Find the speed of this frame relative to the Sun–Earth rest frame.

9–10. A flare occurs on the star α-Centauri, 4 light-years from Earth. If I fall over a chair only 3 years after the flare, the two events are separated by a spacelike interval. In that case there can be no causal interaction between the two events; the flare cannot have caused me to fall. Furthermore, there must exist a reference frame in which the two events are simultaneous. Find the speed of this frame relative to the mutual rest frame of α-Centauri and Earth.

9–11. In the "paradox" of twins Al and Bertha, Al stays home and Bertha travels to a distant star, then turns around and comes home. In spacetime, their world lines are as shown. Using our sign convention, (a) Are the spacetime intervals of Al and Bertha positive or negative? (b) Which has the larger magnitude? (c) Which has experienced the larger proper time?

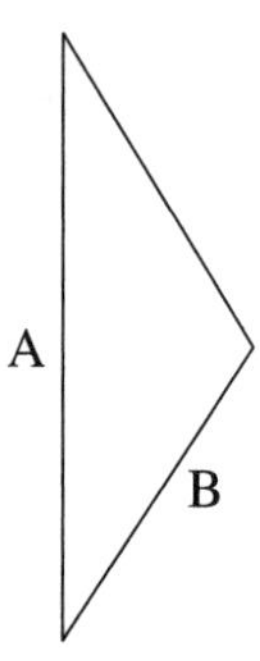

9–12. A "transformation diagram" showing primed and unprimed x and ct coordinate axes was constructed in Sample Problem 1. (a) In terms of the relative velocity V/c, find the angle θ between the x and x' axes (which is the same as the angle between the ct and ct' axes). (b) Suppose that a new diagram is begun, in which the x' and ct' axes are perpendicular to one another. Find the corresponding x and ct axes, and draw them on the new diagram. (c) What is the angle between the x and x' axes in this diagram, in terms of V/c?

9–13. The primed-frame axes in the transformation diagram of Sample Problem 1 are easy to locate for any relative frame velocity V/c, because on the orthogonal x, ct coordinate diagram, the ct' axis corresponds to the line $x = (V/c)ct$ and the x' axis corresponds to the line $ct = (V/c)x$. It is trickier to measure locations x' and times ct' on the diagrams, however, because the primed coordinate axes are nonorthogonal. One way to calibrate these primed coordinates is to draw "invariant hyperbolas" on the diagram, which are the curves $-(ct')^2 + x'^2 = -(ct)^2 + x^2 = \pm a^2$, where $a =$ some constant length. (The equality $-(ct')^2 + x'^2 = -(ct)^2 + x^2$ is a consequence of the Lorentz transformation; see Eq. 9.9.) For example, if we choose $a = 1$ meter and the *plus* sign, this curve intersects the x axis (where $ct = 0$) at $x = 1$ m, and it also intersects the x' axis (where $ct' = 0$) at $x' = 1$ m, as shown on the next page. To find the x, ct coordinates of this latter point, note that the x' axis corresponds to the line $ct = (V/c)x$, so eliminating t we have $-(V/c)^2x^2 + x^2 = (1\text{ m})^2$, from which we find that this hyperbola intersects the x' axis at

$$x = \frac{1\text{ m}}{\sqrt{1 - V^2/c^2}} \equiv \gamma \cdot 1\text{ m}$$

in terms of the usual gamma factor between the two frames. Similarly, the curve with $a = 1$ m and the *minus* sign intersects the ct axis at $ct = 1$ m and the ct' axis at $ct' = 1$ m. This hyperbola is also drawn. (a) Show that all the above-mentioned hyperbolas asymptotically approach the line $x = ct$ (or $x = -ct$) for large values

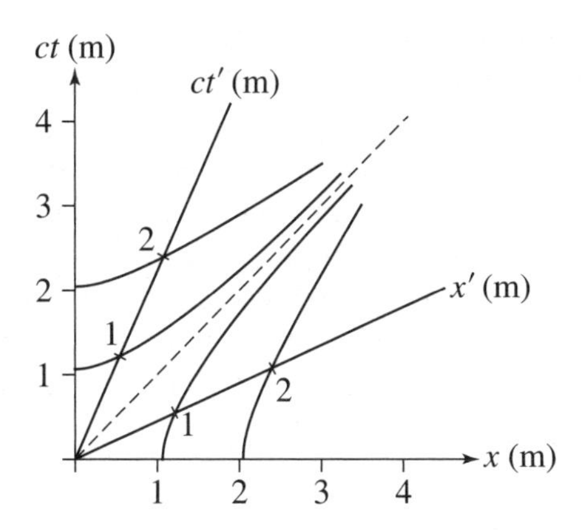

of any one of the four coordinates. (b) Construct a careful transformation diagram for the particular case $V = (3/5)c$, locating on the x' axis the numbers $x' = 1$ m, 2 m, and 3 m, and on the ct' axis the numbers $ct' = 1$ m, 2 m, 3 m. (c) Now that the primed coordinates have been calibrated, you can draw a set of dashed coordinate lines as illustrated below. The lines parallel to the ct' axis correspond to constant values of x', and lines parallel to the x' axis correspond to constant values of ct'. Draw these on your $V = (3/5)c$ diagram for $ct' = 1$ m, 2 m, and 3 m, and for $x' = 1$ m, 2 m, and 3 m. Then from your diagram determine the approximate values of ct' and x' corresponding to an event at $ct = x = 2$ m. (d) From the Lorentz transformation, find the *exact* values ct' and x' corresponding to this same event, and compare with your values found from the diagram. [Such diagrams are also a nice way to illustrate time dilation, length contraction, and the relativity of simultaneity, including their reciprocal natures. See Problems 9–14, 9–15, and 9–16.]

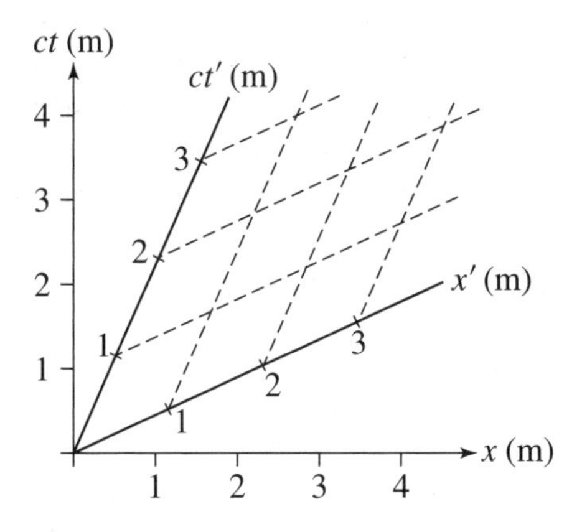

9–14. *Leading clocks lag* (follow-on to Problem 9–13). Redraw the transformation diagram of Problem 9–13 *with calibrated primed-frame axes corresponding to* $V = (3/5)c$, and place two dots on it indicating the spacetime positions of two clocks at rest in the primed frame (and synchronized in that frame) when each reads the same time (say, 1:00 pm). Note that these dots must both lie on a line that is parallel to the x' axis, since all points on such a line have the same value of t'. (a) Show from the diagram that these clocks are *not* synchronized in the unprimed frame. (b) Then show that leading clocks lag; that is, that the clock with the larger value of x (which is leading the other clock in space as they move together in the unprimed frame) lags the other clock in time. (c) Now place two additional dots on the diagram, indicating the events when two clocks at rest in the *unprimed* frame (and synchronized in that frame) both read some definite time (say, 1:00 pm). Show from the diagram that these two clocks are not synchronized in the primed frame. (d) Then show that according to observers in the primed frame, the leading clock lags. *Hint:* As seen in the primed frame, is it the clock with the *larger* value of x' or the *smaller* value of x' that leads the other clock as they move along together?

9–15. *Time dilation* (follow-on to Problem 9–13). Carefully redraw the transformation diagram of Problem 9–13 with calibrated primed-frame axes corresponding to $V = (3/5)c$. (a) Then draw the world line of a clock at rest in the unprimed frame (which will be a vertical line on the diagram), and place two dots on that line, indicating two particular times read by that clock. Show that the clock is moving in the primed frame. Is it moving to the right or to the left? (b) Then show that if the clock ticks through a time interval Δt in the unprimed frame, the corresponding time interval $\Delta t'$ observed in the primed frame is *greater* than Δt. Therefore the moving clock has run slow. (c) Now draw the world line of a second clock at rest in the *primed* frame, and show that if it ticks through a time interval $\Delta t'$ in its rest frame, the corresponding time interval Δt observed in the unprimed frame is greater than $\Delta t'$, so this moving clock has run slow as well.

9–16. *Length contraction* (follow-on to Problem 9–13). The transformation diagram of Problem 9–13 is shown, with calibrated primed-frame axes corresponding to $V = (3/5)c$. A meterstick at rest in the unprimed frame at time $t = 0$ is shown as well, along with the subsequent world lines of its two ends. If observers in the primed frame want to measure the length of the stick (which moves in their frame) they must measure the locations of both ends at the *same time* t' in *their* frame. If this is done at $t' = 0$, for example, the length of the stick to them is the distance $\Delta x'$ along the x' axis, as shown, which is clearly *less* than 1 meter. This illustrates the fact that the stick is Lorentz contracted in the primed frame. Now using the same diagram, show that a stick at rest in the *primed frame* is Lorentz contracted according to observers in the unprimed frame.

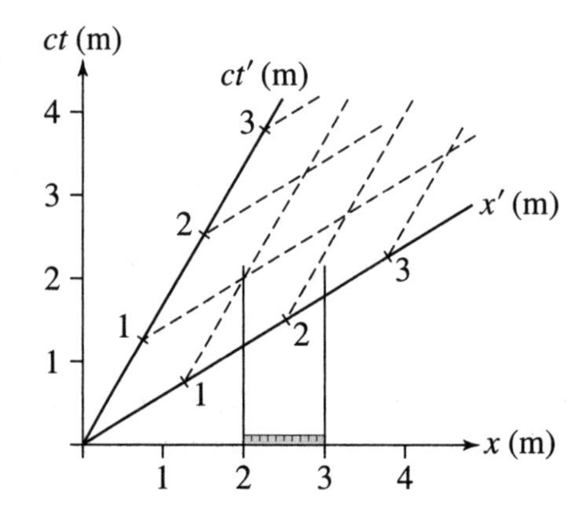

CHAPTER 10

Momentum

IF RELATIVITY HAS greatly changed our understanding of space and time, which form the *arena* of physics, it seems likely that the behavior of *actors* in the arena might be changed as well. In particular, how about the important actors called *particles*? How does relativity influence our understanding of the motion of particles?

10.1 *Classical* Momentum

The *momentum* of a particle is central to our understanding of its motion. That is because in nonrelativistic, classical physics the momentum $\mathbf{p}$ of the particle stays constant unless there is a net force acting on it, in which case Newton's second law states that the force causes a time rate of change of momentum according to $\mathbf{F} = d\mathbf{p}/dt$. Equivalently, since the momentum of a nonrelativistic particle is

$$\mathbf{p} = m\mathbf{v}, \tag{10.1}$$

the product of its mass and velocity, it follows that $\mathbf{F} = m(d\mathbf{v}/dt) = m\mathbf{a}$, where $\mathbf{a}$ is its acceleration.[1]

It is also true that the total momentum of a *system* of particles is conserved (that is, stays constant in time) if no net external force acts. An important property of momentum conservation is that *if the total momentum of a collection of particles is conserved in one inertial frame, it is conserved in all other inertial frames.*

As in Chapter 1, we can assume that the total momentum of a pair of pool balls A and B is conserved during a collision, as illustrated in Fig. 10.1, even if there are

1. As in Chapter 1, boldface symbols represent three-component vectors, such as $\mathbf{F} = (F_x, F_y, F_z,)$, and a vector equation like $\mathbf{F} = d\mathbf{p}/dt$ is short for three equations, one for each component.

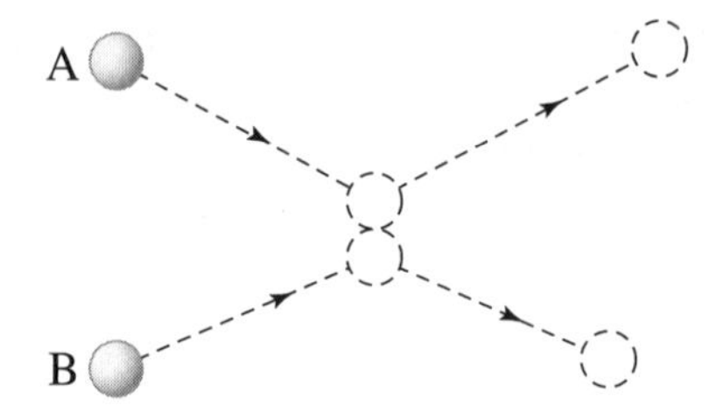

FIGURE 10.1
A collision of two pool balls.

modest external forces like gravity acting on the balls. So in the rest frame of the pool table, we have

$$m_A \mathbf{v}_{Ai} + m_B \mathbf{v}_{Bi} = m_A \mathbf{v}_{Af} + m_B \mathbf{v}_{Bf} \tag{10.2}$$

where $\mathbf{v}_{Ai}$ is the initial velocity of ball A, $\mathbf{v}_{Af}$ is the final velocity of ball A, and so forth, measured in the rest frame of the pool table. Now suppose somebody rides by the pool table at velocity $\mathbf{V}$ on a bicycle. Is momentum conserved as seen by the rider, at rest in the primed frame S' of the bicycle?

Using the classical, "Galilean" velocity transformation $\mathbf{v} = \mathbf{v}' + \mathbf{V}$ for each of the velocities in Eq. 10.2, we have[2]

$$m_A(\mathbf{v}'_{Ai} + \mathbf{V}) + m_B(\mathbf{v}'_{Bi} + \mathbf{V}) = m_A(\mathbf{v}'_{Af} + \mathbf{V}) + m_B(\mathbf{v}'_{Bf} + \mathbf{V}). \tag{10.3}$$

The four terms with $\mathbf{V}$ in them cancel out, so

$$m_A \mathbf{v}'_{Ai} + m_B \mathbf{v}'_{Bi} = m_A \mathbf{v}'_{Af} + m_B \mathbf{v}'_{Bf}, \tag{10.4}$$

showing that momentum is conserved in the bicycle frame as well. In classical physics, if the total momentum of a system of particles is conserved in *one* inertial frame, it is conserved in *any* inertial frame. Naturally, we use the Galilean transformation, including the velocity transformation of Eq. 1.5, to relate quantities between the two frames. We have not *forced* the momentum to be conserved in the S' frame. It is conserved *automatically* if $\mathbf{p} = m\mathbf{v}$ is the momentum and we use the Galilean velocity transformation between the two frames. If momentum were conserved only in one frame it wouldn't be a very useful quantity, because it would be conserved only for a very special set of observers.

2. If $\mathbf{V}$ is in the x direction, the vector equation $\mathbf{v} = \mathbf{v}' + \mathbf{V}$ is short for the three transformation equations $v_x = v'_x + V$, $v_y = v'_y$, $v_z = v'_z$, as given in Eqs. 1.5.

10.2 Momentum in Relativity

Now suppose some particles are moving at *relativistic* speeds. Is their total momentum still conserved if we continue to calculate each individual momentum using the nonrelativistic formula $\mathbf{p} = m\mathbf{v}$?

The problem is that in relativity we have to use the relativistic velocity transformation

$$v_x = \frac{v'_x + V}{1 + v'_x V/c^2} \tag{10.5}$$

for the x components of $\mathbf{v}$ between frames (as derived in Chapter 8) instead of the nonrelativistic Galilean transformation $v_x = v'_x + V$ used in Eq. 10.3.

For example, suppose we have a one-dimensional collision in the x direction between two relativistic pool balls. If $p = mv$ is the momentum of each ball along that direction, then in the S frame the momentum conservation equation is

$$m_A v_{Ai} + m_B v_{Bi} = m_A v_{Af} + m_B v_{Bf}. \tag{10.6}$$

If we transform all velocities to the S' frame using Eq. 10.5, note that the terms proportional to V no longer cancel out, because the denominators in Eq. 10.5 are generally all different. Therefore

$$m_A v'_{Ai} + m_B v'_{Bi} \neq m_A v'_{Af} + m_B v'_{Bf} \tag{10.7}$$

in general, so even if momentum is conserved in one frame it is not conserved in others.[3] Something must be wrong! If conservation of momentum is to be a physical law, Einstein's first postulate requires that it be true in all inertial frames. Equations 10.6 and 10.7 show that the classical definition of momentum flunks this test!

10.3 Momentum in Four-Dimensional Spacetime

To find the *correct* expression for momentum, we need a thoroughly relativistic point of view. Instead of looking at three-dimensional space alone, let us look at physical quantities in the full four-dimensional *spacetime.*

10.3.1 Four-Scalars and Four-Vectors

Ordinary *three*-dimensional space is populated with all kinds of "real" and more-or-less abstract physical quantities that can be classified according to how they change when the coordinates are rotated. For example, a *scalar* (like the temperature at some

3. An example is given in Problem 10–5.

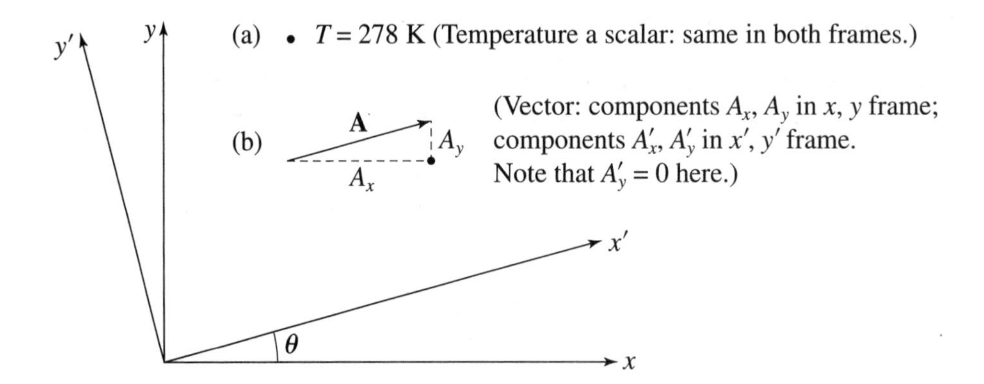

FIGURE 10.2
Two reference frames, one rotated through angle θ relative to the other. (a) A scalar quantity T, the same in both frames. (b) A vector quantity $\mathbf{A}$, with different components in the two frames.

point, or the kinetic energy of a particle) can be represented by a single number that remains unchanged when a rotated coordinate system is used. Take temperature T, for example, as illustrated in Fig. 10.2(a). If a thermometer reads $T = 278$ K (that is, 278 degrees Kelvin) at some point (x, y) using a set of unprimed coordinates in a plane, then obviously $T = 278$ K at the same point, even if we use a rotated (primed) set of coordinates in which the point is labeled as (x', y').

A *vector* (like a particle's position vector $\mathbf{r}$ or momentum vector $\mathbf{p}$) is represented by *three* numbers, corresponding to the x, y, and z components of the vector. Upon rotation to a new coordinate system, some or all of the components change as a function of the rotation angle. If, for example, we want to know the components of an arbitrary vector $\mathbf{A}$ in a primed frame obtained by rotating the coordinate system through an angle θ about the z axis, then

$$\begin{aligned} A'_x &= A_x \cos\theta + A_y \sin\theta \\ A'_y &= -A_x \sin\theta + A_y \cos\theta \\ A'_z &= A_z. \end{aligned} \tag{10.8}$$

Although the components change, the magnitude of the vector does not change. It is independent of the coordinate system from which it is viewed; that is, $|\mathbf{A}'| = |\mathbf{A}|$, or equivalently, the sum of the squares of the components is invariant,

$$A'^2_x + A'^2_y + A'^2_z = A^2_x + A^2_y + A^2_z. \tag{10.9}$$

This invariance is easily verified in the special case of the rotation about the z axis just mentioned, the sort of rotation illustrated in Fig. 10.2(b).

In going over to four-dimensional spacetime, quantities have to be re-examined to see how they transform. We want to identify those quantities that do not change under a

Lorentz transformation (called *four-scalars*) and those that change like the coordinates themselves (called *four-vectors*). That is, the Lorentz transformation between inertial frames takes the place of a rotation in three dimensions.

The speed of light c is an example of a four-scalar; it has the same value in all frames. We will also define the mass m of a particle to be the same in all frames—the same in its rest frame as in a frame in which it is moving.[4] The proper time interval $\Delta\tau$ of a clock, as discussed in Chapter 9, is also a four-scalar; observers in all frames of reference agree upon what a particular clock actually reads, whether it is at rest with respect to them or not.

The simplest four-vector is the position vector of a particle, with components (ct, x, y, and z), as already discussed in Section 9.2. An arbitrary four-vector is specified by its four components (A_0, A_x, A_y, A_z), which are its components in the t, x, y, and z directions. We call these components collectively A_μ, where the index $\mu = 0$, 1, 2, or 3, so in the case of the position four-vector x_μ, for example, the components are (ct, x, y, and z). We require that the arbitrary four-vector $A_\mu = (A_0, A_x, A_y, A_z)$ transform according to the Lorentz transformation; that is, in going from an unprimed to a primed frame with translation at velocity V in the x direction, the components should change according to the equations

$$\begin{aligned} A'_0 &= \gamma(A_0 - VA_x/c) \\ A'_x &= \gamma(A_x - VA_0/c) \\ A'_y &= A_y \\ A'_z &= A_z. \end{aligned} \tag{10.10}$$

Just as for the position four-vector, it follows from the transformation law that

$$-A'^2_0 + A'^2_x + A'^2_y + A'^2_z = -A^2_0 + A^2_x + A^2_y + A^2_z. \tag{10.11}$$

That is, this peculiar way of "summing" squares, where we must include a minus sign in front of the square of the zeroth (that is, time) component, is the same in all inertial frames. Therefore the quantity $-A_0^2 + A_x^2 + A_y^2 + A_z^2$ is a four-scalar, because its value is invariant (that is, does not change) under Lorentz transformations.

10.3.2 The Momentum Four-Vector

Now we can define a *momentum* four-vector, a candidate for the momentum of a relativistic particle. In nonrelativistic physics the momentum $\mathbf{p}$ of a particle of mass m can be written in terms of components $p_i = mv_i$, where $i = 1$, 2, or 3, corresponding to the x, y, and z components

---

4. See the discussion in Appendix F.

$$(p_i) = (p_1, p_2, p_3) \equiv \left(m\frac{dx}{dt}, m\frac{dy}{dt}, m\frac{dz}{dt} \right), \qquad (10.12)$$

with one component for each direction in ordinary three-dimensional space. By analogy, we can define a *four-vector momentum*

$$p_\mu = m u_\mu \qquad (10.13)$$

of a particle of mass m, where u_μ is the *four-vector velocity*

$$u_\mu \equiv \frac{dx_\mu}{d\tau}, \qquad (10.14)$$

with $\mu = 0$, 1, 2, or 3. The four-vector velocity is the derivative of the position four-vector x_μ of a particle with respect to the *proper time* read by a clock that moves along with the particle. We *have* to use the proper time in this definition; we would *not* get a four-vector if we differentiated the position with respect to the time t read by some stationary clock, because t is not a four-scalar, but rather one of the four components of the position four-vector. The four-vector momentum therefore has components

$$(p_\mu) = \left(mc\frac{dt}{d\tau}, m\frac{dx}{d\tau}, m\frac{dy}{d\tau}, m\frac{dz}{d\tau} \right). \qquad (10.15)$$

In the preceding chapter (Eq. 9.15) we found that the proper time $\Delta\tau$ on a clock moving with speed v is related to the time Δt of a stationary clock by the usual time-dilation formula $\Delta\tau = \Delta t\sqrt{1 - v^2/c^2}$. Therefore, taking the limit of infinitesimal time intervals, it follows that

$$\lim_{\Delta\tau \to 0} \frac{\Delta t}{\Delta\tau} \equiv \frac{dt}{d\tau} = \frac{1}{\sqrt{1 - v^2/c^2}}. \qquad (10.16)$$

Using the chain rule, we find that

$$u_x \equiv \frac{dx}{d\tau} = \frac{dt}{d\tau}\frac{dx}{dt} = \frac{1}{\sqrt{1 - v^2/c^2}} v_x \qquad (10.17)$$

where $v_x = dx/dt$ is the ordinary x component of velocity in the unprimed frame. The four components of the four-vector momentum are therefore

$$(p_\mu) = \left(\frac{mc}{\sqrt{1 - v^2/c^2}}, \frac{m\mathbf{v}}{\sqrt{1 - v^2/c^2}} \right) \qquad (10.18)$$

where $m\mathbf{v}/\sqrt{1 - v^2/c^2}$ is a three-vector shorthand for the three spatial components

$$\mathbf{p} = \frac{m\mathbf{v}}{\sqrt{1 - v^2/c^2}} = \left(\frac{mv_x}{\sqrt{1 - v^2/c^2}}, \frac{mv_y}{\sqrt{1 - v^2/c^2}}, \frac{mv_z}{\sqrt{1 - v^2/c^2}} \right). \qquad (10.19)$$

This quantity

$$\mathbf{p} = \frac{m\mathbf{v}}{\sqrt{1 - v^2/c^2}} \tag{10.20}$$

is the correct expression for the momentum of a particle, no matter how fast it is moving! Before relativity, we thought that the momentum was $\mathbf{p} = m\mathbf{v}$, but now we know that the momentum is really

$$\mathbf{p} = \gamma m\mathbf{v} \tag{10.21}$$

where

$$\gamma = \frac{1}{\sqrt{1 - v^2/c^2}} \tag{10.22}$$

in terms[5] of the particle's speed v.

Note that $\mathbf{p}$ reduces to the classical expression $\mathbf{p} = m\mathbf{v}$ in the nonrelativistic limit[6] $v/c \ll 1$. Note also that the numerator has only a single component of $\mathbf{v}$, but the denominator contains the square of the total velocity, $v^2 = v_x^2 + v_y^2 + v_z^2$. That is, the momentum in the x direction depends not only on the velocity in the x direction, but the velocities in the y and z directions as well!

The gamma factor in $\mathbf{p} = \gamma m\mathbf{v}$ means that the momentum becomes indefinitely large as $v \to c$. If the classical form $m\mathbf{v}$ were correct, no momentum component could ever even reach the value $p = mc$, because no massive particles can reach the speed of light. But in fact, the momentum is unbounded, in such a way that a very small change in the velocity of a highly relativistic particle corresponds to a very large change in its momentum. Figure 10.3 sketches the nonrelativistic momentum and the true momentum of a massive particle for speeds less than c, which are the only speeds possible for such a particle.

---

5. Note that the gamma factor $\gamma = 1/\sqrt{1 - v^2/c^2}$ in the definition of momentum, where v is the speed of the particle in the reference frame we are using, is *different* from the gamma factor $\gamma = 1/\sqrt{1 - V^2/c^2}$ in the Lorentz transformation used to help transform quantities between two frames of reference moving at speed V relative to one another. We don't often have to use both gammas in the same problem, so there is seldom any confusion. An example where we *do* have to keep track (of *three* different gamma factors!) is given in Sample Problem 4 of Chapter 11.

6. The mass m in the expression $\mathbf{p} = m\mathbf{v}/\sqrt{1 - v^2/c^2}$ is the ordinary "rest mass" of the object, which we can find by weighing the object on a scale while it is at rest in our frame of reference. It is a constant quantity for any given object, and does not depend upon velocity. Some people introduce a "relativistic mass" m_R instead that increases with velocity according to the rule $m_R = m/\sqrt{1 - v^2/c^2}$ where m is the unchanging rest mass we use. Then the momentum can be written in the classical form $\mathbf{p} = m_R\mathbf{v}$. However, the disadvantages of introducing such a relativistic mass far outweigh the advantages: See Appendix F. (When we use boldface type such as $\mathbf{p}$ for a vector in relativity, we always mean a *three*-vector, only the spatial components of the full four-vector.)

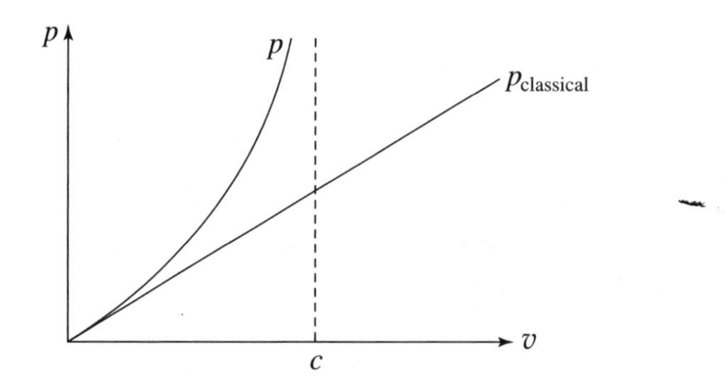

FIGURE 10.3
The classical, nonrelativistic momentum and true momentum of a massive particle, as functions of its velocity. The true momentum is unbounded as $v \to c$.

Now comes the crucial test. Does the four-vector momentum $p_\mu = mu_\mu$ guarantee that if momentum is conserved in *one* inertial frame it is conserved in *all* inertial frames? If so, it has the universal importance we seek. If *not*, it is not worth considering any longer, because we would never know whether such a momentum is conserved or not.

Suppose the total momentum P_μ of a *collection* of particles is conserved in the S frame,[7] that is,

$$(P_\mu)_{\text{initial}} = (P_\mu)_{\text{final}}. \tag{10.23}$$

We can transform this expression into a primed frame by using the Lorentz transformation for the four-vector,

$$P'_0 = \gamma(P_0 - VP_x/c) \tag{10.24a}$$

$$P'_x = \gamma(P_x - VP_0/c) \tag{10.24b}$$

$$P'_y = P_y, \quad P'_z = P_z. \tag{10.24c, d}$$

Then since the four-vector momentum components P_0, P_x, P_y, and P_z are each the same before and after (from Eq. 10.23), it immediately follows from Eqs. 10.24 that P'_0, P'_x, P'_y, and P'_z are also the same before and after. That is, the Lorentz transformation guarantees that if the momentum components are conserved in the unprimed frame, they are conserved in any other inertial frame as well!

The four-vector momentum $p_\mu = mu_\mu$ of a particle has passed two important tests:

1. The spatial components of p_μ reduce to the classical momentum in the limit of small velocities.

---

7. As in Chapter 1, we use lowercase p for the momentum of a single particle and capital P for the total momentum of a collection of particles.

2. If the total momentum of a system of particles is conserved in one inertial frame, it is conserved in all other inertial frames as well, using the Lorentz transformation between frames.

There is one more essential test: *Is momentum actually conserved in collisions of real relativistic particles*? The answer is definitely *yes*; many thousands of experiments have been carried out showing that momentum really is conserved, if we use the correct expression $\mathbf{p} = \gamma m\mathbf{v}$.

Mysteries remain! We have shown that the *spatial* components of the momentum four-vector behave like the momentum of a particle. What is the physical meaning of the *time* component p_0? And what is the meaning of the invariant quantity $-p_0^2 + p_x^2 + p_y^2 + p_z^2$? These are topics for the next chapter.

Sample Problems

1. A particle of mass m moves with speed $(3/5)c$. Find its momentum if it moves (a) purely in the x direction, (b) at an angle of 45° to the x and y axes.

Solution: (a) If it moves purely in the x direction, its momentum is

$$\mathbf{p} = \left(\frac{mv_x}{\sqrt{1 - v^2/c^2}}, 0, 0\right) = \left(\frac{m \cdot (3/5)c}{4/5}, 0, 0\right) = \left(\frac{3}{4}mc, 0, 0\right),$$

larger by a factor 5/4 than it would be if the classical formula $\mathbf{p} = m\mathbf{v}$ were correct. (b) If the particle moves with $v = (3/5)c$ at an angle of 45° to the x and y axes, then $\sqrt{1 - v^2/c^2} = 4/5$ and $v_x = v_y = v/\sqrt{2}$, so

$$\mathbf{p} = \left(\frac{5}{4}m\left(\frac{3c}{5\sqrt{2}}\right), \frac{5}{4}m\left(\frac{3c}{5\sqrt{2}}\right), 0\right) = \left(\frac{3}{4\sqrt{2}}mc, \frac{3}{4\sqrt{2}}mc, 0\right).$$

2. A particle at rest decays into a particle of mass m moving at $(12/13)c$ and a particle of mass M moving at $(5/13)c$. Find m/M, the ratio of their masses.

$(5/13)c$ ← M m → $(12/13)c$

Solution: Conserve momentum. The initial particle is at rest and so has no momentum at all. The final particles must therefore move off in opposite directions with equal but opposite momenta, as shown. The momentum magnitudes are

$$p_m = \frac{mv}{\sqrt{1 - v^2/c^2}} = \frac{m(12/13\,c)}{\sqrt{1 - (12/13)^2}} = \frac{m(12/13\,c)}{5/13} = \frac{12}{5}mc$$

and

$$p_M = \frac{Mv}{\sqrt{1 - v^2/c^2}} = \frac{M(5/13\,c)}{\sqrt{1 - (5/13)^2}} = \frac{M(5/13\,c)}{12/13} = \frac{5}{12}Mc,$$

which must be equal: That is, $(12/5)mc = (5/12)Mc$, so

$$\frac{m}{M} = \left(\frac{5}{12}\right)^2 = \frac{25}{144}.$$

3. Starship *HMS Pinafore* detects an alien ship approaching at $(3/5)c$. The aliens have just launched a 1-tonne (that is, 1000 kg) torpedo toward the *Pinafore,* moving at $(3/5)c$ relative to the alien ship. The *Pinafore* immediately erects a shield that can stop a projectile if and only if the magnitude of the projectile's momentum is less than 6.0×10^{11} kg m/s. (a) Find the torpedo's momentum four-vector in the alien's rest frame. (b) Find the torpedo's momentum four-vector in *Pinafore's* frame. (c) Does the shield stop the torpedo? (d) Find the torpedo's velocity relative to the *Pinafore.*

A ⟶ (3/5)c   t ⟶   P

Solution: (a) In the picture, the alien ship moves to the right (the positive x direction) relative to the *Pinafore,* so the alien's frame is "primed" and the *Pinafore's* frame is "unprimed." The components of the torpedo's momentum four-vector are

$$\left(p'_\mu\right) = \left(\frac{mc}{\sqrt{1 - v^2/c^2}}, \frac{m\mathbf{v}}{\sqrt{1 - v^2/c^2}}\right) = \left(\frac{5}{4}mc, \frac{5}{4}m\frac{3}{5}c, 0, 0\right)$$

$$= (5/4\,mc, 3/4\,mc, 0, 0), \text{ where } mc = (1000 \text{ kg})(3 \times 10^8 \text{ m/s}) = 3 \times 10^{11} \text{ kg m/s}.$$

(b) We can find the torpedo's momentum four-vector in the *Pinafore's unprimed* frame using the inverse Lorentz transformation of the four-vector momentum p_μ. This gives

$$p_0 = \gamma(p'_0 + Vp'_x/c) = \frac{5}{4}\left(\frac{5}{4}mc + \frac{3}{5}\cdot\frac{3}{4}mc\right) = \frac{17}{8}mc,$$

$$p_x = \gamma(p'_x + Vp'_0/c) = \frac{5}{4}\left(\frac{3}{4}mc + \frac{3}{5}\cdot\frac{5}{4}mc\right) = \frac{15}{8}mc,$$

$$p_y = p'_y = 0, \qquad p_z = p'_z = 0.$$

Altogether, $(p_\mu) = (17/8\,mc,\ 15/8\,mc, 0, 0)$.

(c) The momentum of the torpedo in *Pinafore's* frame is therefore

$$p_x = 15/8\, mc = (15/8)(3 \times 10^{11}\,\text{kg m/s}) = 5.6 \times 10^{11}\,\text{kg m/s}.$$

This is not quite enough to penetrate the *Pinafore's* shield, so all is well for the *Pinafore,* and the aliens have been frustrated yet again!

(d) The momentum $p_x = 15/8\, mc = mv/\sqrt{1 - v^2/c^2}$, so the torpedo's velocity in *Pinafore's* frame is the solution to $v/\sqrt{1 - v^2/c^2} = 15/8\, c$. That is, $8^2(v/c)^2 = 15^2(1 - v^2/c^2)$, so

$$\frac{v^2}{c^2} = \frac{15^2}{15^2 + 8^2} = \frac{15^2}{17^2},$$

giving $v/c = 15/17$. This result also follows from the relativistic velocity transformation of Section 8.3.

Problems

10–1. An electron in a linear accelerator attains the speed v. Find the ratio of its momentum to its nonrelativistic momentum mv if (a) $v = c/10$, (b) $v = c/2$, (c) $v = c(1 - 10^{-5})$, (d) $v = c(1 - 10^{-10})$.

10–2. A particle moves at speed $0.99c$. How fast must it move to *double* its momentum?

10–3. A spaceship, at rest in some inertial frame in space, suddenly needs to accelerate. The ship forcibly expels 10^3 kg of fuel from its rocket engine, almost instantaneously, at velocity $(3/5)c$ in the original inertial frame; afterwards the ship has a mass of 10^6 kg. How fast will the ship then be moving, valid to three significant figures?

10–4. An incident ball A of mass m and velocity v_0 moves to the right, striking a target ball B of mass $3m$ initially at rest in the lab frame S. The collision is perfectly elastic, one-dimensional, and *nonrelativistic.* Afterward, ball A bounces back with velocity $-v_0/2$ and ball B is pushed ahead with velocity $+v_0/2$. (a) Show that the sum of the classical momenta $p = mv$ is conserved in the lab frame. (b) Using the Galilean velocity transformation, find the velocity of each ball before and after the collision in an S' frame moving to the right at velocity v_0 relative to the lab. (c) Using the results of part (b), show that the total classical momentum is conserved in S'.

10–5. An incident ball A of mass m and velocity v_0 moves to the right, striking a target ball B of mass $3m$ initially at rest in the lab frame S. Afterward, ball A bounces back with velocity $-v_0/2$ and ball B is pushed ahead with velocity $+v_0/2$. (a) Show that the sum of the classical momenta $p = mv$ is conserved in the lab frame. (b) Using the *relativistic* velocity transformation

$$v'_x = \frac{v_x - V}{1 - v_x V/c^2},$$

find the velocity of each ball before and after the collision in an S' frame moving to the right at velocity v_0 relative to the lab. (c) Using the results of part (b), show that the total classical momentum is *not* conserved in S'.

10–6. A photon of momentum p_γ strikes an atomic nucleus at rest and is absorbed. If the mass of the final (excited) nucleus is M, calculate its speed.

10–7. Two particles make a head-on collision, stick together, and stop dead. The first particle has mass m and speed $(3/5)c$, and the second has mass M and speed $(4/5)c$. Find M in terms of m.

10–8. (a) Find the momentum of a distant galaxy moving away from us at speed $v = c/2$ in our frame of reference. The galaxy contains 10^{11} stars of average mass 2×10^{30} kg, the mass of our Sun. This is 10% of the galaxy's mass; the other 90% is in some kind of unseen "dark matter." (b) How fast would a single proton (mass 1.67×10^{-27} kg) have to move to have the same momentum? Express in the form $v/c = 1 - \epsilon$.

10–9. What is the maximum relative speed v/c a particle can have so there is no more than a 1% error in using the nonrelativistic momentum formula $\mathbf{p} = m\mathbf{v}$?

10–10. Starship *Titanic* suddenly finds itself approaching a space-iceberg at velocity $(3/5)c$, so immediately fires a 100 kg missile at speed $(4/5)c$ relative to the ship, in hopes of breaking up the ice before they crash. Find all components of the missile's momentum four-vector (a) in the spaceship frame, assuming all motion is in the x direction; (b) in the iceberg frame. (c) The iceberg will break up if the momentum impacting it is at least 1.0×10^{11} kg m/s. Does the iceberg break up? (d) What is the velocity of the missile in the iceberg frame?

10–11. Rederive the velocity transformation Eqs. 8.14 using the Lorentz transformation for the velocity four-vector u_μ. *Hint:* Two of the Lorentz transformation equations are needed.

10–12. A particle moves in the x, y plane with velocity $v = (4/5)c$, at an angle of 30° to the x axis. (a) Find all four components of the particle's four-vector velocity u_μ and find the invariant square of its components $-u_0^2 + u_1^2 + u_2^2 + u_3^2$. (b) Find all four components of the particle's four-vector velocity in a frame moving to the right at velocity $v = (3/5)c$. (c) Find the invariant square of the components in this frame.

10–13. Use the relativistic velocity transformation of Section 8.3 to confirm that in Sample Problem 3 of Chapter 10 the speed of the torpedo in *Pinafore's* frame is $(15/17)c$.

10–14. Prove that the sign of the zeroth component of a timelike four-vector is invariant under Lorentz transformations.

CHAPTER 11

Energy

NOW WE take up the important topic of *energy* in special relativity. Like momentum, energy is a concept central to classical physics. It can take on many forms—kinetic energy, potential energy, heat energy, energy of deformation, and so on. The *total* energy of an isolated system always stays the same—that is, it is *conserved.* If the kinetic energy decreases, some other form must increase to make up for it. Whatever else happens, the mass of the system remains unchanged; mass is separately conserved in classical physics.

11.1 Energy and Inertia

A few months after publishing his original 1905 paper on special relativity, Einstein published a follow-on paper, in which he first proposed, using a simple thought experiment, that the *inertia* of an object depends upon its *energy content.* By the "inertia" of an object he meant the reluctance of the object to change its state of motion. According to classical physics, as reviewed in Chapter 1, objects naturally move in a straight line at constant speed, and change this behavior only if a force is applied. The more massive the object is, the more inertia it has; that is, if its mass is larger, it is harder to change its motion. What Einstein deduced from his thought experiment is that the energy content of an object is proportional to its mass, as expressed by $E = mc^2$, probably the most famous equation in all of physics.[1]

In this section we describe a somewhat similar thought experiment, invented by Einstein himself a year later, that arrives at the same result in a more intuitive way. Picture a cardboard tube of length L floating in gravity-free empty space, at rest in

1. An English translation of this paper is contained in the book *The Principle of Relativity,* which is a collection of original papers by H. A. Lorentz, H. Minkowski, and H. Weyl, as well as by Einstein himself (Dover Publications, Inc. 1923).

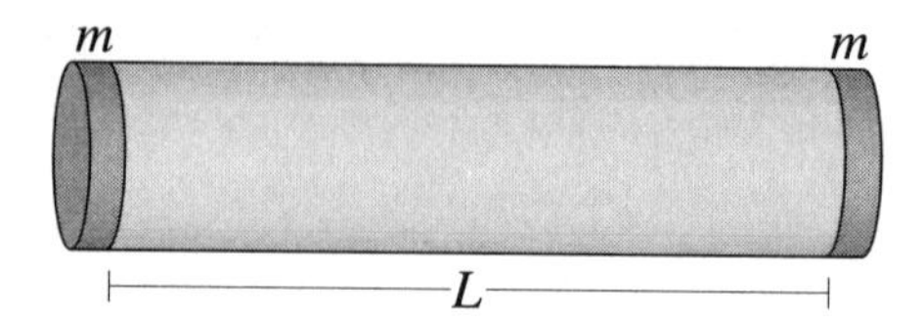

FIGURE 11.1
A tube with endcaps, isolated in space.

some inertial frame, as shown in Fig. 11.1. The tube has a heavy metal cap of mass m at each end. We assume for simplicity that essentially all the mass is tied up in the metal caps; the mass of the tube itself is negligible by comparison, so the mass of the entire assembly is $2m$.

Now a pulse of light is spontaneously emitted by the left-hand cap, directed toward the right, as shown in Fig. 11.2(a). Long before Einstein's theory, it was known that light carries *energy* (which is obvious; you can *feel* the energy from a heat lamp that carries infrared radiation, for example.) In fact, most of the energy we use on Earth can be traced back to energy from light emitted by the Sun.

It was also known that a light beam carries *momentum* $p = E/c$, where E is the energy of the beam. If you lie on the beach in the sun, you are very aware of the energy carried by sunlight because of the heat you feel. But it is also true that the sunlight pushes you down very slightly into the sand, although you don't feel this; the speed of light is so large that the momentum $p = E/c$ in the Sun's rays is not noticeable.

Therefore if the light pulse emitted from the left endcap has a small energy ΔE, it also has a small momentum $\Delta p = \Delta E/c$ directed toward the *right*. To conserve momentum, the tube must therefore recoil toward the left, with an equal but opposite

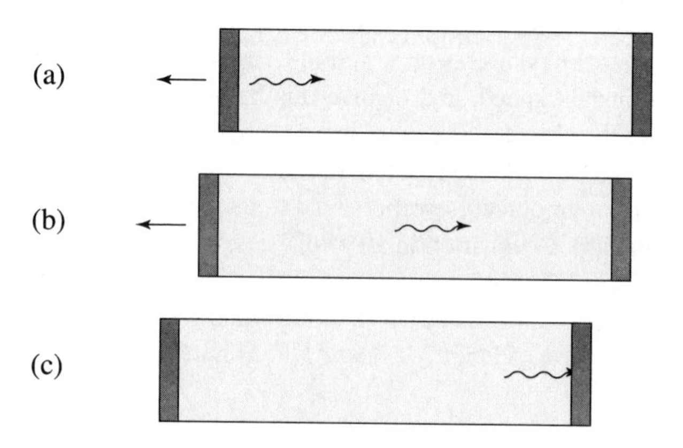

FIGURE 11.2
(a) Light is emitted from the left endcap; the tube recoils. (b) As the light moves from left to right, the tube drifts slightly to the left. (c) The light is absorbed in the right endcap; the tube stops.

small (and therefore nonrelativistic) momentum $\Delta p = \Delta E/c = (2m)\Delta v_{\text{tube}}$, where Δv_{tube} is its velocity. It follows that

$$\Delta v_{\text{tube}} = \frac{\Delta p}{2m} = \frac{\Delta E}{2mc}, \tag{11.1}$$

as shown in Fig. 11.2(b). During the short time it takes the light beam to travel to the right endcap, the tube will drift to the left a little bit.

The time it takes the light to travel is $t = \text{distance/speed} = L/c$, and during this time the tube will recoil the distance

$$\Delta x = \Delta v_{\text{tube}} t = \frac{\Delta E}{2mc} \frac{L}{c} = \frac{\Delta E\ L}{2mc^2}. \tag{11.2}$$

When the light wave reaches the right endcap, it is absorbed, giving its momentum back to the tube. This causes the tube to stop. As shown in Fig. 11.2(c), the net effect of the experiment is to move the tube a distance $\Delta E\ L/2mc^2$ to the left. It started at rest and ended at rest.

The tube, including the light inside it, is completely isolated, and so (as expected) the net momentum, which was zero initially, remains zero throughout the experiment. Yet it seems that *with no outside influence the center of mass of the system has moved!* In fact, the center of mass seems to have moved to the *left* the distance $\Delta x = \Delta E\ L/2mc^2$. This is very odd! Is there some way to prevent this? Is it possible, in spite of appearances, that the center of mass of the system is still at the same location at the *end* of the thought experiment as it was at the *beginning*? Is it possible that the tube can move to the left without moving the center of mass?

Suppose that as the light leaves the left endcap, that endcap loses some mass, dropping from m to $m - \Delta m$. This seems strange, since light has no mass, although it does have energy. Also suppose that as the light is absorbed by the right endcap, the mass of that endcap gains some mass, increasing from m to $m + \Delta m$.

Now it is clear that if we choose Δm just right, it is possible to keep the center of mass fixed, even though the tube has drifted to the left. The reason is that now the tube is lighter on the left end and heavier on the right, so the balancing point (if we imagine a teeter-totter with a fulcrum at the center of mass) can remain at the same point it was originally.

If the center of mass is at the same point, then the mass of the left endcap multiplied by its distance from the center of mass must equal the mass of the right endcap multiplied by its distance from the center of mass. That is, as in Fig. 11.3,

$$(m - \Delta m)(L/2 + \Delta x) = (m + \Delta m)(L/2 - \Delta x). \tag{11.3}$$

Multiplying out both sides, we have

$$mL/2 - \Delta m\ L/2 + m\Delta x - \Delta m\ \Delta x = mL/2 + \Delta m\ L/2 - m\Delta x - \Delta m\ \Delta x. \tag{11.4}$$

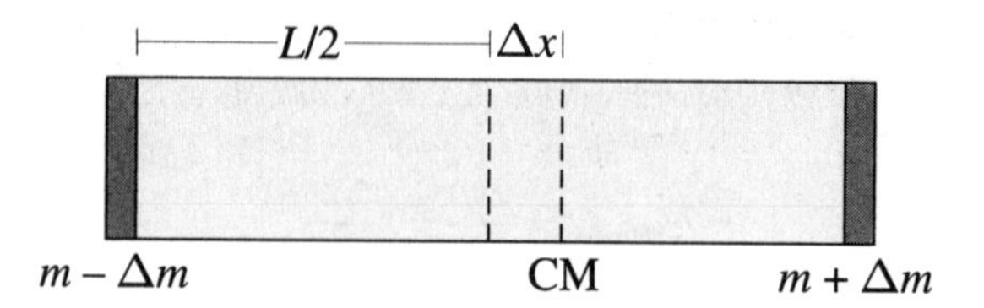

FIGURE 11.3
The tube and endcaps, with the endcap masses affected by the light beam.

Two terms on the left cancel with two terms on the right, leaving

$$2m\ \Delta x = \Delta m\ L. \tag{11.5}$$

Solving for the unknown mass change Δm gives us

$$\Delta m = \frac{2m\ \Delta x}{L} = \frac{2m}{L}\left(\frac{\Delta E\ L}{2mc^2}\right) = \frac{\Delta E}{c^2}, \tag{11.6}$$

substituting in Eq. 11.2 for Δx.

What we have found is that *if the energy of an object changes by* ΔE, its mass changes by $\Delta m = \Delta E/c^2$. Equally well, we can say that *if the mass of an object changes, the energy of the object changes by* $\Delta E = \Delta m\ c^2$.

This thought experiment shows that if energy is removed from an object (even if the energy is in the form of massless light!), the mass of the object is reduced.[2]

11.2 The Energy–Momentum Four-Vector

Now we need to understand $\Delta E = \Delta m\ c^2$ in particular, and relativistic energy in general, in a broader context. We will start with collisions, as we did for momentum in Chapter 10. In classical physics, a collision is *elastic* if the total kinetic energy of the colliding objects stays constant. In that case no other forms of energy are created or destroyed. In *inelastic* collisions some or even all of the kinetic energy is converted into heat energy or some other form of energy. How is this concept of energy changed in special relativity?

In answering this question, we will also answer some questions left over at the end of Chapter 10. In that chapter we showed that the momentum of a particle is

---

2. The "Einstein box" thought experiment has involved approximations, and also a minor swindle. The swindle is the implicit assumption that the cardboard tube is rigid, in that the right endcap starts to move leftward at the very instant the left endcap recoils from the light pulse emission. The implied infinite signal velocity in the cardboard is not possible in relativity. The swindle is easily undone, however, simply by eliminating the tube entirely, leaving only the endcaps. The resulting thought experiment can be solved without approximations and results in Eq. 11.6. See Problem 11–20.

$\mathbf{p} = \gamma m \mathbf{v} = m\mathbf{v}/\sqrt{1 - v^2/c^2}$, which reduces to the nonrelativistic momentum expression $\mathbf{p} = m\mathbf{v}$ if the particle's speed is small. Both momentum expressions have components in the x, y, and z directions; however, the relativistic momentum encompasses only *three* components of the full *four*-vector

$$p_\mu = m u_\mu \equiv \frac{dx_\mu}{d\tau}, \quad (\mu = 0, 1, 2, 3) \tag{11.7}$$

equivalent to components one, two, and three of

$$(p_\mu) = \left(\frac{mc}{\sqrt{1 - v^2/c^2}}, \frac{m\mathbf{v}}{\sqrt{1 - v^2/c^2}} \right). \tag{11.8}$$

What is the meaning of $p_0 \equiv mc/\sqrt{1 - v^2/c^2}$, the *zeroth* component of this four-vector?

The first clue about the meaning of p_0 comes from Chapter 10. We showed there that if all four components of p_μ are conserved in one inertial frame of reference, then all four are conserved in any inertial frame. So p_0, like $\mathbf{p} = (p_x, p_y, p_z)$, is a quantity that is conserved, at least in some circumstances. The obvious candidate is *energy.* Is p_0 in fact the *energy* of the particle?

An initial *objection* to this idea is that energy has the dimensions (mass)(velocity)^2, while p_0 has dimensions (mass)(velocity). We can easily fix this problem, however, by multiplying p_0 by the constant speed of light c to get the right dimensions. That is, let

$$E = p_0 c = \frac{mc^2}{\sqrt{1 - v^2/c^2}}. \tag{11.9}$$

Does this make sense? Is $E = p_0 c$ *really* the energy of the particle?

First, suppose a particle is at *rest* in our frame. In that case, the energy of the particle according to (11.9) would be

$$E_0 = mc^2. \tag{11.10}$$

Even when a particle is at rest it has energy, according to this formula. This is the same result we got more intuitively by the thought experiment of Section 11.1! We will call $E_0 = mc^2$ the *mass energy* of the particle, since the only property of the particle that E_0 depends upon is its mass.

If the particle is moving it has even more energy according to Eq. 11.9, because the denominator then becomes less than 1. That makes sense. We expect that as a particle moves faster and faster it gains more and more energy. In fact, the *kinetic* energy of the particle is the energy above and beyond any energy it has at rest: kinetic energy is the energy of *motion.* Using Eq. 11.9, the kinetic energy of a particle of mass m and speed v is

$$KE = E - E_0 = \frac{mc^2}{\sqrt{1 - v^2/c^2}} - mc^2 = mc^2 \left(\frac{1}{\sqrt{1 - v^2/c^2}} - 1 \right). \tag{11.11}$$

If $v = 0$, the kinetic energy is zero. If the particle has a small velocity v, with $v/c \ll 1$, we can use the binomial expansion of Appendix A to find the kinetic energy for small velocities. The binomial series is[3]

$$(1+x)^n = 1 + nx + \frac{n(n-1)}{2!}x^2 + \frac{n(n-1)(n-2)}{3!}x^3 + \cdots, \tag{11.12}$$

which converges rapidly if $|x| \ll 1$. Therefore, if $v \ll c$,

$$(1 - v^2/c^2)^{-1/2} = 1 + \frac{1}{2}\frac{v^2}{c^2} + \cdots,$$

so

$$\begin{aligned} KE &= mc^2\left(\frac{1}{\sqrt{1-v^2/c^2}} - 1\right) = mc^2\left(1 + \frac{1}{2}\frac{v^2}{c^2} + \cdots - 1\right) \\ &= \frac{1}{2}mv^2 + \cdots, \end{aligned} \tag{11.13}$$

whose biggest term is the nonrelativistic kinetic energy! Once again we retrieve the classical formula $KE = (1/2)mv^2$ in the nonrelativistic limit $v \ll c$, neglecting smaller terms. The identification $E = p_0c$ has passed an important test.

Figure 11.4 illustrates a particle's energy $E = mc^2/\sqrt{1 - v^2/c^2} \equiv \gamma mc^2$ and its kinetic energy $KE = (\gamma - 1)mc^2$ as functions of velocity. The "total" energy E includes the mass energy $E_0 = mc^2$ as well as the kinetic energy. For small velocities the kinetic energy is $KE \equiv (1/2)mv^2$, but for larger velocities, the kinetic energy grows more quickly, diverging as $v \to c$.

The formula

$$E = \gamma mc^2 = \frac{mc^2}{\sqrt{1 - v^2/c^2}} \tag{11.14}$$

for the energy of a particle has now passed some critical tests:

(i) The kinetic energy (the energy of motion) reduces to the nonrelativistic kinetic energy $KE = (1/2)mv^2$ if the particle moves at nonrelativistic speeds.

(ii) If the total energy of two particles is conserved in a collision when viewed from one inertial frame, it is conserved in all inertial frames.

Why is (ii) correct? Because the energy $E = p_0c$ is the zeroth component of what we will now call the energy–momentum four-vector. So it follows from the Lorentz transformation equations[4]

---

3. The binomial series can be derived from the Taylor series. See Appendix A.

4. See Eqs. 10.23 of Chapter 10.

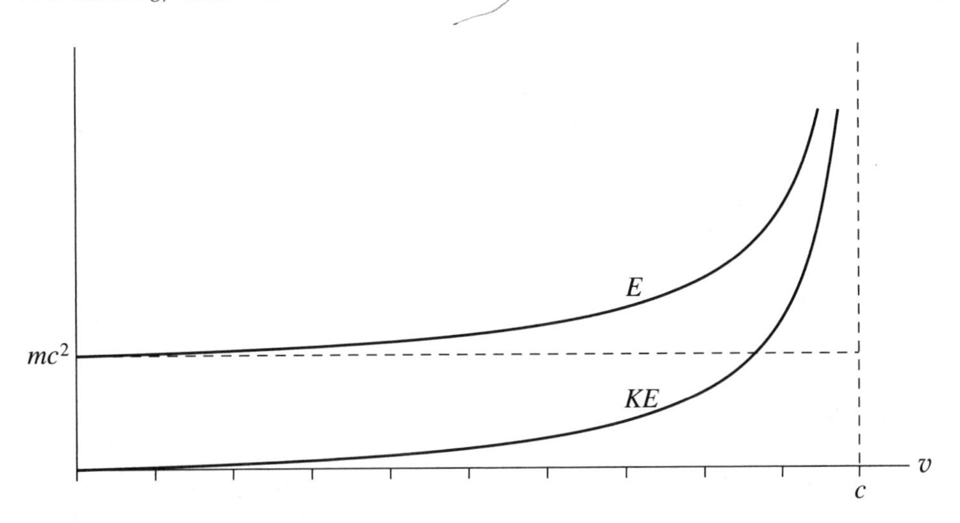

FIGURE 11.4
The relativistic kinetic energy KE and "total" energy $E = mc^2 + KE$.

$$
\begin{aligned}
p'_0 &= \gamma(p_0 - Vp_x/c) \\
p'_x &= \gamma(p_x - Vp_0/c) \\
p'_y &= p_y \\
p'_z &= p_z
\end{aligned}
\tag{11.15}
$$

that if (as experiment shows) the spatial components p_x, p_y, and p_z are conserved in all frames, then the zeroth components p_0 must be conserved in all frames as well.

The final test is to see if E really *is* conserved in the actual collisions of particles. The answer is definitely *yes*; thousands of experiments have confirmed that if we use $E = \gamma mc^2$ for the energy of a particle, the total energy of all the particles is conserved in real collisions.

Now that we have identified the meaning of p_0, there is still one more question left over from Chapter 10. We know that for *any* four-vector A_μ, the particular combination $-A_0^2 + A_x^2 + A_y^2 + A_z^2$ is Lorentz invariant; that is, it has the same value in any inertial reference frame. What is this value for the energy–momentum four-vector? We need to evaluate

$$-p_0^2 + p_x^2 + p_y^2 + p_z^2 \equiv -(E/c)^2 + p_x^2 + p_y^2 + p_z^2. \tag{11.16}$$

Substituting in the expressions $E = \gamma mc^2$ and $\mathbf{p} = \gamma m\mathbf{v}$, we find that

$$
\begin{aligned}
-(E/c)^2 + p_x^2 + p_y^2 + p_z^2 &= -\gamma^2 m^2 c^2 + \gamma^2 m^2 v^2 \\
&= -\gamma^2 m^2 c^2 (1 - v^2/c^2) = -m^2 c^2,
\end{aligned}
\tag{11.17}
$$

which is clearly invariant. We can rewrite this in the slightly different form

$$E^2 = p^2c^2 + m^2c^4, \tag{11.18}$$

which is a very useful relationship between a particle's energy, momentum, and mass that we will use many times in this chapter and beyond. If the momentum of the particle is zero, the energy is $E = mc^2$ as expected; the energy then grows as the momentum increases.

11.3 Example: The Decay of a Particle

Suppose that a particle of mass M breaks up into two identical pieces, each of mass m, as shown in Fig. 11.5. (By now we don't trust any classical "law of nature," so we're not going to automatically assume that $M = 2m$!) This break-up process actually occurs in nature. For example, a particle called the K^0 meson decays spontaneously (in about 10^{-10} s) into two identical π mesons $K^0 \to \pi^0 + \pi^0$. (The "zeros" mean that the particle has no electric charge.) We assume conservation of the momentum four-vector in the decay; that is, we assume that both the spatial (momentum) components and the zeroth (energy) component are conserved.

First conserve *momentum,* that is, the spatial components of the energy–momentum four-vector. The initial momentum is *zero* because M is at rest. Each final particle m has momentum, but the total momentum is zero if the two particles have equal but opposite velocities v. Momentum conservation does not tell us what v is, however.

Now conserve *energy.* The initial energy is Mc^2, since M is at rest. The final energy is $2mc^2/\sqrt{1 - v^2/c^2}$, where v is the speed of each final particle. Therefore conservation of energy tells us that

$$M = \frac{2m}{\sqrt{1 - v^2/c^2}}, \tag{11.19}$$

showing that mass is not conserved in the decay! The initial mass M is *larger* than the total final mass $2m$. Of course, kinetic energy has increased to make up for the decreased mass energy—the final particles do have kinetic energy even though the initial particle had none. There is a trade-off here: Energy can be moved back and forth between mass energy and kinetic energy, as long as the total energy is conserved.

FIGURE 11.5
A particle of mass M breaks up into two particles of mass m.

There is clearly a difference between the classical and relativistic views of mass and energy. Before Einstein, it was thought that mass and energy were separately conserved. No matter what a collection of particles was doing, whether colliding, combining, exploding, or whatever, the total mass always remained the same. Similarly, for an isolated system with no applied external forces doing work on the system, and no escape of energy from the system, energy was conserved. Now we have seen that in relativistic physics mass is not always conserved, so one classical conservation law has been proven false. But if mass energy is included, the total energy of a system is still conserved. *Two* classical conservation laws have been merged by the theory of relativity into a *single* law, the conservation of energy, which includes mass energy as well as kinetic energy.

Conservation of energy gives us the speed of each final particle in terms of the masses M and m. Solving (11.19) for v/c, we find

$$v/c = \sqrt{1 - 4m^2/M^2}. \qquad (11.20)$$

Before evaluating v/c numerically in the $K^0 \to \pi^0 + \pi^0$ decay, it is first convenient to introduce what are called *energy units*.

11.4 Energy Units

In classical mechanics the Système International (SI) unit for energy is the *joule* (J), where $1\,\mathrm{J} = 1\,\mathrm{kg\ m^2/s^2}$. Joules are a convenient unit for classical mechanics, but they are not convenient in typical relativistic collisions or decays, which normally involve subatomic particles of very small mass. The energy of a relativistic electron is likely to be an extremely small fraction of a joule. A more convenient unit of energy in atomic, nuclear, and fundamental particle physics is the *electron volt* (eV).

An electron volt is the energy gained by an electron or proton (or any particle with electric charge e) as it accelerates through a potential difference of 1 volt. An electron accelerating from rest across 12 volts (which might be supplied by a car battery) would end up with 12 eV of kinetic energy. The conversion between eV and joules is

$$1\,\mathrm{eV} = 1.60 \times 10^{-19}\,\mathrm{J}. \qquad (11.21)$$

An electron volt is a good unit to use for electron transitions in atoms, for example, or for the energy of photons emitted in such transitions, which are typically in the visible, infrared, or ultraviolet range of the electromagnetic spectrum. Transitions between inner electrons of heavy elements are typically in the keV range ($1\,\mathrm{keV} = 10^3\,\mathrm{eV}$), where the resulting photons are called x-rays. The unit MeV ($1\,\mathrm{MeV} = 1$ mega electron volt $= 10^6\,\mathrm{eV}$) is more useful in describing energies in nuclear physics, including the energy differences between energy levels in nuclei and the γ-ray (gamma-ray) photons emitted in nuclear transitions. The units GeV ($1\,\mathrm{GeV} = 1$ giga electron volt $= 10^9\,\mathrm{eV}$) or TeV ($1\,\mathrm{TeV} = 1$ tera electron volt $= 10^{12}\,\mathrm{eV}$) are appropriate for describing the energies of fundamental particles in high-energy accelerators such as linear accelerators or synchrotrons.

The mass energies mc^2 of particles are usually given in MeV units: the electron has a mass energy 0.511 MeV (just over half a million electron volts), for example, and the proton has the mass energy 938.3 MeV. (The mass energies of some other particles are given in Appendix I.)

Now that we have the unit "electron volts" for energy, there are also comparable units for momentum and mass that are especially useful for nuclei and fundamental particles. Note from $E^2 = p^2c^2 + m^2c^4$ that if we measure energy in MeV, momentum can be assigned the units MeV/c and mass can be assigned the units MeV/c^2. That would give each term in the equation the same units.

Suppose, for example, an electron (whose mass energy $mc^2 = 0.511\,\text{MeV}$) is moving at speed $(4/5)c$. The mass of the electron is $m = 0.511\,\text{MeV}/c^2$; its energy is

$$E = \frac{mc^2}{\sqrt{1 - v^2/c^2}} = \frac{0.511\,\text{MeV}}{3/5} = 0.852\,\text{MeV},$$

and the magnitude of its momentum is

$$p = \frac{mv}{\sqrt{1 - v^2/c^2}} = \frac{(0.511\,\text{MeV}/c^2)(4/5c)}{3/5} = 0.681\,\text{MeV}/c.$$

The reader can readily verify the fact that the numerical values of E, p, and m in this example are related by $E^2 = p^2c^2 + m^2c^4$.

Now we can also find the speed of each π^0 meson that results from the K^0 decay as given by Eq. 11.20 in Section 11.3. The mass energies are $M_Kc^2 = 498$ MeV and $m_\pi c^2 = 135$ MeV, so the speed of each pion is

$$v/c = \sqrt{1 - 4(135)^2/(498)^2} = 0.840.$$

Having found the speed of each pion, we can find each pion's energy and momentum using $E = \gamma mc^2$ and $\mathbf{p} = \gamma m\mathbf{v}$. It is easier, however, just to use the conservation laws. The initial energy is $M_Kc^2 = 498\,\text{MeV}$ and the initial momentum is zero in the kaon rest frame. The two final pions must therefore have equal but opposite momenta, so from $E^2 = p^2c^2 + m^2c^4$ we see that (since they have equal mass) they must have equal energy as well. From energy conservation, each pion therefore has energy $E_\pi = (498\,\text{MeV})/2 = 249\,\text{MeV}$, and so the magnitude of the momentum of each pion is $p_\pi = (1/c)\sqrt{E_\pi^2 - m_\pi^2c^4} = (1/c)\sqrt{249^2 - 135^2}\,\text{MeV} = 209\,\text{MeV}/c$.[5]

---

5. WARNING: Physicists sometimes leave off (in speech or in writing) the $1/c$ in momentum and the $1/c^2$ in mass, as a kind of shorthand. We might say "The mass of a proton is 938 MeV" or "The momentum of that α-particle is 27 MeV." If we want to keep careful track of units, we can always put the c's back in. In this book the c's will always be kept in.

11.5 Photons

In the middle of the nineteenth century, the Scottish physicist James Clerk Maxwell showed that light waves are *electromagnetic* waves. According to Maxwell's theory, light consists of oscillating electric and magnetic fields that propagate together at the speed of light. Such light waves are typically smooth and extended, something like waves on a string or waves on the surface of water. In 1905, the same year that Einstein published his special theory of relativity, Einstein also published a *different* paper claiming that light sometimes behaves as though it comes in lumps he called energy quanta, not at all the smooth waves of Maxwell. These lumps later came to be called *photons.*

Photons travel at the speed of light and carry discrete amounts of energy $E = h\nu$, where h is known as Planck's constant and ν is the frequency (the number of cycles/second) of the associated light wave. Many experiments confirm that sometimes light does indeed behave as though it is made of such lumps. (A satisfactory understanding of the particle and wave properties of photons comes only from a thorough study of quantum mechanics.)

Suppose we try to apply the usual formulas for relativistic momentum and energy, $\mathbf{p} = \gamma m \mathbf{v}$ and $E = \gamma mc^2$, to *photons.* We immediately see a problem. Associated as they are with light waves, photons travel at the speed $v = c$, so the factor $\gamma \equiv 1/\sqrt{1 - v^2/c^2}$ is *infinite* in each of the above expressions.

However, photons do *not* have infinite momentum and energy! They are easily created in any light bulb. The only way to resolve the difficulty is to let the mass of a photon be *zero.* Then the expressions for $\mathbf{p}$ and E become indeterminate; each involves dividing zero by zero. Even though the usual formulas for $\mathbf{p}$ and E do not make sense for zero-mass photons, the invariant quantity

$$-(E/c)^2 + p_x^2 + p_y^2 + p_z^2 = -m^2c^2$$

does make sense. For photons,

$$-(E/c)^2 + p_x^2 + p_y^2 + p_z^2 = 0, \tag{11.22}$$

so the energy and momentum of a photon are related by

$$E = pc, \tag{11.23}$$

where $p = \sqrt{p_x^2 + p_y^2 + p_z^2}$ is the magnitude of the photon's momentum.[6] Photons are involved in a huge number of particle collisions and decays, and experiments show that photon energies and momenta are indeed related by $E = pc$.

---

6. This is consistent with the relationship $E = pc$ for light beams known even before Einstein, as mentioned in Section 11.1.

Sample Problems

1. A sturdy teakettle has mass 1.0 kg and contains 2 liters of water, with mass 2.0 kg. The teakettle is *tightly sealed.* If the water is heated from 20°C (room temperature) to 100°C, which requires 6.7×10^5 joules of energy, what is the fractional increase in the filled teakettle's mass?

Solution: The mass increases by $\Delta m = \Delta E/c^2 = (6.7 \times 10^5\ \text{J})/(3.0 \times 10^8\ \text{m/s})^2 = 7.44 \times 10^{-12}$ kg, so the fractional increase in mass is

$$\frac{\Delta m}{m} = \frac{7.44 \times 10^{-12}\ \text{kg}}{3.0\ \text{kg}} = 2.5 \times 10^{-12}.$$

This change is so small that it would be *very* difficult to measure!

2. The tau lepton (or *tauon*) has mass $m = 1777\ \text{MeV}/c^2$. A particular tauon decays in time $\tau = 3.0 \times 10^{-13}$ s in its rest frame. If it has energy $100\,mc^2$ in the lab, how far will it move before decaying?

Solution: The tauon's energy is $E = \gamma mc^2 = 100\,mc^2$, so obviously $\gamma = 100$. We know also that the tauon's lifetime in the lab is $t = \tau/\sqrt{1 - v^2/c^2}$. Therefore $t = 100\,\tau = 3 \times 10^{-11}$ s. The distance it moves in that time is

$$d = vt \cong ct = 3 \times 10^8\ \text{m/s} \cdot 3 \times 10^{-11}\ \text{s} = 9.0 \times 10^{-3}\ \text{m} = 9.0\ \text{mm},$$

since with such a large γ, the particle is moving at nearly the speed of light.

(More precisely, $\sqrt{1 - v^2/c^2} = 10^{-2}$, so $1 - v^2/c^2 = 10^{-4}$. Therefore $v/c = (1 - 10^{-4})^{1/2} \cong 1 - (1/2) \times 10^{-4}$ using the binomial expansion of Appendix A. This tells us that $d = 9$ mm to about four decimal places.)

3. A photon of energy 12.0 TeV (1 TeV $= 10^{12}$ eV) strikes a particle of mass M_0 at rest. After the collision there is only a single final particle of mass M, moving at speed $(12/13)c$. Using MeV units, find (a) the momentum of the final particle, (b) the mass M, (c) the mass M_0.

Solution: The before-and-after pictures are as follows:

M_0

M → $(12/13)c$

Energy and momentum are both conserved. Energy conservation gives

$$E_\gamma + M_0c^2 = E_\text{f},$$

where E_γ is the energy of the photon and E_f is the energy of the final particle. Momentum conservation requires that

$$p_\gamma = p_\text{f}$$

since the initial particle of mass M_0 has no momentum. We know that $E_\gamma = 12.0$ TeV, $p_\gamma = E_\gamma/c = 12.0\ \text{TeV}/c$, $E_\text{f} = \gamma Mc^2$, and $p_\text{f} = \gamma Mv$, where γ is the gamma factor of the final particle, which is

$$\gamma = \frac{1}{\sqrt{1 - v^2/c^2}} = \frac{1}{\sqrt{1 - (12/13)^2}} = \frac{1}{5/13} = \frac{13}{5}.$$

(a) The momentum of the final particle is $p_\text{f} = p_\gamma = 12.0\ \text{TeV}/c$.
(b) The mass of the final particle is therefore

$$M = \frac{p_\text{f}}{\gamma v} = \frac{12.0\ \text{TeV}/c}{(13/5)(12/13c)} = \frac{12.0\ \text{TeV}/c}{12/5c} = 5.0\ \text{TeV}/c^2.$$

(c) Using energy conservation, we see that the mass energy of the initial particle is

$$M_0c^2 = \gamma Mc^2 - E_\gamma = \frac{13}{5}(5.0\ \text{TeV}) - 12.0\ \text{TeV} = 1.0\ \text{TeV},$$

so its mass is $M_0 = 1.0\ \text{TeV}/c^2$.

4. Spaceship A fires a beam of protons in the forward direction with velocity $v = (4/5)c$ at an alien ship B fleeing directly away at velocity $V = (3/5)$. Transforming the beam's energy–momentum four-vector using the Lorentz transformation of Eqs. 11.15, find the beam's velocity v' in the frame of ship B.

Solution: Let A and B be at rest in the unprimed and primed frames, respectively. It would be easy to find v' using the velocity transformation of Chapter 8, but here we use the Lorentz transformation instead, partly to illustrate the fact that there are *three different* gamma factors in this problem!

In A's frame, we can write the energy and momentum of the beam as $E = \gamma_m mc^2$ and $p = \gamma_m mv$, where $\gamma_m = 1/\sqrt{1 - v^2/c^2}$, and in B's frame $E' = \gamma'_m mc^2$ and $p' = \gamma'_m mv'$, where $\gamma'_m = 1/\sqrt{1 - v'^2/c^2}$. We use the subscripted γ_m and γ'_m to distinguish these gamma factors from the factor $\gamma = 1/\sqrt{1 - V^2/c^2}$ in the Lorentz transformation between the primed and unprimed frames.

We are given that $\gamma_m = 1/\sqrt{1 - (4/5)^2} = 5/3$, so that $p_0 = E/c = (5/3)mc$ and $p_x = (5/3)m(4/5)c = (4/3)mc$. Then using the first of Eqs. 11.15 with $\gamma = 1/\sqrt{1 - (3/5)^2} = 5/4$, we find that

$$p'_0 = \frac{5}{4}\left(\frac{5}{3}mc - \frac{3}{5}\cdot\frac{4}{3}mc\right) = \frac{5}{4}\left(\frac{25}{15} - \frac{12}{15}\right)mc = \frac{13}{12}mc.$$

Now $p'_0 \equiv E'_0/c = \gamma'_m mc$, so we have found that $\gamma'_m = 1/\sqrt{1 - v'^2/c^2} = 13/12$, which gives the final result $v' = (5/13)c$.

We seldom have to deal with more than one gamma factor in a problem, but when we do, it is safest to use subscripts on some of the γ's to keep them straight!

Problems

11–1. The total energy of a proton that has been accelerated in a synchrotron is 30 times its mass energy mc^2. In terms of m, find the proton's kinetic energy and the magnitude of its momentum. Also find its velocity expressed as a fraction of the speed of light.

11–2. A free neutron at rest decays into a proton, an electron, and an antineutrino. The mass energies of the proton and electron are 938.3 MeV and 0.5 MeV, respectively, and the mass energy of the antineutrino (thought to be a small fraction of an eV) is negligible on the scale of MeV. Find the total kinetic energy of the decay products.

11–3. Calculate the energy required to accelerate a 1.0×10^8 kg spaceship to a speed such that in the Earth frame of reference a passenger inside ages only 10% as fast as we do on Earth. Compare this energy with the mass energy of the ship.

11–4. Should an average proton or electron at 15 million degrees Kelvin (about the temperature at the center of the Sun) be treated by relativistic mechanics? For the purposes of this problem, use as a criterion that the kinetic energy differ from the classical kinetic energy $(1/2)mv^2$ by more than 1%. (The average kinetic energy of a particle at temperature T is $(3/2)\,kT$, where k is Boltzmann's constant.)

11–5. A typical ^{238}U nucleus at rest decays in about 4.5×10^9 years into an α-particle (^{4}He nucleus) and a ^{234}Th nucleus. In atomic mass units (u), defined by $1\,\text{u} = 931.494\ \text{MeV}/c^2$ (corresponding to $M_{^{12}\text{Carbon}} = 12.000000$ u) the atomic masses are ^{238}U : 238.050783, ^{234}Th : 234.043601, and ^{4}He : 4.002603. (a) Find the "decay energy" (in MeV), which is the net loss of mass energy in the decay. (b) The kinetic energy of the emerging α-particle is 4.195 MeV. Why is this different from the decay energy found in part (a)? (c) Explain the discrepancy quantitatively. (Note that kinetic energies in this problem are nonrelativistic, because they are very small compared with the mass energies. Therefore one can use $KE = (1/2)mv^2 = p^2/2m$.)

11–6. A gamma-ray photon emitted from an excited ^{7}Li nucleus has an energy of 0.478 MeV. Calculate its momentum in MeV/c, and its frequency in inverse seconds.

11–7. An unstable particle of mass m decays in time $\tau = 10^{-10}$ s in its own rest frame. If its energy is $E = 1000\,mc^2$ in the lab, how far (in meters) will it move before decaying?

11–8. An antineutrino of energy 1.5 MeV strikes a proton at rest. Is the reaction $\bar{\nu} + p \rightarrow n + e^+$ possible for this antineutrino?

11–9. An electron–positron pair at rest disintegrates into two photons: $e^- + e^+ \rightarrow 2\gamma$. (a) Find the energy (in MeV) and wavelength (in nm, where 1 nanometer (nm) = 10^{-9} m) of each photon. (b) Sometimes an electron–positron pair decays into three photons instead. What is the total kinetic energy of the three photons?

11–10. A particular K^+ meson, with mass energy 494 MeV, decays in 1.2×10^{-8} s in its rest frame. What is its energy in the lab frame, if it travels 10.0 m before decaying?

11–11. A Λ particle of mass energy 1116 MeV has kinetic energy 1116 MeV in the lab and travels 13.7 cm before decaying. How long did it live in its own rest frame?

11–12. The kinetic energy of a particular newly created particle in the laboratory happens to equal its mass energy. If it travels a distance d before decaying, find an expression for how long it lived in its own rest frame.

11–13. A photon of energy $E = 5000$ MeV is absorbed by a nucleus of mass M_0 originally at rest. Afterward, the excited nucleus has mass M and is moving at $(5/13)c$. Using MeV units, find (a) the momentum of M, (b) the mass M, and (c) the mass M_0.

11–14. A particle of mass $M_1 = 1440$ MeV$/c^2$ is moving in the laboratory with speed $(5/13)c$. It decays into two particles with masses M_2 and M_3, where M_2 is at rest in the lab, and M_3 moves at speed $(12/13)c$. Find M_2 and M_3 in units of MeV$/c^2$.

11–15. The "Tevatron" at Fermilab near Chicago accelerates protons to energies of 1 TeV = 10^{12} eV. Find the speed of the protons. (Approximate $mc^2 \approx 1$ GeV = 10^9 eV for protons, and use both the form $v/c = 1 - \epsilon$ and the binomial approximation.)

11–16. The Large Hadron Collider (LHC) at CERN near Geneva, Switzerland, accelerates protons of mass energy mc^2 to an energy $E \gg mc^2$. (a) In terms of mc^2 and E, write down a simple expression for the gamma factor ($\gamma \equiv 1/\sqrt{1 - v^2/c^2}$) of LHC protons. (b) The speed of these protons can be expressed in the form $v/c = 1 - \epsilon$, where ϵ is very small. Using the fact that $\epsilon \ll 1$ (or, equivalently, that $mc^2/E \ll 1$), derive a simple expression for ϵ in terms of mc^2 and E. (c) Each proton has mass energy 938 MeV and travels with energy as large as 7 TeV = 7×10^{12} eV. Find ϵ for these protons. (d) The protons travel in a ring of circumference 27 km. How many microseconds does it take a proton to travel once around the ring?

11–17. A world record cosmic-ray proton energy of 3×10^{20} eV was observed in 1991. How many joules of energy is this? Find the speed and gamma factor ($\gamma \equiv 1/\sqrt{1 - v^2/c^2}$) for this proton. (Express v in the form $v/c = 1 - \epsilon$.) If a baseball of mass 150 g had this much kinetic energy, how fast would it be moving?

11–18. A particle of mass M is moving to the right at $V = (3/5)c$, when it suddenly disintegrates into two particles of mass $m = (2/5)M$, as shown. Find the vertical component of velocity v_y for each particle of mass m.

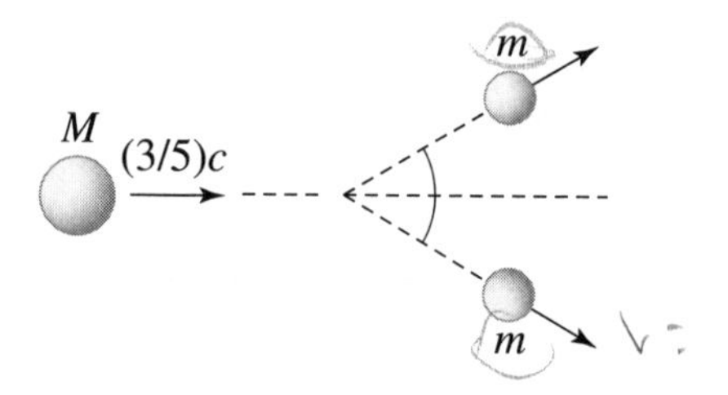

11–19. A cosmic-ray proton moves in the y direction in Earth's frame with velocity components $(0, -(3/5)c, 0)$. Using the Lorentz transformation of Eqs. 11.15 for the energy–momentum four-vector, find the proton's velocity components in the frame of a spaceship moving in the positive x direction with speed $V = (4/5)c$.

11–20. Redo the thought experiment of Section 11.1 by removing the cardboard tube entirely, leaving only the endcaps in place. When the light pulse is emitted the left endcap recoils, drifting leftward indefinitely. Later on the pulse is absorbed in the right endcap, which then drifts rightward indefinitely. Show that if the left endcap loses mass Δm and the right endcap later gains mass Δm, then after the absorption the CM of the system will remain where it was initially if $\Delta m = \Delta E/c^2$, where ΔE is the energy of the pulse.

CHAPTER 12

Applications

IN THIS CHAPTER we describe a very few of the applications of special relativity. In particular, we begin with the concept of *binding energy,* which is intimately involved with Einstein's famous formula $E = mc^2$ and with turning "mass into energy." Then we look at *particle collisions and decays.* Much of what we know about nuclear physics, and almost everything we know about the so-called fundamental particles, has come from studying collision and decay processes at relativistic energies. Finally, we look at *photon rockets,* the highly efficient rockets whose propellant consists entirely of photons.

12.1 Binding Energy in Atoms and Molecules

Atoms have small, heavy, positively charged nuclei, containing nearly all their mass, surrounded by much lighter negatively charged orbiting electrons. The lightest electrically neutral atom is hydrogen, with a single proton as nucleus and a single orbiting electron, bound to one another by their mutual electrical attraction.

Take a hydrogen atom in its ground state; that is, with the electron in its lowest energy level. Rip the atom apart, separating the proton and electron to very large distances from one another by supplying an energy of at least 13.6 eV. This energy is called the *ionization energy* of hydrogen. Atoms can be ionized by firing one or more photons at them. For hydrogen, a single photon of energy 13.6 eV can do the trick, as shown in Fig. 12.1.

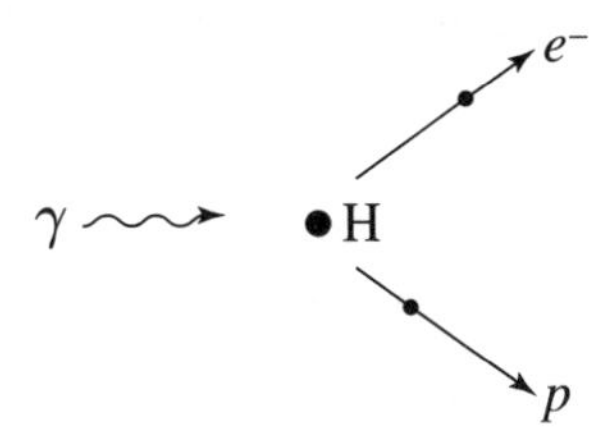

FIGURE 12.1
Ionization of a hydrogen atom.

Before the photon strikes the atom, there are a hydrogen atom and a photon; afterwards there are a proton and an electron widely separated from one another, and no photon at all.

From the results of the previous chapter, this scenario implies that the mass of the hydrogen atom must be *less* than the sum of the masses of the proton and electron, by $\Delta m = 13.6\ \mathrm{eV}/c^2$. Why? The result follows from energy conservation: *Before* the ionization, the total energy is

$$E_{\mathrm{i}} = M_{\mathrm{H}}c^2 + E_\gamma = M_{\mathrm{H}}c^2 + 13.6\ \mathrm{eV}, \tag{12.1}$$

where E_γ is the photon energy. *After* the ionization, the total energy is

$$E_{\mathrm{f}} = m_p c^2 + m_e c^2 + KE, \tag{12.2}$$

including the total kinetic energy of the proton and electron. (The kinetic energy cannot be *zero,* because the final particles must be moving in order to conserve momentum; however, this kinetic energy is some 100,000 times smaller than the ionization energy, so we can safely neglect it in the following calculations.) Energy conservation then gives

$$M_{\mathrm{H}}c^2 = m_p c^2 + m_e c^2 - E_\gamma = m_p c^2 + m_e c^2 - 13.6\ \mathrm{eV}, \tag{12.3}$$

so a hydrogen atom has less mass than the sum of the proton and electron masses by $13.6\ \mathrm{eV}/c^2$. Conversely, if we want to make a hydrogen atom from a proton and an electron, by the time the atom reaches its ground state a net energy of 13.6 eV has to escape, usually in the form of one or more emitted photons.

We can say that in a hydrogen atom, the proton and electron are bound together with a *binding energy* (*BE*) of 13.6 eV. Note that the binding energy is positive, so we write

$$m_p c^2 + m_e c^2 - M_{\mathrm{H}}c^2 = BE_{\mathrm{H}} = 13.6\ \mathrm{eV}. \tag{12.4}$$

The larger the binding energy between two particles, the more tightly they are bound together, and the more energy it takes to rip them apart. To *un*bind the particles, we have to supply at least the binding energy, which is why the binding energy and ionization energy are numerically equal.

Also, the larger the binding energy between two particles, the smaller the mass of the combination.

This too follows from energy conservation. If the binding energy is larger we have to supply more energy to rip the particles apart, so the net mass energy must have been smaller when they were bound together.

What is the *fractional* loss in mass energy when a hydrogen atom is formed from an electron and proton coming together? The mass energy loss is $\Delta E = 13.6\ \mathrm{eV}$, and the sum of the mass energies of the proton and electron is

$$m_p c^2 + m_e c^2 = 938.3\,\text{MeV} + 0.5\,\text{MeV} = 938.8\,\text{MeV}, \tag{12.5}$$

so the fractional loss is

$$\frac{\Delta E}{m_p c^2 + m_e c^2} = \frac{13.6\,\text{eV}}{938.8\,\text{MeV}} = 1.4 \times 10^{-8}, \tag{12.6}$$

which is so small that it is very difficult to measure! It would not be easy to directly weigh hydrogen atoms sufficiently accurately that we could distinguish between their weight and the weight of an appropriate number of free protons plus free electrons. So even though the predicted mass difference is correct, it is not necessarily the most *useful* way to describe the energies involved in combining an electron and proton to form hydrogen.[1]

Chemical reactions similarly involve changes from mass energy into kinetic energy or vice versa. For example, when two oxygen atoms merge to form the molecule O_2, as denoted by $O + O \rightarrow O_2 + \gamma$ (where the symbol γ represents one or more emitted photons), the two individual atoms have a larger mass than the final O_2 molecule. In the process of forming the molecule, some of the initial mass energy has been turned into kinetic energy of the photons.[2] The kinetic energy released in this case is $\Delta E = 5.1$ eV, so the fractional loss in mass energy is

$$\frac{\Delta E}{M_{O_2} c^2} = \frac{5.1\,\text{eV}}{30\,\text{GeV}} = \frac{5.1\,\text{eV}}{30 \times 10^9\,\text{eV}} = 1.7 \times 10^{-10}, \tag{12.7}$$

even smaller than in forming a hydrogen atom. When we burn wood or oil, we also turn some of the original mass energy into kinetic energy. There are various ways to describe these chemical reactions and the sources of the energy released. Regardless of how we describe it, the energy of the light and heat we observe when burning takes

1. There is an alternative way to describe the process: As the proton and electron come together, their electrical *potential energy* is reduced, and this drop in potential energy provides the energy needed to create the kinetic energy of the one or more emitted photons. Both descriptions are correct; when the particles are bound together the total mass is less and the potential energy is less as well. So we can think of the kinetic energy of the photons as arising from the drop in mass energy or from the drop in potential energy. They are equally true.

2. As with changes of electron energy levels in atoms, we can alternatively think of chemical reactions as changing potential energies into kinetic energies or vice versa. When two oxygen atoms come together to form O_2, the electrical potential energy between the two atoms is reduced. So either description is correct: The kinetic energy of emitted photons comes from a reduction in mass energy, or equivalently, the kinetic energy comes from a decrease in potential energy. The two descriptions are equivalent because if two particles have a lower mutual potential energy they have a lower mass energy as well.

place is always associated with a corresponding net reduction in mass energy of the molecules involved in the burning.

12.2 Nuclear Binding Energies

Electrons and nuclei in a single atom are bound together by electrical forces. The two or more atoms that come together to form a molecule are likewise bound together by electrical forces. The protons and neutrons in an atomic nucleus are also bound together, but their binding *cannot* be due to electrical forces, since the neutrons are electrically neutral, and the protons are all positively charged and actually *repel* one another electrically. Some other force must be responsible for nuclear binding. This force is called the *strong nuclear force*—it is very strong and also very short-range. That is, it is a powerful force that acts only if the nucleons (a generic term that includes both protons and neutrons) are very close together.

The simplest composite nucleus is the deuteron, consisting of a single proton and single neutron bound together by the strong nuclear force. The deuteron therefore has the same electrical charge as an ordinary hydrogen nucleus (a single proton) but is about twice as heavy, since neutrons have about the same mass as protons. Like a proton, a deuteron can form a neutral atom by capturing a single electron.

Any atom whose nucleus contains a single proton and any number of neutrons is said to be a hydrogen *isotope*. The nucleus consisting of a proton (p) alone is denoted by ^{1}H, the deuteron (d) by ^{2}H, and the triton (t) (consisting of a proton and *two* neutrons) by ^{3}H, and so on. The "H" tells us that the element is hydrogen, and the superscript indicates the total number of nucleons (protons plus neutrons) in the nucleus. The names of the isotopes corresponding to these nuclei are *hydrogen* (with nucleus ^{1}H), *deuterium* (with nucleus ^{2}H), and *tritium* (with nucleus ^{3}H.)

At the other, upper end of the periodic table, uranium nuclei all contain 92 protons, corresponding to the fact that uranium is the 92nd element. Uranium has many isotopes, whose nuclei differ in their numbers of neutrons. The most common isotope of natural uranium (that is, uranium mined on Earth) has the nucleus ^{238}U, with 92 protons and $238 - 92 = 146$ neutrons, and the next most common isotope has the nucleus ^{235}U, with 92 protons and 143 neutrons.

Returning to the deuteron, its neutron and proton are bound together with energy 2.2 MeV, so it takes the equivalent of a 2.2 MeV photon to rip a deuteron into its constituent parts. This is half a million times larger than the binding energy of the electron in a hydrogen atom! The fractional reduction in mass when a deuteron is made from a neutron and proton is

$$\frac{\Delta E}{m_p c^2 + m_n c^2} = \frac{2.2\,\text{MeV}}{(938.3 + 939.6)\,\text{MeV}} = 0.0012, \tag{12.8}$$

which may seem small but is *huge* compared with the comparable fractions in atomic or molecular binding. Such mass changes in nuclear reactions are substantial, and easily measured.

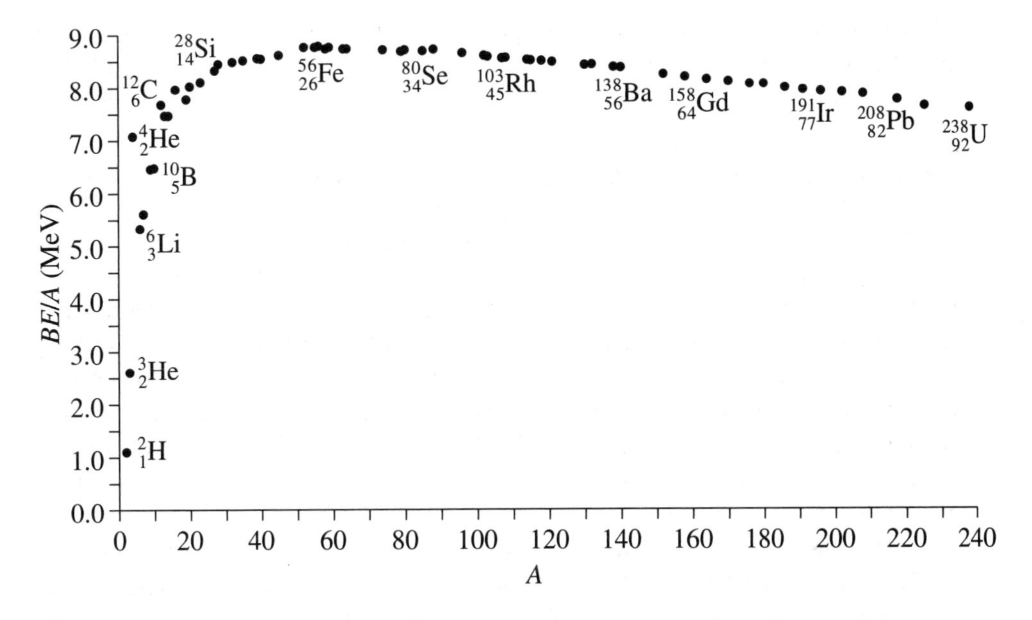

FIGURE 12.2
The curve of binding energy for stable (or nearly stable) atomic nuclei. The horizontal axis represents "A," the number of nucleons in a nucleus, and the vertical axis shows BE/A, the binding energy per nucleon of that nucleus.

The average binding energy of nucleons in heavier nuclei is even greater. Experiments have measured binding energies for a large number of stable nuclei. The results are often presented on the *curve of binding energy* shown in Figure 12.2. The horizontal axis of the graph displays "A," the total number of nucleons in the nucleus. "A" is also called the *atomic weight,* since the weight of the nucleus is (to a good approximation) proportional to the number of nucleons. Thus a deuteron has $A = 2$, and a ^{238}U nucleus (which is nearly stable) has $A = 238$. The vertical axis shows the *binding energy per nucleon, BE/A.* The deuteron has $BE/A = 2.2\,\text{MeV}/2 = 1.1\,\text{MeV}$, for example.

For most light nuclei, like the deuteron ^{2}H, the binding energies per nucleon are lower than for an average nucleus. For somewhat heavier nuclei, BE/A slowly grows, reaching a peak for nuclei of the various isotopes of iron (Fe). For still heavier nuclei, BE/A slowly decreases again, so that uranium nuclei at the far right have a smaller binding energy per nucleon than do those near the middle.

Why do nuclei at both the left and right ends of the curve have smaller binding energies per nucleon than nuclei in the middle? At the left end, this is because nucleons in small nuclei have fewer neighbors than most nucleons in larger nuclei. The proton in the deuteron, for example, bonds with only a single neutron; in a larger nucleus any particular proton bonds with several other neighbor nucleons, thus increasing the average binding energy per nucleon. At the right side of the curve, nuclei contain a great many protons, which repel one another by their electrical forces, reducing the binding energy. A uranium nucleus has 92 protons all repelling one another, which tends to destabilize the nucleus. In nuclei near the middle of the periodic table, most nucleons bond with several neighbors, which adds to their binding energy, while the debonding due to repelling protons is not too great, so the average binding energy can be relatively large. These nuclei excel by compromise.

A few of the nuclei at the far right of the binding-energy curve can be readily caused to split in two, in the process known as *nuclear fission.* The resulting smaller nuclei are nearer the center of the curve, so have larger binding energies per nucleon, which means that kinetic energy is released in the process. There are also nuclei at the far left of the binding-energy curve that can be caused to merge together in what is called *nuclear fusion.* One of the nuclei produced in this merger has larger binding energy per nucleon, leading again to a net increase of kinetic energy. Appendix H provides much more detail about nuclear decays, fission, and fusion, and how increases in binding energy can be associated with enormous releases of kinetic energy, as in fusion reactions in the Sun, fission reactions in nuclear power reactors, and both in nuclear weapons.

In the next few sections, we will switch our attention from binding energies to particle collisions and decays. Analyzing events like these is one of the most important applications of the expressions for the energy and momentum of relativistic particles. Just as in nonrelativistic mechanics, the conservation laws are enormously useful in analyzing the outcome of collisions. In fact, we already used them in Chapter 11 to explore the K-meson decay $K^0 \rightarrow \pi^0 + \pi^0$. We will now go on to exploit the conservation laws in a much wider variety of particle collisions and decays.

12.3 Decays of Single Particles into Two Particles

Many particles decay into just *two* other particles, as illustrated in Fig. 12.3. The initial particle of mass m_0 is shown in its rest frame; it has energy m_0c^2 and momentum zero. It subsequently decays into two particles, with masses m_1 and m_2. These two final particles must move away in opposite directions in order to conserve momentum. As we will show, given the initial and final masses, conservation of energy and momentum are sufficient to determine the energies, momenta, and speeds of each final particle.

In the initial rest frame, conservation of energy and momentum require

$$E_0 = m_0c^2 = E_1 + E_2 \tag{12.9}$$

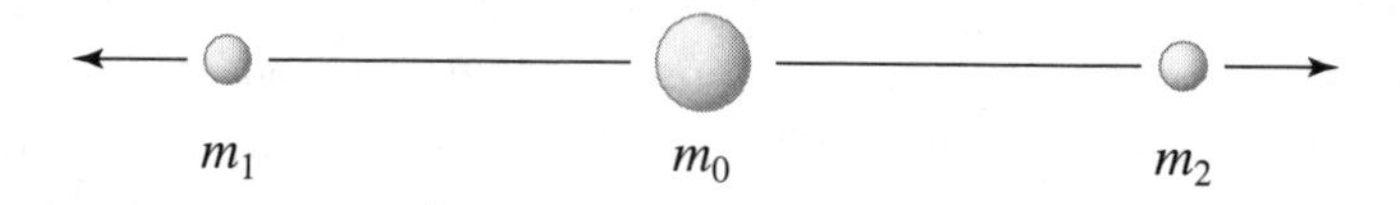

FIGURE 12.3
A particle of mass m_0 at rest decays into two particles with masses m_1 and m_2.

and

$$0 = \mathbf{p}_1 + \mathbf{p}_2. \tag{12.10}$$

We can also use $E^2 - p^2c^2 = m^2c^4$ for each of the three particles.

We will first find the energy of particle 1 in terms of the known masses of all the particles. We know that

$$E_2^2 - p_2^2c^2 = m_2^2c^4 = (m_0c^2 - E_1)^2 - p_1^2c^2, \tag{12.11}$$

using energy conservation for the first term on the left and momentum conservation for the second term (note that momentum conservation ensures that in terms of magnitudes, $p_1 = p_2$.) Now multiply out the right-hand side, giving

$$\begin{aligned} m_2^2c^4 &= m_0^2c^4 - 2m_0c^2E_1 + E_1^2 - p_1^2c^2 \\ &= m_0^2c^4 - 2m_0c^2E_1 + m_1^2c^4. \end{aligned} \tag{12.12}$$

We can solve this last equation for E_1, giving

$$E_1 = \left(\frac{m_0^2 + m_1^2 - m_2^2}{2m_0}\right)c^2. \tag{12.13}$$

Now that we have found E_1 in terms of known quantities, we can also find E_2, both momenta, the particle velocities, and other quantities as well, using the conservation laws along with $E^2 - p^2c^2 = m^2c^4$. A few examples follow.

(a) $\pi^0 \to 2\gamma$ (decay into two massless particles)

A π^0 meson decays in about 10^{-16} seconds, nearly always into two photons, written $\pi^0 \to 2\gamma$, as illustrated in Fig. 12.4. Each photon is massless, so Eq. 12.13 gives

$$E_\gamma = \left(\frac{m_{\pi^0}^2}{2m_{\pi^0}}\right)c^2 = \frac{m_{\pi^0}c^2}{2} = \frac{135\,\text{MeV}}{2} = 67.5\,\text{MeV}. \tag{12.14}$$

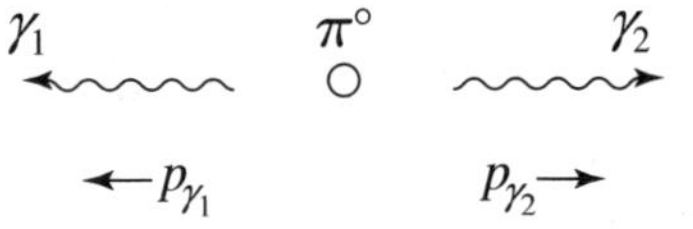

γ_1 π^0 γ_2

$\leftarrow p_{\gamma_1}$ $p_{\gamma_2} \rightarrow$

FIGURE 12.4
The decay $\pi^0 \to 2\gamma$.

That is, since the photons have equal but opposite momenta and the same (zero) mass, each must have the same energy. So each photon carries away half the mass energy of the π^0, and the momentum of each photon is

$$p_\gamma = E_\gamma / c = m_{\pi^0} c/2 = 67.5 \text{ MeV}/c.$$

(b) $\Lambda \to n + \gamma$ (decay into one massive and one massless particle)

The Λ (lambda) particle is unstable, decaying almost always into a proton or neutron and a pi meson. However, about one time in a thousand, a Λ decays into a (massive) neutron and a (massless) photon, $\Lambda \to n + \gamma$. We treat this decay here, and the more common decays in the following section. Our goal is to find as much as we can about the final particles in terms of m_Λ and m_n. In the Λ rest frame, Eq. 12.13 gives

$$E_n = \frac{(m_\Lambda^2 + m_n^2)c^2}{2m_\Lambda}, \tag{12.15}$$

since the photon is massless. Having found E_n, we can then find

$$KE_n = E_n - m_n c^2 \tag{12.16a}$$

$$p_n = \frac{1}{c}\sqrt{E_n^2 - m_n^2 c^4} \tag{12.16b}$$

$$v_n = \text{neutron speed} = p_n c^2 / E_n \tag{12.16c}$$

$$\gamma_n \equiv 1/\sqrt{1 - v_n^2/c^2} = E_n / m_n c^2 \tag{12.16d}$$

$$p_\gamma = p_n \tag{12.16e}$$

$$E_\gamma = p_\gamma c. \tag{12.16f}$$

(c) $\Lambda \to p + \pi^-$ (decay into two massive particles)

Another type of decay is illustrated by the much more common Λ decay $\Lambda \to p + \pi^-$, where the lambda particle decays into two massive particles, a proton and a pi-minus meson. There are many decays of this type, including

$\Lambda \to n + \pi^0$ (an alternative lambda decay mode)

$\Xi^- \to \Lambda + \pi^-$ (a "cascade particle" decay)

$\Omega^- \to \Xi^- + \pi^0$ (an omega-minus decay),

all with lifetimes in the ballpark of 10^{-10} seconds. Equation 12.13 gives

$$E_\pi = \frac{\left(m_\Lambda^2 - m_p^2 + m_\pi^2\right) c^2}{2m_\Lambda}, \tag{12.17}$$

which relates the pion energy to known particle masses. (Alternatively, if the lambda mass m_Λ were unknown, an experimental measurement of E_π could be used to determine it.) Other variables are then easily found:

$$KE_\pi = E_\pi - m_\pi c^2 \tag{12.18a}$$

$$p_\pi = \frac{1}{c}\sqrt{E_\pi^2 - m_\pi^2 c^4} \tag{12.18b}$$

$$v_\pi = p_\pi c^2 / E_\pi \tag{12.18c}$$

$$p_p = p_\pi \tag{12.18d}$$

$$E_p = \sqrt{p_p^2 c^2 + m_p^2 c^4} = m_\Lambda c^2 - E_\pi \tag{12.18e}$$

and so on.

12.4 Decay into Three Particles

In the preceding two-particle decays, the conservation laws are such a powerful restriction that the energy and momentum of each final particle are completely determined by the particle masses. During the 1920s, this seemed to contradict the experimental evidence on the energy of the electrons emitted in what are called *beta decays*.[3]

For example, experimentalists observed only two particles emitted in the beta decay of a ^{14}C nucleus: $^{14}\text{C} \to {}^{14}\text{N} + e^-$. If that were correct, the electrons emitted in the decays of a large sample of ^{14}C nuclei at rest should all have identical energies $E_e = \left(m_{^{14}\text{C}}^2 - m_{^{14}\text{N}}^2 + m_e^2\right) c^2 / \left(2m_{^{14}\text{C}}\right)$, a result similar to that in Eq. 12.17. On the contrary, experimentalists found that the electrons have a continuous range of energies, as shown in Fig. 12.5.

It was the Vienna-born physicist Wolfgang Pauli who first provided the basic explanation: there are not *two* final particles but *three.* In a famous letter written on 4th December, 1930 to a group of nuclear physicists, Pauli wrote:

> Dear Radioactive Ladies and Gentlemen:
>
> The continuous beta spectrum would then become understood by the assumption that in beta decay a neutron is emitted in addition to the electron such that the sum of the energiés of the neutron and electron is constant. . . .

Pauli's "neutrons" are now called *neutrinos* or *antineutrinos,* depending upon the particular beta decay (the entirely different particle we call the neutron was not

3. Beta decays are described more fully in Appendix H.

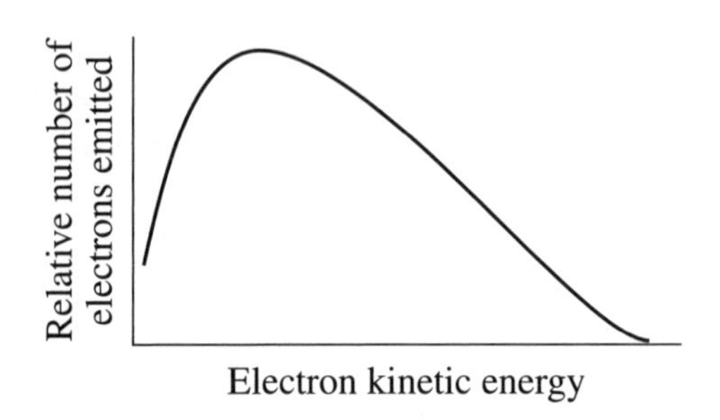

FIGURE 12.5
Energy spectrum of β-decay electrons.

discovered until 1932.)[4] Also, strictly speaking, it is the sum of the energies of *all three* final particles that is constant. So, for example, ${}^{14}\mathrm{C} \to {}^{14}\mathrm{N} + e^- + \bar{\nu}$, where (in the rest frame of the ^{14}C nucleus) the sum of the energies of all three final particles is equal to the mass energy of ^{14}C. Other three-particle beta decays are given in Appendix H.

In all these cases, the conservation laws cannot specify the energy or momentum of each final particle—there is a wide range of possibilities. For example, the emitted electron in the ^{14}C decay has its maximum possible energy in Fig. 12.5 when the emitted antineutrino has no kinetic energy at all. More commonly, all final particles have nonzero momentum and kinetic energy, with their sums constrained by the conservation laws.

12.5 Photoproduction of Pions

The collision of two stable particles is the usual method for producing unstable particles, and is a way of studying the stable particles themselves. Frequently, one of the initial particles is essentially at rest, so it is called the "target" particle. The "incident" particle may have come from cosmic rays or a particle accelerator such as a linear accelerator or a synchrotron. As an example of a collision process, consider the production of pi mesons by bombarding stationary protons with γ-ray (that is, high-energy) photons. We can produce either π^0 or π^+ mesons this way, by means of the reactions[5]

$$\gamma + p \to \pi^0 + p \quad \text{and} \quad \gamma + p \to \pi^+ + n.$$

4. Neutrinos and antineutrinos were initially extraordinarily difficult to detect in the lab; they were not observed directly until the 1950s. We now know that neutrinos (and antineutrinos) come in three "flavors": electron type, muon type, and tauon type. The antineutrino in the ^{14}C decay is electron type, and the neutrino and antineutrino in the muon decay $\mu^- \to e^- + \nu + \bar{\nu}$ are muon type and electron type, respectively. Neutrinos are now routinely studied in laboratories around the world.

5. π^- mesons can be produced by the reaction $\gamma + n \to \pi^- + p$.

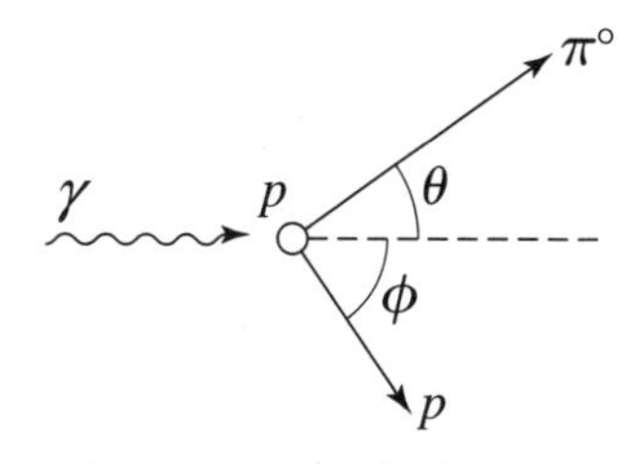

FIGURE 12.6
The photoproduction of a pion.

In the laboratory the target proton is at rest (for example as a hydrogen nucleus in a liquid hydrogen bubble chamber). The first reaction is shown schematically in Fig. 12.6.

In this π^0 photoproduction process, energy conservation gives

$$E_\gamma + m_p c^2 = E_\pi + E_p. \tag{12.19}$$

Momentum conservation, in terms of the angles defined in Fig. 12.6, gives

$$\mathbf{p}_\gamma = \mathbf{p}_\pi + \mathbf{p}_p, \tag{12.20}$$

which in component form becomes

$$p_\gamma = p_\pi \cos\theta + p_p \cos\phi \quad \text{(in the incident direction)}$$

$$0 = p_\pi \sin\theta - p_p \sin\phi \quad \text{(in the transverse direction).}$$

We may not be interested in the proton recoil angle ϕ, so we can eliminate it between these two equations for momentum conservation. The easiest way to do this is to draw the "momentum conservation triangle" of Fig. 12.7, which shows that $\mathbf{p}_\gamma = \mathbf{p}_\pi + \mathbf{p}_p$.

By the law of cosines, it follows that

$$p_p^2 = p_\pi^2 + p_\gamma^2 - 2p_\pi p_\gamma \cos\theta. \tag{12.21}$$

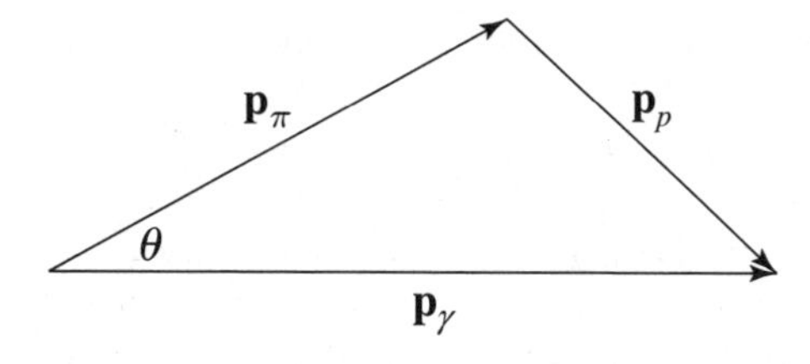

FIGURE 12.7
Momentum conservation triangle.

In addition to energy and momentum conservation, we need to make use of the formula $E^2 = p^2c^2 + m^2c^4$ for each particle involved. Then the energy conservation equation (12.19) can be written

$$p_\gamma c + m_p c^2 = \sqrt{p_\pi^2 c^2 + m_\pi^2 c^4} + \sqrt{p_p^2 c^2 + m_p^2 c^4}. \tag{12.22}$$

We know the particle masses, and we may also know the momentum of the incoming photon. If so, there are three unknowns, p_π, p_p, and $\cos\theta$, and two equations (12.21) and (12.22) involving them. Therefore these quantities are not determined, but if one unknown is measured, the others can be calculated. For example, an experiment can be set up with the pion detector at some fixed angle θ from the incident photon direction. If a pion is detected at this angle, from the conservation laws we can deduce both its momentum and the momentum of the recoil proton. If our principal interest is in the pion momentum at various angles, the proton momentum p_p can be eliminated between the two equations, resulting in a single relation for p_π as a function of the angle θ. A similar type of experiment is described in the next section.

12.6 Compton Scattering

One of the first experiments showing that light has a particle aspect (that is, that light should be thought of as consisting of photons) was performed and analyzed by the American physicist A. H. Compton in 1922. He fired a beam of x-rays at a block of graphite and measured the wavelength of the scattered x-rays as a function of the angle at which they emerge. If we assume that electrons in the graphite are causing the scattering, the collision would look as shown in Fig. 12.8.

Energy conservation gives

$$h\nu + m_e c^2 = h\nu' + E_e, \tag{12.23}$$

where ν and ν' are the initial and final x-ray frequencies, and the momentum conservation triangle along with the law of cosines gives

$$p_e^2 = \left(\frac{h\nu}{c}\right)^2 + \left(\frac{h\nu'}{c}\right)^2 - 2\frac{h\nu}{c}\frac{h\nu'}{c}\cos\theta. \tag{12.24}$$

The electron momentum and energy are related by

$$\begin{aligned} p_e^2 c^2 &= E_e^2 - m_e^2 c^4 \\ &= (h\nu - h\nu' + m_e c^2)^2 - m_e^2 c^4 \\ &= h^2(\nu - \nu')^2 + 2h\, m_e c^2(\nu - \nu') \end{aligned} \tag{12.25}$$

using the energy conservation equation.

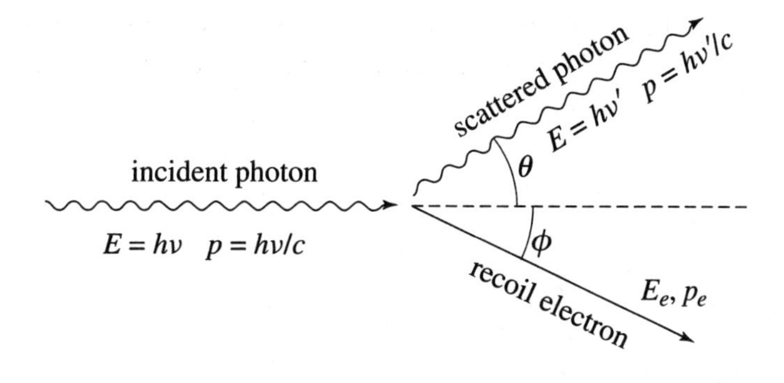

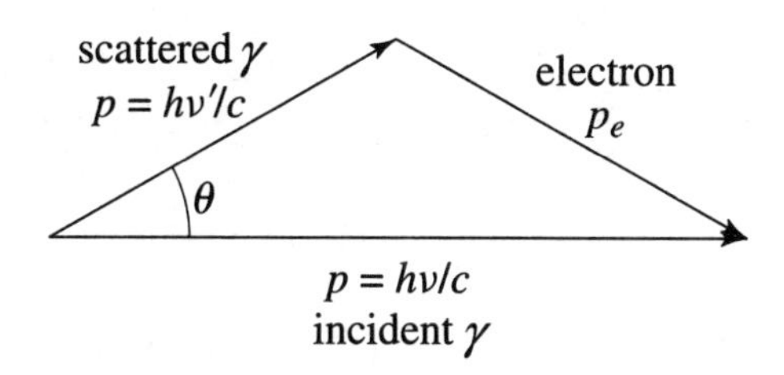

FIGURE 12.8
An incident photon scattering off an electron initially at rest.

Eliminating the electron momentum between Eqs. 12.24 and 12.25 gives

$$h^2(\nu - \nu')^2 + 2h\, m_e c^2(\nu - \nu') = h^2\nu^2 + h^2\nu'^2 - 2h\nu\, h\nu' \cos\theta, \qquad (12.26)$$

which can be simplified by cancelling terms to

$$m_e c^2(\nu - \nu') = h\nu\nu'(1 - \cos\theta). \qquad (12.27)$$

The result can be further simplified by substituting the frequency–wavelength relations $\nu = c/\lambda$ and $\nu' = c/\lambda'$ to get the final Compton scattering formula

$$\lambda' - \lambda = \frac{h}{m_e c}(1 - \cos\theta). \qquad (12.28)$$

Obviously $\lambda' > \lambda$, so the photon has reddened upon scattering, since it has transferred energy to the electron.[6] Compton observed photons obeying this formula! He *also* found scattered photons that *were not* shifted in wavelength. These are photons that scattered off an entire atom (or even the crystal lattice as a whole) instead of a single

6. If light consisted of classical waves rather than photons, there would be no appreciable wavelength shift when light is scattered off electrons.

electron. In this case the mass that should be used in the Compton formula is so large that the change in wavelength is extremely small and is not detected.

12.7 Forbidden Reactions

There are conceivable processes that are nevertheless forbidden because they violate the conservation of energy and momentum. A case in point is the decay of a particle into a single final particle of different mass. It is also easy to see that an electron and a positron (which is an antielectron) cannot annihilate one another in such a way that only *one* photon is produced; that is, $e^- + e^+ \not\rightarrow \gamma$. This can be proved in any reference frame, but it is most easily shown in the frame in which there is no net momentum.[7] In this zero-momentum frame the electron and positron approach one another with equal speeds. By energy conservation, the resulting photon energy would be $E_\gamma = 2E_e = 2(m_e c^2 + KE_e)$, or twice the total energy of each initial particle. The energy and momentum of a photon are related by $E_\gamma = p_\gamma c$, so the photon momentum could not be zero, which contradicts the zero momentum required by momentum conservation. Having proved the reaction cannot occur when viewed from the zero-momentum frame, it is evident that it cannot occur when viewed from *any* frame.[8] In exactly the same way you can show that a perfectly isolated electron cannot radiate ($e^- \not\rightarrow e^- + \gamma$).

Other illegal reactions include those in which a particle decays into two or more particles having total mass greater than that of the original particle, which follows from energy conservation alone. Thus a proton cannot decay into a neutron and something else, because protons are lighter than neutrons. On the other hand, a neutron can (and does) decay into a proton, an electron, and an antineutrino. A K meson can decay into either two or three pi mesons, but never into four, because the mass of four π's is greater than the mass of a K.

12.8 Photon Rockets

A spaceship of mass M_0 is at rest in some initial frame in outer space. It suddenly starts shooting out photons from an exhaust nozzle at the rear of the ship. The photons have momentum, so the spaceship recoils in the forward direction, becoming a *photon rocket.* It is a particularly *efficient* rocket, because the photon propellant moves as fast as any propellant can. After some period of time, the spaceship has ejected a stream of photons with total energy E_{photons} in the spaceship's initial rest frame, and the spaceship's mass has been reduced to M, as shown in Fig. 12.9.

---

7. For relativistic collisions or decays the zero-momentum frame is called the *center of momentum frame,* or *CM frame.* See the footnote on page 181.

8. The reactions $e^- + e^+ \rightarrow 2\gamma$ and $e^- + e^+ \rightarrow 3\gamma$ *are* allowed by the conservation laws, and they happen.

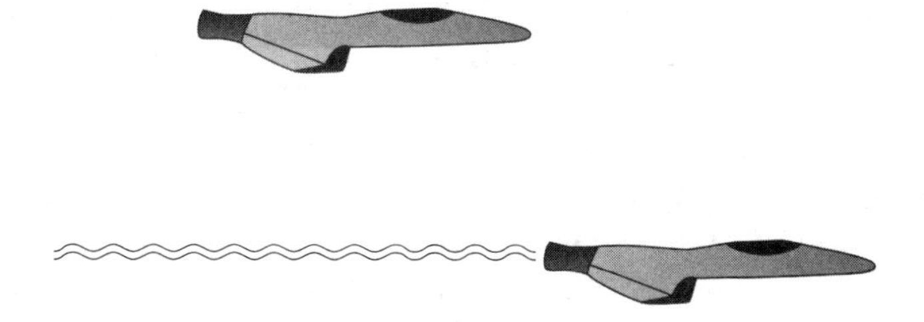

FIGURE 12.9
A photon rocket. A spaceship of mass M_0, originally at rest in some inertial frame in outer space, emits photons from an exhaust nozzle at the rear. The spaceship now has a smaller mass M, and it moves in the forward direction in order to conserve overall momentum.

Conserving energy and momentum for the spaceship and the stream of photons, we have

$$M_0 c^2 = E_{\text{photons}} + E_{\text{ship}} \tag{12.29}$$

$$0 = \mathbf{p}_{\text{photons}} + \mathbf{p}_{\text{ship}}. \tag{12.30}$$

This should look vaguely familiar! In Section 12.3, example (b), we discussed the decay of a Λ particle into a neutron and a photon. Before the decay, all the energy of the system was tied up in the mass energy of the Λ particle. The emitted photon had energy, consistent with the fact that a neutron has less mass than a Λ. The only important differences between the Λ decay and spaceship problems are that the spaceship emits a *stream* of photons instead of only one, and not all of its photons are emitted at the same time. However, this makes no difference in the analysis, because no matter when the photons are emitted, and no matter how energetic they are, their total energy and momentum are related by $E_{\text{photons}} = p_{\text{photons}}c$. Therefore all the results of example (b) are valid for the photon rocket, if we replace the Λ particle by the initial ship (with subscript zero), and the neutron by the ship at a later time (with no subscript at all). Equation 12.15 then becomes

$$E = \frac{\left(M_0^2 + M^2\right) c^2}{2M_0} \tag{12.31}$$

for the energy of the ship at any time after ignition, and Eq. 12.16d becomes

$$\gamma = \frac{1}{\sqrt{1 - v^2/c^2}} = \frac{E}{Mc^2} = \frac{\left(M_0^2 + M^2\right) c^2}{2M_0 M c^2} = \frac{1 + (M/M_0)^2}{2(M/M_0)} \tag{12.32}$$

for the ship's γ-factor. (Note that $M/M_0 = 1$ at ignition corresponds to $\gamma = 1$, since the ship is then at rest.)

When we design an interstellar spaceship, we will want it to be highly relativistic, meaning that it should reach large values of γ. Note from Eq. 12.32 that this will happen for a photon rocket if $M/M_0 \ll 1$, since we can then neglect the term $(M/M_0)^2$ in the numerator, so $\gamma \cong M_0/2M$. For example, suppose the ship has a payload that is 10% of the initial mass, with the other 90% taken up by fuel. Then $M/M_0 = 1/10$; therefore, by the time the fuel has been used up, the ship has achieved $\gamma \cong 5$, and so travelers (human or robotic) will age only 1/5 as fast as those who remain in the original rest frame, according to people who stayed behind.

How could a photon rocket be achieved? One method would be to use onboard fuel to pump up huge lasers on the ship, which would emit beams of photons out the exhaust nozzle. It is likely that such a rocket could be built, but it is much less likely that it could achieve relativistic speeds, because it is not terribly efficient. Another method would be to carry separate(!) matter and antimatter onboard fuel chambers. The matter and antimatter particles would be allowed to annihilate one another at a controlled rate, and since the annihilation would be largely into photons, the photons could be made to emerge from the exhaust. Needless to say, such a rocket would be quite difficult to build and use. Antihydrogen can be made by combining positrons (that is, antielectrons) and antiprotons, each made in accelerators on Earth. It would be technically challenging and enormously expensive to make and store the (probably mostly frozen) antihydrogen in sufficient quantities for a long-range voyage. Antihydrogen atoms would have to be confined in electromagnetic bottles so they would not touch ordinary atoms until needed. Needless to say, any onboard confinement failure would lead to a sudden, spectacular conclusion to the voyage. For these reasons and more, interstellar matter–antimatter photon rockets would be extraordinarily difficult to build, fuel, and operate. But not impossible!

Sample Problems

1. Suppose that a single ^{238}U nucleus, at the far right of the curve of binding energy, spontaneously fissions into two identical nuclei. Using Fig. 12.2, give a rough estimate of the kinetic energy released.

Solution: Each final nucleus has $92/2 = 46$ protons with $238/2 = 119$ nucleons overall, corresponding to the palladium nucleus ^{119}Pd. From the curve of binding energy, the *BE/A* of ^{238}U is roughly 7.5 MeV, while the *BE/A* of ^{119}Pd is roughly 8.5 MeV. Therefore the change in *BE/A* is roughly 1 MeV, so the total reduction in *BE* in this case is $238 \times 1\,\text{MeV} \sim 200$ MeV (where we have rounded to a single decimal place because our estimate of $\Delta(BE) = 1$ MeV was quite rough). A reduction of *BE* by 200 MeV then corresponds to an increase in kinetic energy of 200 MeV from a single fissioning of ^{238}U. That is a lot of energy!

2. The sigma-minus particle decays, with a half-life of 1.5×10^{-10} s, into a neutron and a negative pion: $\Sigma^- \to n + \pi^-$. Find the ratio of kinetic energies KE_{π^-}/KE_n of the two final particles in the rest frame of the Σ^-. (The mass energies are Σ^-: 1197.4 MeV; n: 939.6 MeV; and π^-: 139.6 MeV.)

Solution: From Eq. 12.13 we know that $E_{\pi^-} = \left(m_{\Sigma^-}^2 + m_{\pi^-}^2 - m_n^2\right) c^2/2m_{\Sigma^-}$, so

$$KE_{\pi^-} = \left(\frac{m_{\Sigma^-}^2 + m_{\pi^-}^2 - m_n^2}{2m_{\Sigma^-}}\right) c^2 - m_{\pi^-}c^2$$

$$= \left(\frac{m_{\Sigma^-}^2 + m_{\pi^-}^2 - m_n^2 - 2m_{\Sigma^-}m_{\pi^-}}{2m_{\Sigma^-}}\right) c^2 = \left(\frac{(m_{\Sigma^-} - m_{\pi^-})^2 - m_n^2}{2m_{\Sigma^-}}\right) c^2,$$

where in the last step we combined the first, second, and fourth terms of the previous step. By exchanging the neutron and pion labels, it follows that

$$KE_n = \left(\frac{(m_{\Sigma^-} - m_n)^2 - m_{\pi^-}^2}{2m_{\Sigma^-}}\right) c^2.$$

The ratio of kinetic energies is therefore

$$\frac{KE_{\pi^-}}{KE_n} = \frac{(m_{\Sigma^-} - m_{\pi^-})^2 - m_n^2}{(m_{\Sigma^-} - m_n)^2 - m_{\pi^-}^2} = \frac{(1197.4 - 139.6)^2 - 939.6^2}{(1197.4 - 939.6)^2 - 139.6^2} = 5.026.$$

That is, the *less* massive final particle (the pion) takes away most of the kinetic energy released in the decay. This can be proven in general: in the frame of the initial particle, in two-particle decays the less massive final particle emerges with most of the kinetic energy. See Problem 12–9.

3. In the photoproduction of a pion $\gamma + p \to p + \pi^0$ from a proton initially at rest, is it possible for the final proton to be at rest as well?

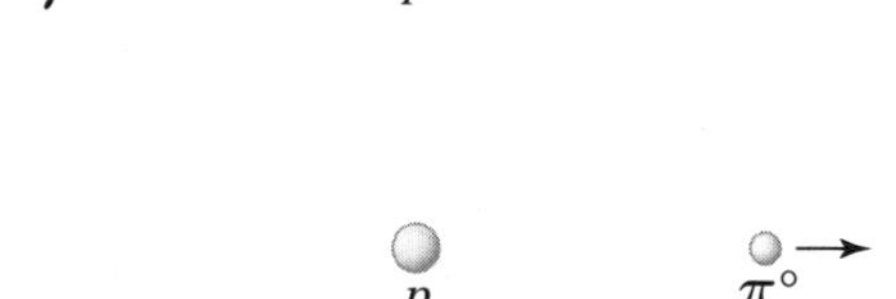

Solution: If the proton is at rest before and after, energy conservation gives $m_p c^2 + E_\gamma = m_p c^2 + E_{\pi^0}$, so $E_\gamma = E_{\pi^0}$. Squaring this equation gives $E_\gamma^2 = E_{\pi^0}^2$, so $p_\gamma^2 c^2 = p_{\pi^0}^2 c^2 + m_{\pi^0}^2 c^4$ (using the relation $E^2 = p^2c^2 + m^2c^4$ for the photon and pion). However, momentum conservation gives $p_\gamma = p_{\pi^0}$, so $p_\gamma^2 c^2 = p_{\pi^0}^2 c^2$. Combining the energy and momentum conservation equations gives $m_{\pi^0}^2 c^4 = 0$, which is *false*. So the proton *cannot* remain at rest.

4. The bright star Arcturus is 36 light-years from the Sun. A photon-rocket trip to Arcturus is planned so that when the ship reaches its final velocity, the distance between Arcturus and the Sun will be only 12 light-years from the traveler's point of view. What fraction of the initial ship mass will remain at that point?

Solution: The distance between Arcturus and the Sun will be Lorentz-contracted to 1/3 their rest-frame distance; that is, $\sqrt{1 - v^2/c^2} = 1/3$, so $\gamma = 1/\sqrt{1 - v^2/c^2} = 3$. From Eq. 12.32 it follows that $1 + (M/M_0)^2 = 6(M/M_0)$, so M/M_0 is a solution of the quadratic equation

$$(M/M_0)^2 - 6(M/M_0) + 1 = 0.$$

That is, $M/M_0 = (6 \pm \sqrt{36 - 4})/2 = 3 \pm \sqrt{8}$, where we must choose the minus sign because we know that $M/M_0 < 1$. Therefore the fraction of the initial mass remaining when the final velocity is achieved is $M/M_0 = 3 - \sqrt{8} = 3 - 2\sqrt{2} = 0.1716$. About 17% of the initial mass remains.

Problems

12–1. Two wooden blocks are each measured to have mass m. The blocks are then brought close together with a spring (of negligible mass) compressed between them, as shown. The potential energy stored in the spring is $(1/2)k(\Delta x)^2$, where k and Δx are the force constant and compression of the spring. The assembly is held together by a string (of negligible mass) opposing the outward force of the spring. If the string is cut, the spring expands to its natural length (so stores no more potential energy), and the blocks fly apart with total kinetic energy equal to the potential energy that had been stored in the spring. In terms of given quantities, what was the mass of the assembly before the string was cut?

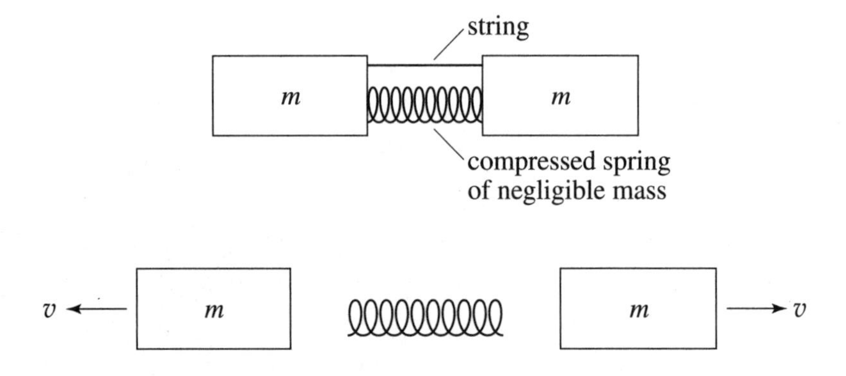

12–2. Hydrogen-like atoms are atoms with only a single electron. They include neutral hydrogen, singly ionized helium (with a helium nucleus and one electron, so the atom has a net positive charge), doubly ionized lithium, and so on. The ionization energy of a hydrogen-like atom is proportional to the square of the electric charge on the nucleus. Find the binding energy of the electron (in eV) in (a) singly ionized helium and (b) doubly ionized lithium, assuming in each case that the electron is in its ground state.

12–3. The binding energy per nucleon of the ^{3}He nucleus is 2.6 MeV, as read from the binding energy curve of Fig. 12.2, but the binding energy per nucleon of tritium (^{3}H) is not shown. Deduce the binding energy of tritium to two significant figures, from the following information: (*i*) The mass energies of isolated protons, neutrons, and electrons are 938.3 MeV, 939.6 MeV, and 0.5 MeV, respectively. (*ii*) Tritium decays by the reaction $^3\text{H} \rightarrow {}^3\text{He} + e^- + \bar{\nu}$, where the kinetic energy of the final particles is 0.0186 MeV. (*iii*) The mass energy of the antineutrino $\bar{\nu}$ is negligible on the MeV scale.

12–4. A *positronium* atom consists of an electron (e^-) and positron (e^+) bound together with a binding energy of a few electron volts. The mass energies of the electron and positron are each 0.511 MeV. Suppose that a particular positronium atom, at rest in the lab, annihilates into two photons. Find the energy and the magnitude of the momentum of each photon, to three significant figures.

12–5. A π^- meson most often decays into a muon and antineutrino: $\pi^- \rightarrow \mu^- + \bar{\nu}$. The mass energies are π^-: 139.6 MeV; μ^-: 105.7 MeV; and $\bar{\nu}$: unknown, but a small fraction of an electron volt. Using energy units, and in the π^- rest frame, find the muon's (a) total energy, (b) kinetic energy, (c) momentum, and (d) velocity, as a multiple of c.

12–6. A π^+ meson sometimes decays into a positron and neutrino: $\pi^- \rightarrow e^+ + \nu$. The mass energies are π^+: 139.6 MeV; e^+: 0.5 MeV; ν: unknown, but a small fraction of an electron volt. Using energy units, and in the π^+ rest frame, find the positron's (a) total energy, (b) kinetic energy, (c) momentum, and (d) velocity, as a multiple of c, to three significant figures.

12–7. In the decay $\Lambda \rightarrow n + \pi^0$, find v/c of each final particle in the rest frame of the Λ particle. (Particle masses are given in Appendix I.)

12–8. An Ω^- particle sometimes decays into a Ξ^0 hyperon and a pion, written $\Omega^- \rightarrow \Xi^0 + \pi^-$. The mass energies are Ω^-: 1676 MeV; Ξ^0: 1311 MeV; and π^-: 140 MeV. Using energy units, in the rest frame of the Ω^-, find the pion's (a) total energy, (b) kinetic energy, (c) momentum, and (d) speed, as a multiple of c.

12–9. A particle decays into two particles of unequal mass. (a) Prove that in the rest frame of the initial particle, it is the *less* massive final particle that carries away most of the final kinetic energy (see Sample Problem 2). (b) Then show that in the nonrelativistic limit, in which mass is nearly conserved, the ratio of the final particle kinetic energies is equal to the inverse ratio of the final particle masses.

12–10. The Sun's energy comes from nuclear fusion reactions, in which mass energy is converted into kinetic energy. The first step in the so-called *proton–proton chain* is the collision of two protons to form a deuteron, a positron, and a neutrino: $p + p \rightarrow d + e^+ + \nu$. (a) Find the total kinetic energy of the final particles in the zero-momentum (CM) frame, in terms of m_p, m_d, m_e, and c. (Neglect the mass of the neutrino and the initial kinetic energy of the protons.) (b) Suppose that in a particular reaction the final deuteron is at rest. Find the energy of the electron in terms of the same parameters.

12–11. The uranium nucleus ^{239}U is created in nuclear reactors in the collision of neutrons with the common nucleus ^{238}U as follows: $n +^{238} \text{U} \rightarrow (^{239}\text{U})^*$, where the star means that the ^{239}U nucleus is in an excited state (with an energy higher than its ground-state energy). One way for the excited nucleus to lose its excess energy is to emit a gamma-ray photon, $(^{239}\text{U})^* \rightarrow {}^{239}\text{U} + \gamma$. (a) Why is it true that $E_\gamma < (m^* - m)c^2$, assuming that the excited nucleus was at rest before the photon was emitted? (b) Find the gamma-ray photon energy E_γ in terms of the speed of light c and the masses m^* and m of the excited and ground-state nuclei only.

12–12. Calculate the frequency of the photons emitted in the decay $\pi^0 \rightarrow \gamma + \gamma$, assuming the pion is at rest.

12–13. A particular neutron at rest decays so that the final proton is at rest, while the electron and antineutrino have equal and opposite momenta. Find the energy of the electron, in MeV. (Note that the neutron, proton, and electron mass energies are 939.565 MeV, 938.272 MeV, and 0.511 MeV, respectively. The antineutrino mass is unknown, but it is very much less than the electron mass.)

12–14. A particular negative muon at rest decays in such a way that the final antineutrino is at rest, and the final electron and neutrino move in opposite directions. Find the kinetic energy of the electron, in MeV. (The neutrino and antineutrino masses are unknown, but are both very much less than the electron mass. The muon and electron masses are given in Appendix I.)

12–15. In the photoproduction of a π^0 meson from a proton at rest, is it possible that if we select the photon energy just right, the produced pion will be at rest in the initial proton rest frame? If so, find the right photon energy in terms of the particle masses. If not, show why it can't be done.

12–16. K^+ mesons can be photoproduced by the reaction $\gamma + p \rightarrow K^+ + \Lambda$ where the initial proton is at rest in the lab. From the conservation laws, discover if it is possible for either the K^+ or the Λ to be at rest in the lab, and for what photon energy (in terms of the particle masses) this could happen.

12–17. It is possible to create antiprotons ($\bar{p}$) by the reaction

$$p + p \rightarrow p + p + p + \bar{p}$$

where one of the initial protons comes from a high-energy accelerator and the other is at rest in the lab. Find the "threshold energy," the minimum energy of the incident proton needed to make the reaction go. It is helpful to see that for this energy all of the final particles move in the forward direction at the same velocity.

12–18. The *tachyon,* a hypothetical particle that travels faster than light, has imaginary energy and momentum using the traditional formulas. (a) Show that these quantities can be made *real* by assigning tachyons the imaginary mass im_0, where m_0 is real. How then do the resulting real momentum and energy depend upon velocity? Sketch both quantities for $v > c$. In terms of m_0 and c, what is the quantity $E^2 - p^2c^2$ for tachyons? Now suppose an ordinary particle of mass M at rest decays into an ordinary particle of mass m and an unseen particle that may or may not be a tachyon. Knowing M and m, show how you could tell from measurements of the energy of the final ordinary particle m, whether the unseen particle is a tachyon, a massless particle, or an ordinary massive particle. Assume that energy and momentum are conserved.

12–19. A sodium "D-line" photon (wavelength $\lambda = 589$ nm) is scattered through an angle $\theta = 90^\circ$ by a free electron at rest. Find the change in wavelength of the scattered photon, and the kinetic energy of the recoil electron.

12–20. The quantity $\lambda_C \equiv h/m_e c$ is called the "Compton wavelength" of the electron. (a) If a photon scatters off an electron at rest with scattering angle $\theta = 90^\circ$, what is the photon's change of wavelength $\lambda' - \lambda$ in terms of λ_C? (b) At what scattering angle is $\lambda' - \lambda$ a maximum? What then is $\lambda' - \lambda$?

12–21. Prove that two colliding particles cannot transform into a single photon.

12–22. Explain why a photon that strikes a free electron cannot be absorbed: $\gamma + e \nrightarrow e$. Such a reaction *does* take place if the electron is bound in an atom, and causes atomic excitation or even ejection of the electron from the atom. Why can absorption occur in this case?

12–23. A 10^6-kilogram spaceship fires a burst of laser light involving 10^{33} photons of wavelength $\lambda = 400$ nm. Find the recoil velocity of the ship.

12–24. We are designing a photon rocket that will achieve a final speed corresponding to $\gamma = 2$. (a) How fast will the ship then be moving, compared with c? (b) What will be the final fraction M/M_0 of the ship's mass remaining?

12–25. (a) Show that in terms of the fraction M/M_0 of the ship's mass remaining, the speed v achieved by a photon rocket is given by

$$v/c = \frac{[1 - (M/M_0)^2]}{[1 + (M/M_0)^2]}.$$

(b) If the fraction of the original ship mass remaining is $M/M_0 = 1/2$, what speed has the ship achieved? (c) Invert the result of part (a) to find a formula for the mass fraction remaining in terms of v/c. (d) If the ship achieves $v = (4/5)c$, what fraction of its initial mass remains?

12–26. Departing Earth, a photon rocket achieves a large $\gamma = \gamma_1$ if its final mass ratio is M_1/M_0. Later, it uses its photon rocket to bring it back to rest at its destination. Later still, it starts back home, again achieving the same $\gamma = \gamma_1$, and finally it uses its photon rocket to bring it back to rest at Earth. (a) Find its ultimate mass ratio M_f/M_0 in terms of M_1/M_0. (b) If $\gamma_1 = 5$, find M_1/M_0 and M_f/M_0. If $M_0 = 100$ metric tons ($= 10^5$ kg), what is the payload mass M_f?

12–27. In classical, nonrelativistic physics, a rocket starting from rest in gravity-free empty space achieves the velocity $v = u \ln(M_0/M)$ if it ejects its propellant at constant speed u relative to the rocket. This is called the *rocket equation.* (Here M_0 is the initial total mass of the rocket, and M is the mass remaining at any time later.) If we naively apply this equation to a photon rocket (where the exhaust speed is highly relativistic!), we predict $(v/c)_{NR} = \ln(M_0/M)$ rather than the correct result $v/c = [1 - (M/M_0)^2]/[1 + (M/M_0)^2]$ quoted in Problem 12–25. (a) Find the ratio $(v/c)/(v/c)_{NR}$ if $M/M_0 = (i)1/3, (ii)1/2, (iii)2/3$. (b) Suppose that a photon rocket has not yet achieved a relativistic speed. That is, suppose $M = M_0 - \Delta M$ where $\Delta M/M_0 \ll 1$. Then show that the two expressions for v/c give the *same* result through terms of order $(\Delta M / M_0)^2$. *Hint:* Use the binomial expansion of Appendix A together with the natural logarithm expansion $\ln(1 + x) = x - x^2/2 + \cdots$ valid for $x \ll 1$. Neglect powers $(\Delta M/M_0)^3$ and higher. This problem shows that the classical rocket equation remains valid for photon rockets before they get moving very fast.

CHAPTER 13

Transforming Energy and Momentum

THE MOMENTUM AND ENERGY of a particle obviously depend upon the frame of reference in which they are measured. In the particle's rest frame, for example, its only energy is mass energy mc^2, and it has no momentum at all. In any *other* frame, the particle has nonzero momentum and nonzero kinetic energy. In general, suppose we know the momentum and energy in some particular frame: How can they be found in any other frame?

The answer to this question is most important in translating information about particle collisions or decays between the *laboratory* and *center of momentum* (CM) frames of reference. The laboratory frame is the rest frame of the laboratory in which the experiment is performed and observed, and the CM frame (by definition!) is the frame in which the total momentum of the system of particles is zero.[1] The decay problems of Chapter 12 used the CM frame, for example: The initial particle was at rest, so had zero momentum. We may need to know the results in a laboratory frame instead, in which the initial particle is generally moving. How do we do that?

---

1. In *classical* physics, the "center of mass," or "CM," frame is a frame in which the center of mass is at rest, and also the frame in which the total momentum of a system of particles is *zero,* as explained in Chapter 1. In *relativistic* physics, however, these two properties no longer go together: that is, a frame in which the total momentum is zero is no longer a frame in which the center of mass is necessarily at rest. Take, for example, the decay $N^* \rightarrow N + \gamma$ of an excited nucleus N^* into the final less-excited nucleus N and a gamma-ray photon. A frame in which N^* is at rest is a CM frame, since in such a frame the momentum is zero. (Note that the center of mass also happens to be at rest in this frame before the decay.) *After* the decay, the total momentum is still zero in this frame by momentum conservation, since the N and γ have equal but opposite momenta. However, the photon is massless, so the center of mass is now moving along with the final nucleus N. Therefore in relativistic physics, a zero-momentum frame is still called a "CM" frame, but now CM is taken to mean "center of momentum" rather than "center of mass."

13.1 The Energy–Momentum Transformation

We already came close to figuring out how to transform energy and momentum from one frame to another back in Chapters 10 and 11. The relativistic energy E and momentum $\mathbf{p}$ both contribute to the energy–momentum four-vector

$$p_\mu = (E/c,\ p_x,\ p_y,\ p_z), \tag{13.1}$$

and since they are components of a four-vector, they must transform according to the Lorentz transformation, as described in Chapter 10. These equations are

$$E' = \gamma(E - Vp_x) \tag{13.2a}$$

$$p'_x = \gamma(p_x - VE/c^2) \tag{13.2b}$$

$$p'_y = p_y, \quad p'_z = p_z \tag{13.2c, d}$$

where $\gamma \equiv 1/\sqrt{1 - V^2/c^2}$. We will call Eqs. 13.2 the *energy–momentum transformation*. If p and E are known, p' and E' can be found right away from these equations.

The transformation is easy to remember, because it is very similar to the Lorentz transformation for t, x, y, and z. The only difference is the number of c's in the first two equations, which can always be found by checking dimensions. As a simple example, suppose that a particle of mass m is at rest in the unprimed frame, so the momentum and energy of the particle are $p_x = p_y = p_z = 0$ and $E = mc^2$. What are the momentum and energy of the particle in a *primed* frame moving in the positive x direction at velocity V? According to Eqs. 13.2,

$$E' = \gamma(E - 0) = \frac{mc^2}{\sqrt{1 - V^2/c^2}} \tag{13.3a}$$

$$p'_x = \gamma(0 - VE/c^2) = -\frac{mV}{\sqrt{1 - V^2/c^2}} \tag{13.3b}$$

while p_y and p_z are zero in both frames. We could have written down these equations without using the transformation! The particle is moving to the left at speed V in the primed frame, so the derived formulas $E' = \gamma mc^2$ and $p'_x = -\gamma mV$ are clearly correct.

We can also write out the inverse transformation (from the primed to the unprimed frame) just as we did for the Lorentz transformation, by interchanging primes and unprimes and changing the sign of V. This gives

$$E = \gamma(E' + Vp'_x) \tag{13.4a}$$

$$p_x = \gamma(p'_x + VE'/c^2) \tag{13.4b}$$

$$p_y = p'_y, \quad p_z = p'_z. \tag{13.4c, d}$$

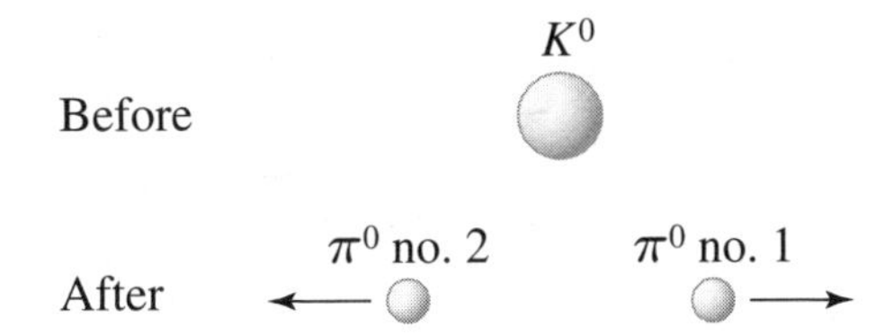

FIGURE 13.1
K^0 decay in the center-of-momentum frame.

As an example, we can apply the energy–momentum transformation to the neutral kaon decay $K^0 \to \pi^0 + \pi^0$ already discussed in Chapter 11. Suppose the K^0 is moving at velocity $v = (3/5)c$ in the lab, and suppose also that it decays into one π^0 moving forward (in the direction the K^0 was moving), and another π_0 moving backward. We would like to know the momentum and energy of each pion in the lab. One way to solve this problem would be to analyze it completely in the lab frame, using conservation of momentum and energy. However, it is easier to begin by analyzing the decay in the CM frame, and then finding the results in the lab frame using the energy–momentum transformation.

A diagram of the decay in the CM frame is shown in Fig. 13.1. By convention, the primed frame moves to the right, so the CM frame is the primed frame. For this frame we found the momentum and energy of each particle in Section 11.3. The results are summarized in Table 13.1.

The energies and momenta in the (unprimed) laboratory frame can now be found by applying the inverse energy–momentum transformation of Eqs. 13.4. The relative velocity of the two frames is $V = (3/5)c$ and $\gamma \equiv 1/\sqrt{1-(3/5)^2} = 5/4$, so

$$E = \frac{5}{4}\left(E' + \frac{3}{5}cp'_x\right) \quad \text{and} \quad p_x = \frac{5}{4}\left(p'_x + \frac{3}{5}E'/c\right). \tag{13.5}$$

Evaluating E and p_x for each particle, we obtain the laboratory values as given in Table 13.2.

As a good check, notice that energy and momentum are conserved in this frame as well. The results are summarized in Fig. 13.2.

TABLE 13.1
CM-frame momenta and energies in the decay $K^0 \to \pi^0 + \pi^0$.

	Momentum p'	Energy E'
K^0 meson	0	498 MeV
π^0 no. 1	209 MeV/c	249 MeV
π^0 no. 2	−209 MeV/c	249 MeV

TABLE 13.2
Momenta and energies in the $K^0 \to \pi^0 + \pi^0$ decay in a lab frame with $V = (3/5)c$.

	Momentum p'	Energy E'
K^0 meson	374 MeV/c	623 MeV
π^0 no. 1	448 MeV/c	468 MeV
π^0 no. 2	−75 MeV/c	155 MeV

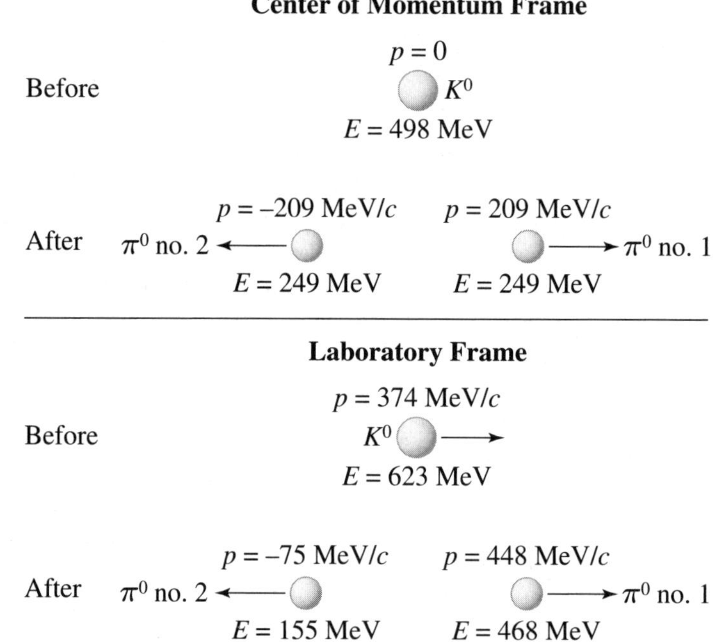

FIGURE 13.2
K^0 decay in the CM frame and in the laboratory frame.

13.2 Light Aberration and the Relativistic Doppler Effect

The energy–momentum transformation can also be used to find the Doppler shift of light. Everyone is familiar with the Doppler shift of sound waves—the apparent frequency of a source depends upon its velocity with respect to the observer. The siren of an approaching ambulance has a higher pitch, which changes to a lower pitch as the ambulance speeds away. The Doppler effect has long been known to hold for light waves also. A radiating source emits light that appears to be shifted to higher frequencies if the source is moving toward the observer, and to lower frequencies if the source is moving away. The shift is quite small in most circumstances, but can become large if the relative velocity is high.

The best-known situation in which the Doppler effect for light is observed is in the shift in line-spectra of stars. Atoms of a given element emit a characteristic set of discrete wavelengths, which we can measure in the laboratory. We can identify which elements are present in a star by these spectral lines. Most stars or galaxies are moving with respect to us, however, so the wavelengths in their spectra are somewhat shifted toward the blue or red. In the visible spectrum blue light has high frequency (small wavelength) and red light has low frequency (large wavelength). Light from an approaching star is shifted toward the blue, while light from a receding star is shifted toward the red. We know the frequencies of lines in the spectrum of hydrogen, for example, so if the same set of hydrogen lines is shifted toward the red in a particular star, we know that star is moving away from us, and we can figure out how fast it is moving from the size of the redshift.

The light from some stars changes frequency in a periodic way, changing from redder to bluer over the course of several days. This is how we have discovered planets around distant stars. Because of their gravitational pull on their sun, they cause their sun to wobble slightly back and forth, so that its light is Doppler-shifted in a periodic way as we observe it. From the size and period of the Doppler shift we can learn much about the planet and its orbit.

The light from very distant galaxies (those outside our local group of galaxies) is always *red*shifted, which is the main reason we know that the universe is expanding. The size of the redshift increases with the distance of the galaxy from us, so the more distant a galaxy, the faster it is moving away. This is what we would expect if the visible universe resulted from a "big bang" explosion at some time in the past—faster moving objects would naturally be farther from us by this time than slower objects. Some of the most distant objects we observe are "quasars," or quasi-stellar objects. The redshifts in their spectral lines indicate that some of them are moving away at appreciable fractions of the speed of light. Figure 13.3 shows the line spectra of four actual quasars whose redshifts vary from $z = 0.178$ up to $z = 0.389$, where $z \equiv \Delta\lambda/\lambda$, which is the increase in wavelength due to the Doppler effect divided by the wavelength emitted in the quasar rest frame.[2]

The Doppler effect is also one cause of line broadening in the spectra of hot gases, which has been observed in the laboratory as well as in the spectra of the Sun and stars. At any given time some molecules approach and some recede from an observer, so the net effect is to broaden each spectral line to include a range of wavelengths,

2. Cosmologists have an alternative way of understanding the redshifts of quasars and distant galaxies. They depend on Einstein's *general theory of relativity* of 1915, rather than his earlier special theory of relativity, to study the expanding universe on large scales. It is then natural to use an *expanding reference frame* in which all galaxies are more or less at rest. Then since distant galaxies are at rest in this frame, just as we are, there is no Doppler redshift of the light propagating from them to us. However, as space expands, the wavelengths of light expand along with it, to generate the "cosmological redshift" we observe.

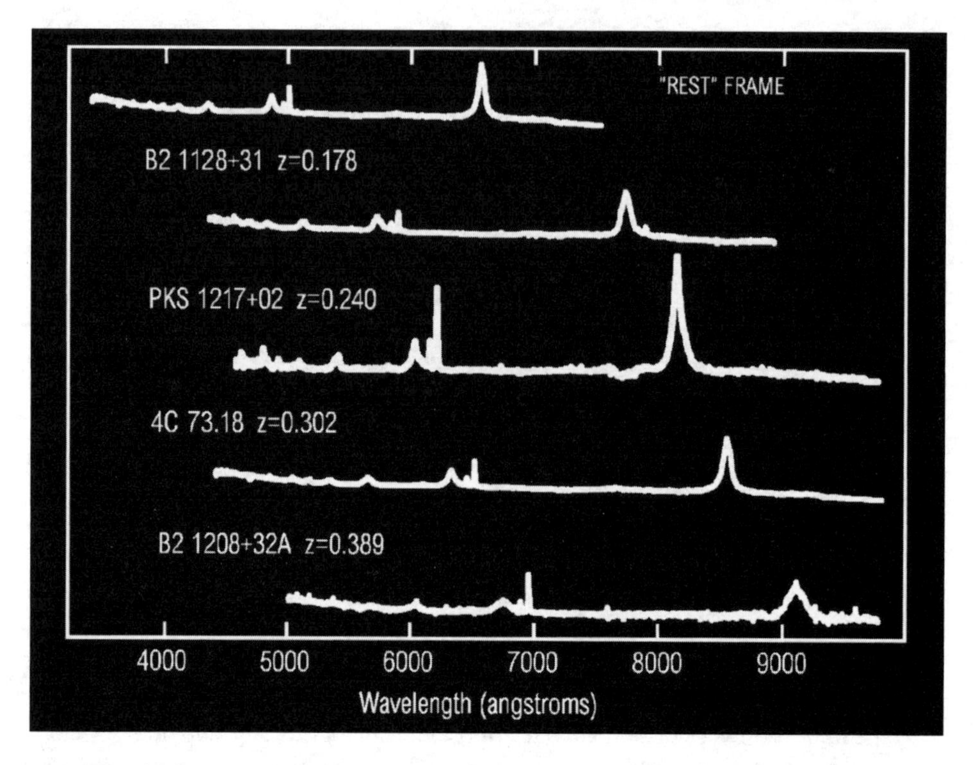

FIGURE 13.3
Spectra of four quasars (or QSOs, for "quasi-stellar-objects"), together with the spectrum of a hypothetical quasar (at the top) that is at rest relative to us. The horizontal axis represents the wavelength of light in Ångstroms (1 Å $= 10^{-10}$ m), and the vertical axes are proportional to the light intensity. The second spectrum, from the actual quasar B2 1128+31, has the lowest redshift of the four; and the lowermost spectrum, from the quasar B2 1208+32A, has the highest redshift. The rather broad spectral lines have wavelengths 4340 Å, 4860 Å, and 6532 Å from the quasar at rest relative to us, and increasingly larger wavelengths from the redshifted quasars. These three spectral lines are all produced by electron transitions in hydrogen. The prominent sharper line and another smaller line nearby, both just to the right of the 4860 Å line, are produced by electron transitions in doubly ionized oxygen. Figure courtesy of C. Pilachowski and M. Corbin/NOAO/AURA/NSF.

some redder and some bluer. The average energy of the molecules in a hot diffuse gas is proportional to the temperature, so the hotter the gas, the more Doppler shift we would expect in the radiation emitted from the molecules.

A relativistically correct formula for the Doppler shift, in terms of the relative velocity of the source and observer, can be found from the transformation laws of energy and momentum. Consider a source that is at rest at the origin of the primed coordinate system and emits a photon at some angle θ' with respect to the x' axis. We can assume that this photon moves in the x', y' plane, as shown in Fig. 13.4. If the photon frequency is ν' in the frame of the source, its energy is $E' = h\nu'$, and its momentum is $p' = E'/c = h\nu'/c$. In terms of components,

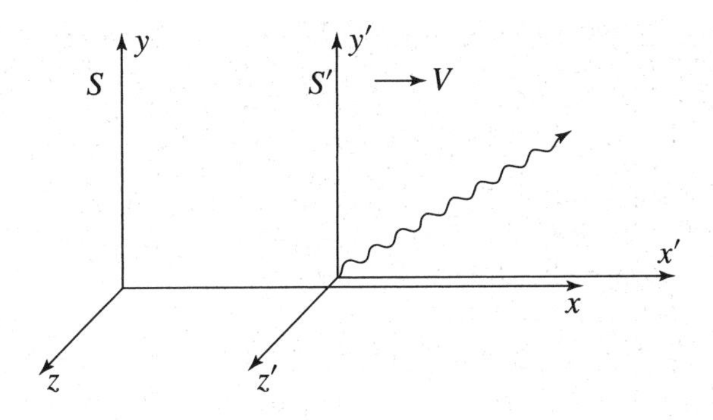

FIGURE 13.4
A photon from a source at the origin of the S' frame moves in the x', y' plane.

$$E' = h\nu' \tag{13.6a}$$

$$p'_x = (h\nu'/c)\cos\theta' \tag{13.6b}$$

$$p'_y = (h\nu'/c)\sin\theta' \tag{13.6c}$$

$$p'_z = 0. \tag{13.6d}$$

Now suppose that the observer is at rest in the unprimed frame, and that as usual the primed frame is moving to the right at speed V, as in Fig. 13.4. The observer finds that the photon has energy E, momentum components p_x, p_y, p_z, and angle θ with respect to the x axis. Each of these quantities can be calculated from the inverse transformations of Eqs. 13.4. We will need to calculate only E and p_x:

$$E = h\nu = \gamma(E' + Vp'_x) = \gamma(h\nu' + (V/c)h\nu'\cos\theta')$$

$$\text{so} \quad \nu = \gamma\nu'(1 + (V/c)\cos\theta') \tag{13.7a}$$

$$\text{and } p_x = (h\nu/c)\cos\theta = \gamma(p'_x + VE'/c^2) = \gamma((h\nu'/c)\cos\theta' + V\,h\nu'/c^2)$$

$$\text{so} \quad \nu\cos\theta = \gamma\nu'(\cos\theta' + V/c). \tag{13.7b}$$

The frequencies can be eliminated by dividing Eq. 13.7b by Eq. 13.7a, giving

$$\cos\theta = \frac{\cos\theta' + V/c}{1 + (V/c)\cos\theta'}, \tag{13.8a}$$

which expresses the photon angle in the unprimed frame (the observer's frame) in terms of its angle in the primed frame (the frame of the source). These angles are obviously different (except when $\theta' = 0$ or π), which is simply the aberration of light effect discussed in Chapter 2. The result can be inverted to give $\cos\theta'$ in terms of $\cos\theta$:

$$\cos\theta' = \frac{\cos\theta - V/c}{1 - (V/c)\cos\theta}. \tag{13.8b}$$

Using this equation, the primed angle in Eq. 13.7a can be eliminated, to obtain

$$\nu = \gamma\nu'\left\{1 + \frac{V}{c}\frac{\cos\theta - V/c}{1 - (V/c)\cos\theta}\right\} = \frac{\gamma\nu'(1 - V^2/c^2)}{1 - (V/c)\cos\theta}, \tag{13.9}$$

so

$$\nu = \frac{\nu'\sqrt{1 - V^2/c^2}}{1 - (V/c)\cos\theta}, \tag{13.10}$$

which is the *relativistic Doppler equation,* giving ν, the frequency observed, in terms of ν', the frequency emitted by the source in its own rest frame, and the angle θ at which the observer sees the photon.

What does the equation *mean*? First, suppose that $\theta' = 0$, so the photon is moving to the right along the x' axis. To observe it, we must be situated somewhere along the x axis, and it is clear from Fig. 13.4 that the source is *approaching* us. Then (since $\cos\theta' = 1$ for $\theta' = 0$), it follows from Eq. 13.8a that

$$\cos\theta = \frac{\cos\theta' + V/c}{1 + (V/c)\cos\theta'} = \frac{1 + V/c}{1 + V/c} = 1, \tag{13.11}$$

so the photon is also moving at $\theta = 0$ from our point of view, as expected. The Doppler equation then gives

$$\nu = \frac{\nu'\sqrt{1 - V^2/c^2}}{1 - (V/c)\cos\theta} = \frac{\nu'\sqrt{1 - V^2/c^2}}{1 - V/c} = \nu'\sqrt{\frac{1 + V/c}{1 - V/c}} \tag{13.12}$$

for the frequency of a source coming directly *toward* us. Note that $\nu > \nu'$, corresponding to a *blue*shift. If, for example, a spaceship is coming toward us at $(3/5)c$ and broadcasts a signal at frequency ν' from its point of view, we will observe the higher frequency $\nu = \nu'\sqrt{(1 + 3/5)/(1 - 3/5)} = 2\nu'$.

Now suppose the photon is emitted with $\theta' = \pi$, as shown in Fig. 13.5. In that case the source is moving directly *away* from us, as shown; and since $\cos\theta' = -1$ we find from Eq. 13.8a that

$$\cos\theta = \frac{\cos\theta' + V/c}{1 + (V/c)\cos\theta'} = \frac{-1 + V/c}{1 - V/c} = -1, \tag{13.13}$$

so the angle of the light from the source is also $\theta = \pi$ from our point of view. The Doppler equation then gives

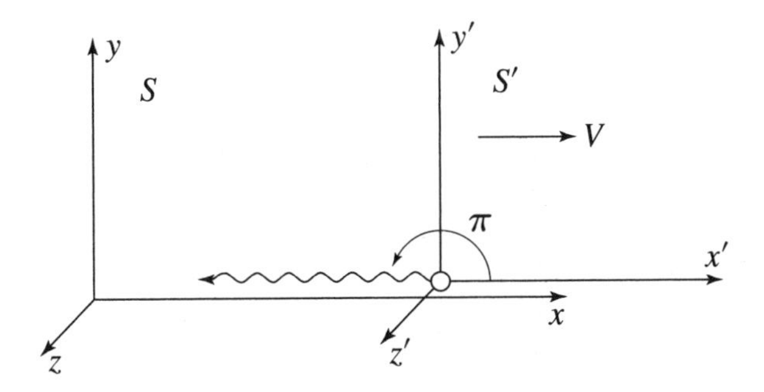

FIGURE 13.5
A photon from a source at the origin of the S' frame is emitted with $\theta' = \pi$.

$$\nu = \frac{\nu'\sqrt{1 - V^2/c^2}}{1 - (V/c)\cos\theta} = \frac{\nu'\sqrt{1 - V^2/c^2}}{1 + V/c} = \nu'\sqrt{\frac{1 - V/c}{1 + V/c}} \tag{13.14}$$

for a source moving directly away from us, so $\nu < \nu'$, corresponding to a *red*shift. If, for example, a distant galaxy is moving away at $(3/5)c$, we will observe that each spectral line has a much lower frequency: in fact, $\nu = \nu'\sqrt{(1 - 3/5)/(1 + 3/5)} = \nu'/2$.

From Eqs. 13.12 and 13.14, the Doppler shift for sources directly approaching or directly receding from us is given by

$$\nu = \nu'\sqrt{\frac{1 \pm V/c}{1 \mp V/c}} \tag{13.15}$$

where upper signs mean approaching and lower signs mean receding.[3]

Another interesting special case is that in which the source emits a beam of light with $\theta' = \pi/2$, perpendicular to the relative frame translation direction, as shown in Fig. 13.6(a). Then $\cos\theta' = 0$, so from Eq. 13.8a we see the light beam at angle θ such that $\cos\theta = V/c (= 3/5$, say, corresponding to $\theta = 53^\circ$), as illustrated in Fig. 13.6(b). This is the aberration of light effect described in Chapter 2, which is very similar to watching snowflakes fall while in a moving car: They seem to come from the front, more and more as the car moves faster and faster. The Doppler shift in this case is

$$\nu = \frac{\nu'\sqrt{1 - V^2/c^2}}{1 - (V/c)\cos\theta} = \frac{\nu'\sqrt{1 - V^2/c^2}}{1 - V^2/c^2} = \frac{\nu'}{\sqrt{1 - V^2/c^2}}, \tag{13.16}$$

---

3. The wavelength λ of a beam of light is related to the light's frequency by $\lambda = c/\nu$, so the Doppler shift in terms of wavelength, with upper signs for approaching and lower signs for receding, is $\lambda = \lambda'\sqrt{(1 \mp V/c)/(1 \pm V/c)}$. Also, a more physically transparent derivation of the Doppler formula is outlined in Problem C–3 of Appendix C.

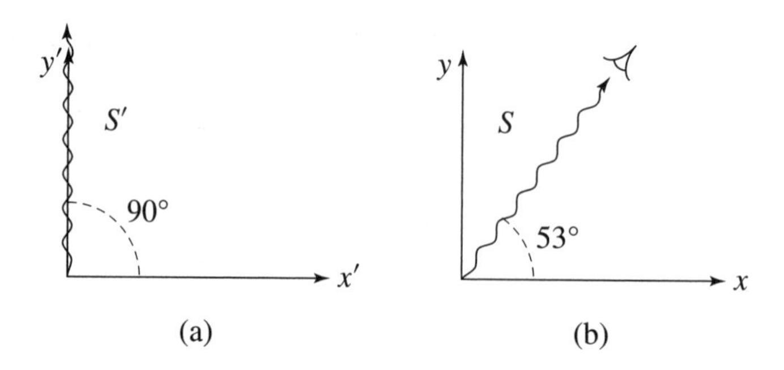

FIGURE 13.6
(a) Light beam emitted by the source at angle $\theta' = 90^\circ$ in the primed frame. (b) The beam is seen by an observer in the unprimed frame to be moving at angle $\theta = 53^\circ$, if the primed frame is moving to the right with $V = (3/5)c$.

which is a *blue*shift. If the relative velocity between the source and observer is $V = (3/5)c$, for example, then the observed frequency is $\nu = (5/4)\nu'$.

Now suppose the *observer,* who is at rest in the unprimed frame, sees that the source, which is at rest in the primed frame, is moving perpendicular to the line of sight; namely, suppose that $\theta = \pi/2$, as shown in Fig. 13.7(b). Then $\cos\theta = 0$, so from Eq. 13.8b we learn that $\cos\theta' = -V/c$. If $V/c = 3/5$, for example, the beam of light was emitted at angle $\theta' = 127^\circ$, as shown in Fig. 13.7(a). In this case the Doppler shift is

$$\nu = \frac{\nu'\sqrt{1 - V^2/c^2}}{1 - (V/c)\cos\theta} = \nu'\sqrt{1 - V^2/c^2} = (4/5)\nu', \qquad (13.17)$$

which is a *red*shift! This particular special case is called the *transverse* Doppler effect. The observer sees a source moving at right angles relative to the line of sight. In

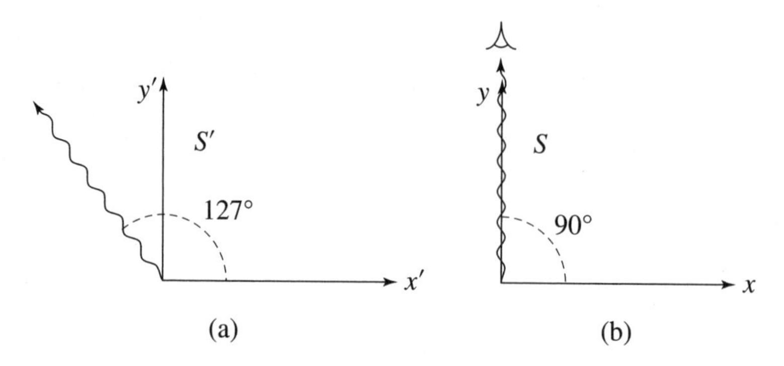

FIGURE 13.7
(a) Light beam emitted by the source at angle $\theta' = 127^\circ$ in the primed frame, which moves to the right at $(3/5)c$. (b) The observer, at rest in the unprimed frame, sees the beam come from the direction $\theta = 90^\circ$, perpendicular to the source's velocity direction.

classical physics there would be no Doppler effect at all in this case, since the source was moving neither toward nor away from the observer when the light was emitted. The transverse Doppler effect is easily understood in relativity using time dilation. Everything on a moving source will seem to the observer to be running slow in time, including the atoms responsible for the radiation. Therefore the frequency of the emitted light, which serves as a kind of clock demonstrating the time rate on the source, will be decreased by the time-dilation factor $\sqrt{1 - V^2/c^2}$.

13.3 The Appearance of Stars to a Fast-Moving Spaceship

A spaceship is at rest in the center of a star cluster, with an equal density of stars in all directions. All of the stars have the same yellow color. What does the cluster look like to spaceship passengers if the ship accelerates up to relativistic speeds? If we suppose that they can look out in all directions, what do they see?

There are three effects, two of which were already discussed in the preceding section. One of these is the Doppler effect, and the other is the aberration of light. The Doppler effect tells us that the light from the stars *ahead* of the ship will be shifted toward higher and higher frequencies to ship passengers as the ship travels faster and faster, from yellow to green to blue to violet and even into the ultraviolet if the ship moves fast enough. Stars to the *rear* of the ship will have lower and lower frequencies, shifting from yellow to orange to red and then into the infrared at very high speeds.

Light aberration tells us that as the ship moves faster and faster, a particular star appears to be closer and closer to the forward direction. A star dead ahead stays in the same apparent position, while a star directly to the rear also stays directly to the rear. But all other stars shift somewhat toward the forward direction. Stars that are really halfway from the front to the rear will now appear to be in the forward hemisphere, and even some stars that are really in the rear hemisphere will now appear to be in the forward hemisphere. At very high speeds there will appear to be a large concentration of stars nearly straight ahead, and very few in the rear hemisphere.

Fig. 13.8 shows what the front and rear hemispheres look like when the ship (a) is at rest in the cluster and (b) is moving at $(4/5)c$. The figure is only in black and white, so remember also that when moving the forward stars are shifted toward the blue (the star dead ahead the most of all), while the rear stars are shifted toward the red.

There is a *third* effect we have not discussed so far, but which is implicit in the Doppler effect. The energy $E = h\nu$ of each photon from stars in the forward direction is larger than when the ship is at rest, and the ship also runs into more photons per second while it is moving, so as long as the light is still more or less in the middle of the visible spectrum, it will not only be shifted toward the blue, but it will also tend to be *brighter.* Light from stars to the rear will not only be redder than before, but it will also tend to be *dimmer,* both because each photon has less energy and because fewer photons are received by the ship each second.

Summarizing all three effects, the forward sky is bluer, brighter, and packed with stars; the backward sky is redder, dimmer, and nearly devoid of stars.

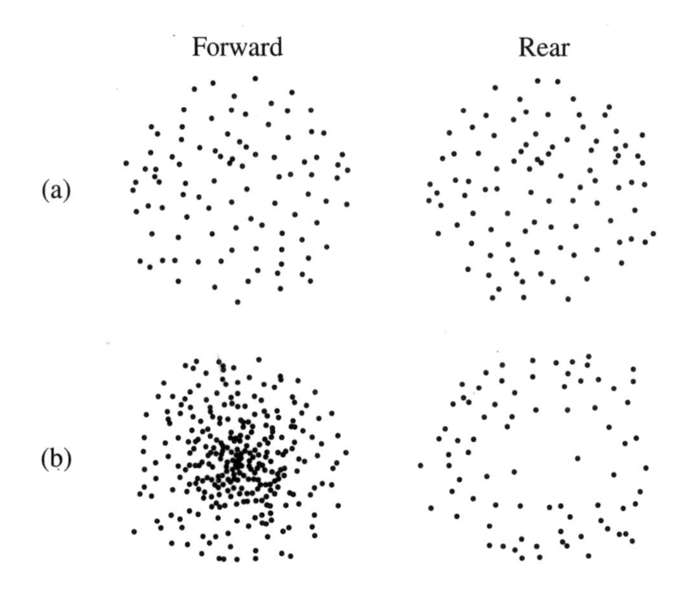

FIGURE 13.8
Stars in the forward hemisphere and rear hemisphere (a) as seen when the ship is at rest and (b) when the ship is moving at $(4/5)c$. Stars in the front are blueshifted and those in the rear are redshifted.

13.4 Threshold Energies

High-energy physicists often want to create new particles in the collisions of other particles. If the total mass of the new particles exceeds that of the initial particles, the collision requires that the initial particles have enough *kinetic* energy to create the additional *mass* energy. What is this initial kinetic energy? More precisely, what is the minimum initial kinetic energy required to create the final particles? Often it is very easy to figure this out in the center-of-momentum frame. That is because in the CM frame the final particles can all be at rest, since by definition the CM frame is the frame in which there is no net momentum. In this case *all* the final energy is tied up in mass energy. Therefore the initial kinetic energy required is simply the difference between the total final mass energy and the total initial mass energy.

For example, consider the collision in which a pion strikes a proton, to form a Λ hyperon and a K meson: $\pi^- + p \to \Lambda + K^0$. In the CM frame, where the pion and proton bang into one another with equal but opposite momenta, the minimum kinetic energy of the pion and proton required to just barely create the final particles is such that

$$(KE_\pi + M_\pi c^2) + (KE_p + M_p c^2) = M_\Lambda c^2 + M_K c^2. \tag{13.18}$$

The total initial kinetic energy $KE_\pi + KE_p$ is obviously the difference between the final and initial mass energies.

However, in the laboratory we may want to create the Λ and K by firing pions at a *stationary* target of protons. What is the minimum energy a pion must have to make the

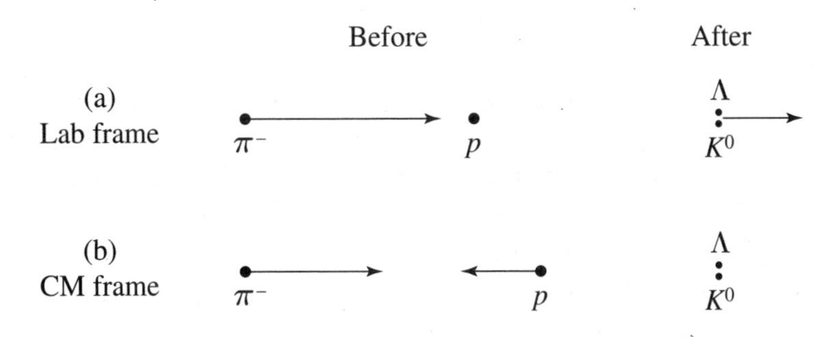

FIGURE 13.9
The initial and final particles in (a) the lab frame and (b) the CM (zero-momentum) frame.

reaction go? This minimum energy is called the *threshold energy*. An energy less than this is not enough to create the final particles. An energy *greater* than the threshold energy is fine, because then the final particles can have just enough kinetic energy to conserve the total energy.

So here is the question: Given the masses of all the particles, what is the threshold energy of the incident particle in the lab, in which the target particle is at rest? Sketches of the collision are shown in Fig. 13.9 in the lab frame (which is the unprimed frame) and in the CM frame (the primed frame). In answering this question, it is very helpful to transform the *total* energy and *total* momentum between the lab and CM frames. For example, we would like to relate the *total* energy and *total* momentum of the π^- and p before the collision (or of the Λ and K after the collision) in the lab (stationary-target) frame and in the CM frame.

For each individual particle, we know that the energy and momentum in the two frames are related by

$$E' = \gamma(E - Vp_x)$$
$$p'_x = \gamma(p_x - VE/c^2)$$
$$p'_y = p_y \quad p'_z = p_z \tag{13.19}$$

where the CM frame is primed and moves to the right at speed V relative to the unprimed lab frame. The *total* energy and momentum components of the initial two particles (or the final two particles) are simply the sums

$$E_T = E_1 + E_2 \qquad p_{Tx} = p_{1x} + p_{2x}$$
$$p_{Ty} = p_{1y} + p_{2y} \quad p_{Tz} = p_{1z} + p_{2z}. \tag{13.20}$$

Therefore, since the coefficients $\gamma \equiv 1/\sqrt{1 - V^2/c^2}$ and γV in Eqs. 13.19 are the same for each particle, it follows that *the total energy and momentum transform exactly the same way as the energy and momentum of the individual particles.* In other words,

Eqs. 13.19 are as valid for the *total* energy and momentum as they are for each individual particle.

This leads to a very elegant and useful result. Since the sum transforms the same way as each individual piece, and each piece transforms like a four-vector, it follows that the *total* energy and momentum of a system of particles form a *four-vector* P_{T} with components $(P_{\mathrm{T}\mu}) = (E_{\mathrm{T}}/c,\, p_{\mathrm{T}x},\, p_{\mathrm{T}y},\, p_{\mathrm{T}z})$ that transform according to Eqs. 13.19. Also, like any other four-vector, the combination $E_{\mathrm{T}}^2 - p_{\mathrm{T}}^2c^2$ must be invariant, the same in any inertial frame, where $p_{\mathrm{T}}^2 = (p_{\mathrm{T}x})^2 + (p_{\mathrm{T}y})^2 + (p_{\mathrm{T}z})^2$. That is,

$$E'^2_{\mathrm{T}} - p'^2_{\mathrm{T}}c^2 = E^2_{\mathrm{T}} - p^2_{\mathrm{T}}c^2 \tag{13.21}$$

for *any* primed and unprimed inertial frames. We are especially interested in the case where the primed frame is the CM frame and the unprimed frame is the lab frame. In the lab frame, the target particle, which is initial particle 2, is at rest, so $p_2 = 0$ and $E_2 = m_2c^2$. In the CM frame, the total momentum is by definition zero; that is, $p'_{\mathrm{T}} = p'_1 + p'_2 = 0$, where we have implicitly chosen the x direction as the direction of motion. Then the invariant equation $E'^2_{\mathrm{T}} - p'^2_{\mathrm{T}}c^2 = E^2_{\mathrm{T}} - p^2_{\mathrm{T}}c^2$ becomes

$$E'^2_{\mathrm{T}} = E^2_{\mathrm{T}} - p^2_{\mathrm{T}}c^2 = (E_1 + m_2c^2)^2 - p_1^2c^2. \tag{13.22}$$

We can use $p_1^2c^2 = E_1^2 - m_1^2c^4$, the usual relationship for a single particle, so

$$E'^2_{\mathrm{T}} = \left(E_1 + m_2c^2\right)^2 - E_1^2 + m_1^2c^4. \tag{13.23}$$

The E_1^2 terms cancel, which makes it easy to solve for E_1. The result is

$$E_1 = \frac{E'^2_{\mathrm{T}} - \left(m_1^2 + m_2^2\right)c^4}{2m_2c^2}, \tag{13.24}$$

which gives the incident particle energy in the lab frame in terms of the particle masses and the total energy of the two particles in the CM frame.

So far we have discussed particles 1 and 2 before a collision takes place. If there are the same two particles after the collision as before, the collision is said to be *elastic*; otherwise it is *inelastic*. In an elastic collision, such as $\pi^- + p \to \pi^- + p$, the mass is the same before and after, so the total kinetic energy must also be the same before and after. In an inelastic collision, the total mass afterward may either be greater or less than before, and the total kinetic energy makes up for any change in mass so as to keep the total energy constant. The collision $\pi^- + p \to \Lambda + K^0$ is inelastic, with greater mass and less kinetic energy after the collision than before. Whether the collision is elastic or inelastic, the total energy in the lab frame must be the same before and after, and the total energy in the CM frame must also be the same before and after. Therefore the quantity E'_{T} in Eq. 13.24, which is the total *initial* energy in the CM frame, is also the total *final* energy in the CM frame.

Now we seek to find the *minimum* laboratory energy E_1 of an incident particle in the laboratory frame (that is, the stationary-target frame) required to create a particular new particle; this minimum energy E_1 is the threshold energy. The easy way to find it is to first find the minimum energy in the CM frame, and then find the corresponding minimum energy in the lab frame.[4] In the CM frame, because the total momentum is zero, it is possible that all final particles are at rest. The minimum total energy in the CM frame is simply the sum of the mass energies of all the final particles. Any additional energy in the CM frame beyond the minimum energy goes into kinetic energy of the final particles. The minimum energy in the CM frame is therefore

$$E'_{\mathrm{T}} = \sum_{\substack{\text{final}\\ \text{particles}}} mc^2,$$

the sum of the final particle mass energies. We have therefore found that the threshold lab frame energy of the incident particle can be found in terms of the particle masses alone:

$$(E_1)_{\text{threshold}} = \frac{\left(\sum\limits_{\substack{\text{final}\\ \text{particles}}} mc^2\right)^2 - \left(m_1^2 + m_2^2\right) c^4}{2m_2c^2}. \tag{13.25}$$

For example, suppose we want to create an electron–positron pair by firing a positron e^+ (which is an antielectron, with the same mass as an electron e^-) at an electron at rest in the lab:

$$e^+ + e^- \to e^+ + e^- + e^+ + e^-. \tag{13.26}$$

The original positron and electron are still there, and an additional electron–positron pair has been created. What is the minimum energy of the original positron needed to make this reaction go?

The answer is

$$(E_1)_{\text{threshold}} = \frac{(4mc^2)^2 - (m^2 + m^2)c^4}{2mc^2} = 7mc^2. \tag{13.27}$$

The total energy of the incident positron must be at least *seven* times its mass energy, and its kinetic energy must therefore be at least $E - mc^2 = 6mc^2$.

When we create two new particles, each of mass energy mc^2, why isn't the minimum required incident kinetic energy just $2mc^2$ instead of $6mc^2$? The answer is that to conserve momentum the final particles must be moving in the lab frame, so

4. Note from Eq. 13.24 that (since the masses are fixed) if the CM-frame energy is minimized, the lab-frame energy is minimized as well.

they have kinetic as well as mass energy. Of the $6mc^2$ of kinetic energy required, $4mc^2$ goes into kinetic energy of the final particles and $2mc^2$ into the additional mass energy.

13.5 Colliding-Beam Experiments

In the preceding section we found that the threshold energy for electron–positron pair creation from electron–positron (or electron–electron) collisions is $7mc^2$, the required energy of the incident particle in the lab frame, in which the target particle is at rest. This is quite high compared with the mass energy of the created particles. Therefore, it occurred to particle physicists that if they could actually *perform* the experiment in the CM frame, with both initial particles racing toward one another with equal but opposite momenta, much less energy would be needed to create new particles. In the reaction $e^+ + e^- \to e^+ + e^- + e^+ + e^-$, for example, each initial particle needs only $2mc^2$ of energy in the CM frame (since the final particles can all be at rest), so in that frame the minimum required energy at the end is just $4mc^2$. This means that each initial particle then needs a kinetic energy of only mc^2, in contrast with the lab frame, in which the incident particle requires a kinetic energy of $6mc^2$, which is much harder to achieve.

Therefore *colliding-beam* experiments were suggested, in which *both* of the initial particles would be accelerated and then made to crash into one another. At first this suggestion seemed crazy to many people, because it would be like firing bullets at bullets and hoping that some of them would hit each other. Careful analysis showed that it should work, however, and so several colliding-beam accelerators (called *colliders*) have by now been built, and they have worked spectacularly well at providing the very high collision energies that are sometimes needed. In fact, they are so effective that fixed-target accelerators for achieving high energies are now largely a thing of the past.

A good example is the Tevatron at Fermilab, a large complex not far from Chicago. In the Tevatron, bunches of protons circle around through an evacuated ring of 4-mile circumference, while bunches of antiprotons circle in the opposite direction through the same ring. Each proton (or antiproton) can achieve an energy up to about 1 TeV = 1000 GeV, so the total collision energy is up to 2 TeV, which is the energy available to create new particles. A small section of the Tevatron tunnel and ring are shown in Fig. 13.10.

A more recent large colliding-beam machine is the Large Hadron Collider (LHC) at CERN near Geneva, Switzerland.[5] The LHC has *two* rings, one for protons circling clockwise, and one for protons circling counterclockwise (the LHC does not need to use antiprotons). The two underground rings are each 27 km in circumference, housed within a large tunnel crossing under the border of Switzerland and France. Each proton

5. CERN is an acronym for the original organization title *Conseil Européen pour la Recherche Nucléaire.* The acronym was not changed when the organization was renamed *Organization Européen pour la Researche Nucléaire* (European Organization for Nuclear Research).

FIGURE 13.10
The Tevatron ring. Protons and antiprotons circle in opposite directions within the ring. Photo courtesy of Fermilab.

can reach an energy up to 7 TeV, so that the resulting collision energy can be used to create new particles with mass energies up to 14 TeV.

Sample Problems

1. Nearly crippled by a surprise attack from an alien vessel, our starship *Endeavor* has only a single photon torpedo left, which can deliver a sufficient number of 1 MeV photons to provide a total of 1 gigajoule of energy. The aliens are attacking us head-on at speed $(12/13)c$ and will only be knocked out if, from their perspective, our photon torpedo strikes them with at least 4 gigajoules of energy. (a) How many photons are in the torpedo? (b) What is the frequency of the photons in our frame, and in the alien's frame? (c) Are the aliens beaten, or do they go on to destroy civilization as we know it?

Solution: (a) The conversion between electron volts and joules is $1\,\text{eV} = 1.60 \times 10^{-19}$ J, so each photon in the torpedo has energy $1\,\text{MeV} = 10^6\,\text{eV} = 1.60 \times 10^{-13}$ J. The number of photons in the torpedo is therefore

$$\text{No. of photons} = \frac{\text{torpedo energy}}{\text{energy/photon}} = \frac{10^9\ \text{J}}{1.60 \times 10^{-13}\ \text{J}} = 6.25 \times 10^{21}.$$

(b) In the picture, the aliens are shown moving to the *left* from our point of view, so their frame is *unprimed* and our frame is *primed.* The frequency of the 1 MeV photons in our frame is

$$\nu' = \frac{E'}{h} = \frac{1.60 \times 10^{-13}\,\text{J}}{6.63 \times 10^{-34}\,\text{J} \cdot \text{s}} = 2.41 \times 10^{20}\,\text{s}^{-1}.$$

In the alien's frame the frequency is *higher* by the Doppler shift,

$$\nu = \sqrt{\frac{1+V/c}{1-V/c}}\nu' = \sqrt{\frac{1+12/13}{1-12/13}}\nu' = \sqrt{\frac{25/13}{1/13}}\nu' = 5\nu',$$

five times as great.

(c) The number of photons received by the aliens is the same as the number sent in our frame, but in their frame each photon has five times the energy that it has in our frame, so they will receive a total of 5 gigajoules, more than enough to destroy their ship. We can also find this result directly from the energy–momentum transformation: in our frame the energy and momentum of the photon torpedo are $E' = 10^9$ J and $p'_x = E'/c = 10^9$ J/c, so the torpedo's energy in the unprimed alien's frame is

$$E = \gamma(E' + Vp'_x) = \frac{13}{5}\left[10^9\,\text{J} + \frac{12}{13}c(10^9\,\text{J}/c)\right] = \frac{13}{5}\left[\frac{25}{13}\right] 10^9\,\text{J} = 5 \times 10^9\,\text{J}.$$

2. A synchrotron accelerates a beam of protons and then directs them at a target of stationary protons (the nuclei in liquid hydrogen.) Find the threshold incident-proton energy required to produce neutral pi mesons by means of the reaction $p + p \to p + p + \pi^0$. (Note that $m_p = 938.3\,\text{MeV}/c^2$ and $m_{\pi^0} = 135.0\,\text{MeV}/c^2$.)

Solution: The threshold energy is given by Eq. 13.25,

$$(E_1)_{\text{threshold}} = \frac{\left(\sum_{\substack{\text{final}\\\text{particles}}} mc^2\right)^2 - \left(m_1^2 + m_2^2\right)c^4}{2m_2c^2}$$

where particle 1 is the incoming proton and particle 2 is the target proton. So here

$$(E_1)_{\text{threshold}} = \frac{\left((2m_p + m_{\pi^0})c^2\right)^2 - \left(m_p^2 + m_p^2\right)c^4}{2m_pc^2}$$

$$= \frac{(2 \cdot 938.3 + 135.0)^2 - 2 \cdot (938.3)^2}{2 \cdot 938.3}\text{MeV} = 1218.0\,\text{MeV}.$$

The threshold *kinetic* energy of the incident proton is

$$(E_1)_{\text{threshold}} - m_p c^2 = (1218.0 - 938.3)\ \text{MeV} = 279.7\ \text{MeV}.$$

Note that this threshold kinetic energy has to be greater than the mass energy 135.0 MeV of the newly created pion, because the final particles must have some kinetic energy in order to satisfy overall momentum conservation.

Problems

13–1. We observe an electron of mass $0.511\,\text{MeV}/c^2$ moving in the $+x$ direction at speed $v = (3/5)c$. Find its momentum and energy in our frame, and also in a frame moving in the $+x$ direction at speed $V = (4/5)c$.

13–2. We observe a proton of mass $938.3\,\text{MeV}/c^2$ moving in the $-x$ direction at speed $v = (4/5)c$. Find its momentum and energy in our frame, and also in a frame moving in the $+x$ direction at speed $V = (3/5)c$.

13–3. We observe a neutron of mass $939.6\,\text{MeV}/c^2$ moving in the $+y$ direction at speed $v = (3/5)c$. Find its momentum components and energy in our frame, and also in a frame moving in the $+x$ direction at speed $V = (3/5)c$.

13–4. The decay of the Λ particle into a proton and pion was discussed in Chapter 12, and the momentum and energy of the final particles were calculated in the CM frame, in terms of all the particle masses. (a) Find the numerical values in MeV units of the energy and momentum of the pion in the CM frame, using the masses found in Appendix I. (b) Now suppose the initial Λ moves to the right in the laboratory with speed $V = (4/5)c$, and also suppose the decay pion moves to the right in the CM frame. Find the energy and momentum of the pion in the laboratory frame.

13–5. A comet approaches Earth at speed $v = 3.0 \times 10^5$ m/s. Calculate the fractional Doppler shift $\Delta\nu/\nu$ as we observe this comet.

13–6. A cautionary tale: A physicist is in court, defending herself on the charge of running a red light. She testifies that because of the Doppler effect, as she approached the intersection the relatively low-frequency red light was shifted to higher-frequency green light, from her point of view. The judge fined her instead for speeding, 1 dollar for every meter/second she was driving over the speed limit of 10 m/s (36 km/hr). How large was the fine? (The wavelength of red light is about 700 nm $= 700 \times 10^{-9}$ m, and the wavelength of green light is about 550 nm.) How much would the fine have been if the speed limit had been 20 m/s?

13–7. In observing a distant galaxy with a spectrometer, astronomers find that the wavelength of the "Lyman alpha" hydrogen line in the spectrum of its stars has shifted from its laboratory value of $\lambda = 121.6$ nm to three times that value, $\lambda = 364.8$ nm. According to the Doppler effect, how fast is the galaxy receding from us? That is, what is v/c of this galaxy relative to us?

13–8. The radius of the Sun is 7×10^8 m, and its equatorial rotation period is 24.6 days. Find the shift in wavelength of 500 nm light due to the Doppler shift, between parts of the Sun moving (a) toward us and (b) away from us, as a result of its rotation. Is the relativistic factor $\sqrt{1 - v^2/c^2}$ significantly different from unity in this case, if we want 1% accuracy?

13–9. Astronomers and cosmologists use the symbol "z" for the redshift of distant galaxies or quasars, where $z \equiv \Delta\lambda/\lambda = (\lambda_{ob} - \lambda_{em})/\lambda_{em}$. Here, λ_{ob} is the wavelength of a spectral line as observed on Earth, and λ_{em} is the wavelength emitted by the distant object in its own rest frame. (a) Show that, in terms of z, according to the Doppler effect the recessional velocity of a distant quasar is given by

$$v/c = \frac{(1+z)^2 - 1}{(1+z)^2 + 1}.$$

(b) Find the recessional velocities v/c of all four quasars described in Fig. 13.3, according to the Doppler effect.

13–10. A light source glows uniformly in all directions, in a frame at rest relative to the light source. Show that if the source is moving at speed v in our frame, half of the emitted photons are radiated into a forward cone whose half-angle is $\theta = \cos^{-1}(v/c)$. This is called the "headlight effect."

13–11. *Watching a receding photon rocket* (See Chapter 12). A photon rocket has departed Earth with total mass M_0. Astronomers watch it recede through a telescope—they can observe the photons emitted in the rocket exhaust. What redshift $\nu_{\text{observed}}/\nu_{\text{emitted}}$ do they observe in terms of M_0 and the total mass M of the rocket when the photons were emitted? (The result is very simple!)

13–12. Show that for a system of particles, the invariant $E_T^2 - p_T^2c^2$ is equal to the square of the total energy in the CM frame of the system, and that this is *not* necessarily equal to $m_T^2c^4$, where m_T is the total mass of the system.

13–13. Antiprotons $\bar{p}$ have been produced in accelerator experiments by the reaction $p + p \rightarrow p + p + p + \bar{p}$. If one of the initial protons (the target proton) is at rest in the lab, what is the minimum energy needed for the other initial proton (the incident proton) to initiate this reaction?

13–14. The particles carrying the weak nuclear force are called W^+, W^-, and Z. The Z particle, in particular, has been created in the collision $e^- + e^+ \rightarrow Z$ of electrons e^- with positrons e^+, in which the initial particles had equal energies. (Electrons and positrons have equal mass energies $m_ec^2 = 0.5$ MeV.) (a) The Z was found to have a mass energy $M_Zc^2 = 91.2$ GeV. What was the energy of the e^- in the experiment, in GeV units? (b) What was the momentum of the e^- in the experiment in units GeV/c? (c) If the Z had been created in an experiment in which the initial e^- was *at rest* in a target, what then would have been the energy E of the e^+, in GeV units?

13–15. Electron–positron pairs can be formed in collisions of photons and electrons: $\gamma + e^- \to e^- + e^+ + e^-$. (Electrons and positrons have the same mass m_e.) Suppose the initial electron is at rest in the lab. (a) For the reaction to proceed, is the minimum initial energy of the photon *less than, greater than,* or *equal to* $2m_ec^2$? Explain. (b) Find the actual minimum initial photon energy required, expressed as a multiple of m_ec^2.

13–16. Researchers wish to create an electron–positron pair from the head-on collision of a high-energy ("gamma-ray") photon γ and a low-energy visible-light photon γ_0. The desired reaction is $\gamma + \gamma_0 \to e^+ + e^-$. In the lab, the photon energies are E_γ and E_{γ_0}, and the electron and positron each has mass m_e. (a) In terms of m_e and c, what is the minimum total photon energy E'_{T} required in the CM (that is, zero-momentum) frame, to create the pair? (b) What then is the minimum energy E_γ required of the high-energy photon in the *lab* frame, in terms of m_e, c, and E_{γ_0}? (c) Find E_γ numerically in GeV, given that $m_ec^2 = 0.511$ MeV and $E_{\gamma_0} = 2.35$ eV (γ_0 is a green photon emitted by a neodymium:glass laser). (d) In fact, the reaction described above is difficult because the required energy E_γ is very large. However, in experiments carried out at the Stanford Linear Accelerator Center (SLAC) during the late 1990s by a team from several institutions, electron–positron pairs were created for the first time in photon–photon collisions when a 29.2 GeV photon struck N green photons (each of $E_{\gamma_0} = 2.35$ eV), essentially simultaneously. What was the minimum value of N that allowed the reaction to proceed?

13–17. The J/ψ meson was discovered in the 1970s at MIT and at SLAC, using different reactions. SLAC used $e^- + e^+ \to J/\psi$, in which the initial electron and positron had equal but opposite velocities, and each had total energy $E = 1.55$ GeV. (a) What is the mass energy of the J/ψ meson? (b) What is the velocity of the original electron, expressed in the form $v/c = 1 - \epsilon$? (c) Suppose the SLAC researchers had planned to create the J/ψ meson by firing a positron at an electron at *rest*. What must the total energy of the positron have been in that case?

13–18. The Large Hadron Collider (LHC) at CERN accelerates protons up to 7 TeV. If two such beams are made to collide head-on, energies of up to 14 TeV are available to create new particles. Suppose a single new particle X of mass $M_X = 12\ \mathrm{TeV}/c^2$ is created in the LHC. If we wanted to create an X in a different accelerator instead, in which a very high-energy beam of protons is fired at a *stationary* target of protons, what beam energy would be required to create an X?

13–19. A team of particle physicists wants to create a new particle Q in the collision $p + p \to Q$. Each initial proton has mass m, and particle Q has a mass M that is unknown, except one expects that $M \gg m$. Suppose that one or both of the protons can be accelerated up to energy E_0. (a) Show that if *both* protons achieve energy E_0 in a colliding-beam experiment, the largest mass energy that can be created is $Mc^2 = 2E_0$. That is, in a colliding-beam experiment, Mc^2 increases *linearly* with E_0. (b) Show that if only *one* of the protons achieves E_0, while the other is a stationary target, the largest mass energy that can be created is $Mc^2 \cong \sqrt{2mc^2E_0}$.

That is, in a stationary-target experiment, Mc^2 increases only as the *square root* of E_0. (c) If $E_0 = 72mc^2$, find the maximum mass that can be created in each type of experiment.

13–20. Pion photoproduction via the reaction $\gamma + p \rightarrow p + \pi^0$ was discussed in Section 12.5. (a) Find the threshold photon energy for this reaction in the rest frame of the initial proton, in terms of m_{π^0} and m_p. The result is greater than the added mass energy $m_{\pi^0}c^2$. Why? (b) Evaluate the threshold energy in MeV (particle masses are given in Appendix I). (c) Ultra high-energy cosmic rays consist primarily of protons that may have originated in far-away active galactic nuclei. As they zip through space they will inevitably encounter low-energy photons in the cosmic background radiation (CBR) that was set loose in the early universe. CBR photons have a wide range of wavelengths, peaked at approximately 1 mm. These photons have energies that are way below the threshold to cause pion photoproduction off protons that are at rest in the CM frame of our galaxy, but can have very high energies in the rest frame of the cosmic-ray protons themselves. If these energies exceed the threshold, pions will be produced and the proton energy in the frame of our galaxy will be reduced, leading to an upper limit in the cosmic-ray proton energies we can observe. This is called the GZK limit (for the physicists Greisen, Kuzmin, and Zatsepin who predicted it). Estimate the GZK limit for the cosmic-ray proton energy (in eV) by pretending that the CBR consists entirely of photons with wavelength 1 mm in our galactic frame of reference. [*Hint:* Let the unprimed frame be the rest frame of the cosmic-ray proton and the primed frame be the frame of our galaxy. Then show that the gamma-factor $\gamma = (1 - V^2/c^2)^{-1/2}$ between these two frames is given by $\gamma = E'_p/m_pc^2 \cong E_\gamma/(2E'_\gamma)$ where E_γ and E'_γ are the photon energies in the unprimed and primed frames. To derive the last (approximate) equality, we can use the energy–momentum transformation in the limit where the relative frame velocity is $V/c = 1 - \varepsilon$, where $\varepsilon \ll 1$. (Also $E'_\gamma = h\nu' = hc/\lambda_{\text{CBR}}$ where we are taking $\lambda_{\text{CBR}} = 1$ mm.)] The actual limiting proton energy obtained by GZK was 6×10^{19} eV. Nevertheless, cosmic-ray protons with energies of up to 3×10^{20} eV have apparently been observed. The reason for the discrepancy is unclear.

13–21. A particle of mass M is moving to the right at $V = (3/5)c$, when it suddenly disintegrates into two particles of mass $m = (2/5)M$, as shown. Find the vertical component of velocity v_y for each particle of mass m. Use the energy-momentum transformation. (This problem was previously given in Chapter 11, since it can also be solved without the use of four-vectors.)

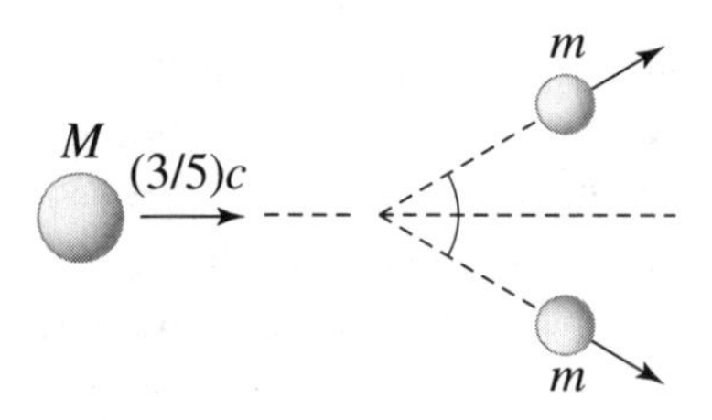

CHAPTER 14

Gravitation

EINSTEIN'S THEORY OF SPECIAL RELATIVITY is powerful in scope, because it affects the very foundations of physics. *Every* part of physics has to be looked at to make sure it conforms to relativistic requirements. The electromagnetism of James Clerk Maxwell passes the test, and is actually partially justified and strengthened by the relativistic point of view. On the other hand, classical mechanics has to be revised in order to satisfy Einstein's demand that the fundamental laws be invariant under the Lorentz transformation. We will now examine Newton's law of gravitational attraction, summarized by the central force

$$F = \frac{GMm}{r^2}, \tag{14.1}$$

which varies as the product of the two masses involved and inversely as the square of the distance between them. Does this law pass the test of relativity? Or is a different, relativistic theory of gravity required?

The questioning of Newtonian gravitation seems at first a waste of time! After all, calculations made on planetary motion using Newton's formula agree with observations to a high degree of accuracy. In fact, two planets, Neptune and Pluto, were discovered by the deviations they caused in the calculated orbits of other planets. So Newton's theory has a great deal going for it. Yet from a relativistic point of view, we can see that Newton's law of gravitation cannot be correct!

First of all, how should we interpret the distance "r" between two objects? Who is supposed to measure this distance? Lengths have lost their absolute character in relativity, so an observer on one object may measure a different distance from an observer on the other object. It is also implied in the formula that if you want to know the force between two objects *now*, you should put in the "distance between them" *now*. One problem with this is that we learned in Chapter 6 that *there is no universal now*. Also, how does one object "know" where the other is *now* from its point of view? If no signal can travel with infinite speed, a given object can detect where another was only at an earlier time—at least the time it takes light to travel between them.

Newton's formula also depends on the masses of the objects. Yet in relativity, mass is just one form of energy. Why should gravity act only on that part of the energy contained in mass? Might it not also act upon (and be caused by) kinetic energy and massless photons?

Finally, there is a property of gravitation that doesn't disagree with Newton's theory but isn't explained by it either. This is the fact that in writing the equation of motion of a particle in a gravitational field the mass of the particle cancels out, and so does not influence the motion. If the gravitational force $F = GMm/r^2$ is the only appreciable force acting, then the m's cancel when we write $F = ma$, so the acceleration of a falling object is independent of its mass. An illustration is Galileo's "Leaning Tower of Pisa experiment," in which he supposedly dropped two objects off the tower and showed they take the same time to fall to the ground, regardless of their masses. The mass m in $F = GMm/r^2$ is called the *gravitational mass,* the property of an object upon which gravity pulls, whereas the mass m in $\mathbf{F} = m\mathbf{a}$ is called the *inertial mass,* the property of "sluggishness" of an object, proportional to its reluctance to accelerate. The fact that these masses are equal to one another, or at least proportional through the constant G, is not explained by Newton's theory of gravity.

For these and other reasons, after completing special relativity Einstein set to work to find a new theory of gravitation. This led, after several years of concentrated effort, to his *general theory of relativity* of 1915. The theory overcomes the faults present in Newton's theory and also has the required properties that it reduces to special relativity in the absence of nearby masses, and to Newton's gravitation for nonrelativistic weak fields.

It has been said that[1]

Every successful physical theory eats its predecessors alive.

The predecessors (special relativity and Newtonian gravitation) are still inside general relativity, alive and kicking! In certain limiting cases they work as well as ever, but general relativity is indeed more general, and it predicts striking new effects such as gravitational waves and black holes.

14.1 The Principle of Equivalence

In 1908 Einstein had what he called the "happiest thought of my life"—the *principle of equivalence.* While the proposal that light moves at the same speed in all inertial frames led to the special theory of relativity, the equally simple but powerful equivalence principle led to the general theory of relativity. By using this principle, together with some beautiful mathematics describing curved space invented by Gauss, Riemann, Ricci, and others during the nineteenth century, Einstein proposed a theory of gravity that accounts for gravitational phenomena in terms of a curved "non-Euclidean"

1. From a talk given by Sidney Coleman, a physicist's physicist.

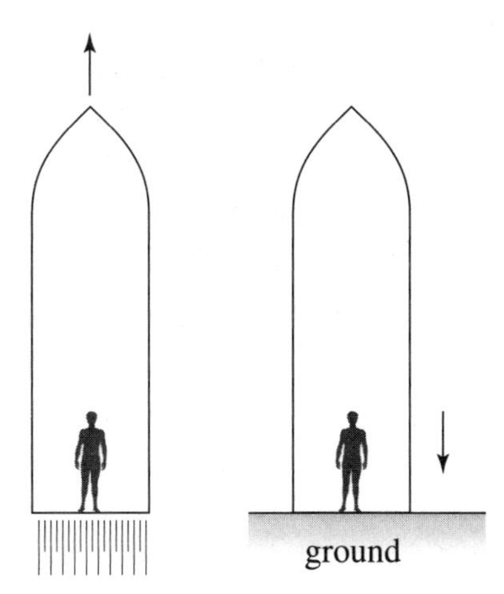

FIGURE 14.1
Two spaceships, one accelerating in space, and the other at rest on the ground.

geometry of spacetime. Although a retracing of the general theory is far beyond the scope of this book, the equivalence principle by itself leads to fascinating insights. We discuss it here for its own interest and because it leads to a gravitational effect on clock rates, which together with special relativistic time dilation has been measured to a high degree of precision.

There are several ways to state the equivalence principle. Here is the approach we will use. Imagine two spaceships, one of which is uniformly accelerating in gravity-free empty space, while the other is standing at rest in a uniform gravitational field, as shown in Fig. 14.1. The equivalence principle then claims that an experiment performed inside the accelerating ship will give the same result as an exactly similar experiment inside the ship at rest in the uniform field, if we choose the acceleration to be numerically equal to the gravitational field g.[2]

A key word here is "inside," since you could clearly distinguish between the two situations by poking your head through a porthole to see if you were standing on some large planet or plummeting through the blackness of space. The principle is a way

2. An alternative form of the equivalence principle is the claim that uniform gravity is eliminated in a freely falling reference frame. If we evacuate the air from an elevator shaft and cut its cables, an elevator will fall freely. People inside will float around, and the experiments they do will verify the law of inertia in the absence of gravity. The elevator will be a gravity-free inertial frame! Similarly, if a hole were drilled through Earth from North to South Pole and a box full of people were dropped into the hole, as long as the box is protected from the rock walls, molten lava, and the extreme heat of the interior, the people inside will feel no gravity, and any experiments they do entirely within the box will not allow them to tell for sure whether they are falling along Earth's axis or drifting in a box in outer space, far from any gravitating stars or planets. In either case, they will be in a gravity-free inertial frame.

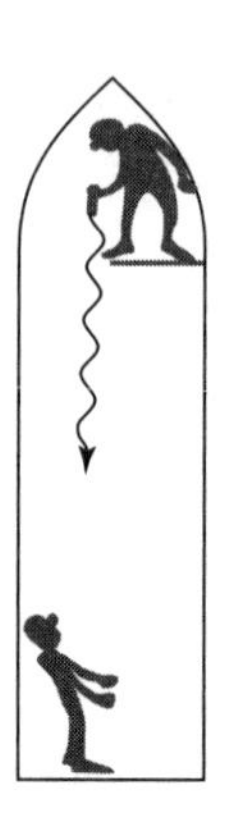

FIGURE 14.2
A laser beam is emitted.

of expressing the observed fact that inertial and gravitational mass are equal. In the accelerating ship, it is a passenger's *inertial* mass that "causes" the passenger to be pressed against the floor; in the stationary ship, it is his or her *gravitational* mass that performs this function. The equivalence of these two situations is quite reasonable for mechanics: intuition and analysis agree that the motion of an object inside is the same in either case. What is not so clear is that the principle applies to experiments with electricity, light, atomic and nuclear physics, as well as mechanics. Yet Einstein decided to pursue this principle, supposing it to be universally valid, to see where it would lead him. We will use the principle here only to deduce two related effects of gravity that are not contained in Newton's theory: the gravitational frequency shift, and the effect of gravity on the rate of clocks.

We start by considering a particular thought experiment with light waves. We first deduce the result of the experiment when performed in the uniformly accelerating ship, and then claim, by virtue of the equivalence principle, that the experiment will give the same result in a ship at rest in a uniform gravitational field.

At the top of the accelerating rocket is an observer who shines a laser beam at another observer at the bottom of the ship, as shown in Fig. 14.2. We assume the laser emits monochromatic light (that is, a single frequency only), and that the distance traveled by the bottom observer while the beam is traveling is very small compared with the length of the ship; this is equivalent to the assumption that the acceleration a is not too large. It follows that the time it takes for the beam to reach the bottom observer is about $t = h/c$, where h is the distance between the two observers. During this light-travel time, the bottom observer has attained a velocity $v = at = ah/c$ with respect to the velocity of the laser when the light was emitted.

The bottom observer is moving *toward* the source, so will observe a blueshift due to the Doppler effect. We've already assumed that this velocity is small compared with the speed of light, so we can use the nonrelativistic Doppler formula

$$\nu_{\text{observed}} = \nu_{\text{emitted}}(1 + v/c) = \nu_{\text{emitted}}(1 + ah/c^2) \qquad (14.2)$$

to compare the observed frequency with the emitted frequency.[3] (Clearly, if instead the bottom observer were to shine a monochromatic beam at the top observer, the top observer would observe a *redshift,* since by the time the top observer received the light, he or she would be moving away from where the source was when the light was emitted.)

Now by means of the equivalence principle we claim that the same effects would be observed in a ship at rest in a uniform gravitational field, if we substitute the acceleration of gravity g for the rocket acceleration a. That is, if the observer at the top of the stationary ship were to shine light with emitted frequency ν_{em} toward the lower observer, the lower observer would see an observed frequency

$$\nu_{\text{ob}} = \nu_{\text{em}}(1 + gh/c^2), \tag{14.3}$$

a blueshift, whereas the top observer would see a redshift if he or she looked at a light beam sent off by the lower observer. But in this case we can hardly blame the shift on Doppler, because neither observer is moving. Instead, the shift in this case is due to a difference in altitude of the two clocks at rest in a uniform gravitational field.[4]

The gravitational frequency shift of Eq. 14.3 was first observed in 1960 for an entirely Earth-based experiment by researchers at Harvard University,[5] observing gamma-ray photons emitted by atomic nuclei. They put emitting and absorbing nuclei at the top and bottom of a 74-foot tower and (within experimental error) found that the frequency shift of the photons was just what the equivalence principle predicts. The emitter and absorber had to be (and were) phenomenally sensitive, because in this experiment the fractional frequency shift was only $\Delta\nu/\nu_{\text{em}} = gh/c^2 = (9.8\ \text{m/s}^2)(22.6\ \text{m})/(3 \times 10^8\ \text{m/s})^2 = 2.5 \times 10^{-15}$.

14.2 Clock Rates

How can we *explain* the blueshift seen by the person at the bottom of the stationary ship? If we think of the laser atoms that radiate the light at the top as clocks whose rate is indicated by the frequency of their emitted light, the observer at the bottom would be

---

3. The relativistic Doppler effect for approaching source and observer (as in Eq. 13.12) becomes

$$\nu_{\text{observed}} = \nu_{\text{emitted}}(1 + v/c)^{1/2}(1 - v/c)^{-1/2} \cong \nu_{\text{emitted}}(1 + v/2c)(1 - v/2c) \cong \nu_{\text{emitted}}(1 + v/c)$$

for small v/c, using the binomial expansion of Appendix A and dropping second and higher terms in v/c. The result is the nonrelativistic Doppler formula.

4. This situation is reminiscent of the muon decay problem of Chapters 4 and 5, where it was necesssary to explain why the muons failed to decay before they had penetrated the atmosphere. The *fact* of penetration held in both the Earth and muon frames, but the *reason* for the fact was different in the two frames. Observers on Earth explained it by saying that the muon clocks ran slower, while the muon explained it by saying that the atmosphere was thinner because of the Lorentz contraction.

5. R. V. Pound and G. A. Rebka, Jr., *Physical Review Letters* 4 (1960).

forced to conclude that these top clocks are running *fast* compared to similar clocks at the bottom of the ship. Bottom clocks radiate a certain frequency, while similar atoms higher up radiate a higher frequency. The observer at the top would agree with this judgment. The top observer sees a redshift when looking at clocks at the bottom, so it would be natural for a person at the top to believe that bottom clocks run slower than top clocks.

If atomic clocks up high run faster, it is of course true that all clocks up high run faster, because they can be continuously compared with one another. For suppose a clock at the top of the ship has a luminous second hand that emits light of frequency $\nu = 5 \times 10^{14}\ \mathrm{s}^{-1}$, corresponding to a yellow color. Then in one complete revolution of the second hand, the number of wavelengths emitted is $60 \times 5 \times 10^{14} = 3 \times 10^{16}$. The observer at the bottom must collect all these wavelengths, 3×10^{16} waves per revolution, since none is created or destroyed in transmission. However, the frequency of the waves increases by the factor $(1 + gh/c^2)$, so it follows that the second hand of the clock at the top appears to complete a revolution in *less* than 60 seconds to the observer at the bottom, by the exact same factor.

If *all* clocks at high altitude run fast compared with clocks at lower altitude, we can conclude that time itself runs fast at higher altitude: that is, for time intervals Δt,

$$\Delta t_{\text{high altitude}} = \Delta t_{\text{low altitude}}(1 + gh/c^2). \tag{14.4}$$

The gravitational effect on clock rates is not as seemingly paradoxical as the time-dilation effect for moving clocks, since *both* the upper and lower observers agree that the upper clocks run faster than the lower clocks. It is easy to calculate the magnitude of the gravitational clock effect at Earth's surface. Using a uniform g of about 9.8 m/s^2, a clock on top of 29,029-foot Mt. Everest would run faster than a similar clock at sea level by about one part in 10^{12}, which amounts to a gain of only 1 second in 30,000 years!

14.3 The Hafele–Keating Experiment

By 1971, human-made atomic clocks had become accurate enough to test the altitude effect. Two researchers, J. C. Hafele of Washington University in St. Louis and R. E. Keating of the U. S. Naval Observatory in Washington, D. C., decided to test *both* the altitude effect and special relativistic time dilation by taking some of these atomic clocks on commercial air flights.[6] In fact, beginning on October 4, they carried a package of four atomic clocks onto a series of scheduled air flights traveling more-or-less eastward around the world. The total trip lasted 65.4 hours, including 41.2 hours in flight. Using information provided by the flight captains, they kept records of their altitude and speed at frequent intervals, so they could calculate the expected

6. Hafele and Keating, *Science* 177 (1972), p. 166 ff.

TABLE 14.1
Observed time-interval differences between the traveling and Earth-based clocks by the time the traveling clocks returned to their starting point. Results are in nanoseconds. A positive sign means that the traveling clocks read a larger time than ground clocks.

Eastward travel:	-59 ± 10
Westward travel:	$+273 \pm 7$

clock rates. And when they returned to their starting point they compared the clock readings with those of similar atomic clocks kept at the U. S. Naval Observatory.

Departing again about a week after their return, Hafele and Keating carried a set of four clocks on flights traveling *westward* around the world and again compared clock readings with those of the clocks left in the lab. This time the trip lasted 80.3 hours, of which 48.6 hours were in flight. The observed time differences between the clocks flown eastward and those on the ground, and between clocks flown westward and those on the ground, are shown in Table 14.1. The table shows how much the traveling clocks *advanced* relative to the ground-based clocks, measured in nanoseconds ($1\text{ ns} = 10^{-9}$ s). It is interesting that the eastward-traveling clocks *lost* time, while the westward-traveling clocks *gained* time on the ground clocks. (The errors shown were found from the spread of the four clock readings in the plane and of the clocks on the ground.)

Hafele and Keating also had to calculate the *theoretical* difference in clock readings, based on special relativistic time dilation due to the motion, and the altitude effect due to gravity. Of course the altitude and speed of the airplanes changed continually, so the calculations had to be made for small time intervals and summed to obtain the total effect. Careful records of altitude and speed had to be kept throughout the flights. The results, including the altitude and motion effects, and the sums of these, are given in Table 14.2.

The primary source of errors here were the altitude and speed data collected during the several flights in each direction it took to circumnavigate the globe. Note that within

TABLE 14.2
Calculated differences between the traveling and Earth-based clocks, due to the altitude and motion effects (in nanoseconds).

Eastward:	altitude (gravitational) effect:	144 ± 14
	motion (special relativity) effect:	-184 ± 18
	total effect predicted:	-40 ± 23
Westward:	altitude (gravitational) effect:	179 ± 18
	motion (special relativity) effect:	96 ± 10
	total effect predicted:	275 ± 21

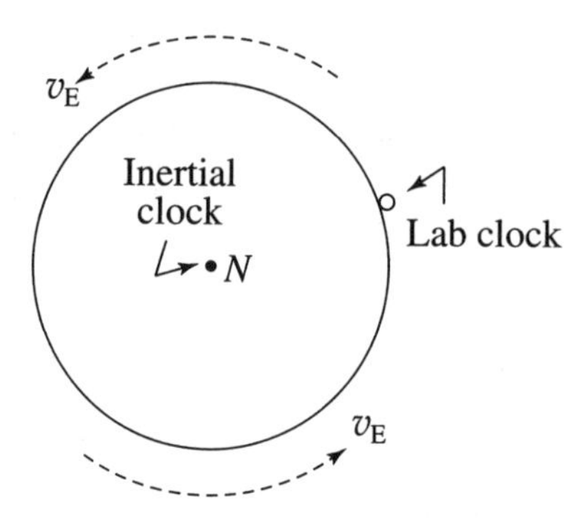

FIGURE 14.3
Clocks observed in an intertial frame.

the errors quoted, the calculated differences are entirely consistent with the observed time interval differences, and that in effect both special relativistic time dilation and the gravitational altitude effect are needed to explain these experiments. Also note that the *altitude* effects are comparable for the eastward and westward flights (although on average the planes flew higher on the westward flights), but that the *time-dilation* effects have opposite signs for the two directions! The net effect is that the clocks flown to the east *lost* 59 ± 10 ns to the ground clocks, while the clocks flown to the west *gained* 273 ± 7 ns on the ground clocks (see Problem 14–7).

How were the theoretical calculations carried out? Consider a simplified model just to get the idea, by supposing that the ground clocks reside in a lab fixed to a point on the equator, and that all the plane flights were directed either due east or due west around the equator. Let Δt_E be a small advance in time of a clock at rest on the ground. Unfortunately, this clock is not inertial, because it accelerates as Earth turns. (If something moves in a circle, even at constant speed, it is accelerating toward the center of the circle, so is not inertial.) To use special relativity, therefore, we need to invent a hypothetical clock, also at ground level but not attached to the surface, that does *not* rotate as Earth turns. This clock stays at rest in an inertial frame fixed to Earth's center.[7] Such an inertial clock might be placed, for example, on the North Pole.

Let this inertial clock advance by Δt_{In} as the lab clock advances by Δt_E. Now according to this inertial clock, the actual clock in the lab is moving with speed v_E, the speed of the surface as Earth turns, as shown in Fig. 14.3. (At the equator, this speed is $v_E = 2\pi R_E/T_E$, Earth's circumference divided by its period of rotation, $T_E = 24$ hours.) Therefore to the inertial clock the lab clock is moving and so runs slow by the usual time-dilation factor. The two times are related by

$$\Delta t_E = \Delta t_{In}\sqrt{1 - v_E^2/c^2} \cong \Delta t_{In}(1 - v_E^2/2c^2), \tag{14.5}$$

where in the last step we used the binomial expansion, valid here since $v_E \ll c$. (We

7. For the purposes of these experiments we can ignore the fact that Earth's center is rotating about the Sun once every year; that is, even Earth's center is not exactly inertial.

have assumed that these two clocks are both at ground level, so there is no altitude effect between them.)

Now what about the airplane clocks? According to the hypothetical ground-level observers in the inertial frame, while their inertial clock is advancing by Δt_{In}, a clock in the airplane runs slow because of its motion, and fast because of its altitude. Altogether,

$$\Delta t_{\text{plane}} = \Delta t_{\text{In}}\sqrt{1 - v^2/c^2}(1 + gh/c^2), \tag{14.6}$$

with factors from both time dilation and the gravitational altitude effect. Of course, the speed v of the airplane is small compared with the speed of light, so using the binomial expansion, we can approximate $\sqrt{1 - v^2/c^2} \cong 1 - v^2/2c^2$. Also, the small terms $v^2/2c^2$ and gh/c^2 turn out to be comparable in size, so to first order in the small quantities, we can write

$$\Delta t_{\text{plane}} = \Delta t_{\text{In}}(1 - v^2/2c^2 + gh/c^2). \tag{14.7}$$

Now we can eliminate the hypothetical clock reading Δt_{In} between Eqs. 14.5 and 14.7, so we can compare directly the observed times Δt_{E} and Δt_{plane} with one another. We can also set v (the speed of the plane relative to the inertial frame) to be $v = v_{\text{E}} \pm v_{\text{g}}$, the speed of Earth's surface v_{E} *plus* the speed of the plane relative to the ground (for the eastern flights) or *minus* the speed of the plane relative to the ground (for the western flights), as shown in Fig. 14.4. The result is

$$\Delta t_{\text{plane}} = \Delta t_{\text{E}}\left[1 + gh(t)/c^2 - v_{\text{g}}(t)(v_{\text{g}}(t) \pm 2v_{\text{E}})/2c^2\right]. \tag{14.8}$$

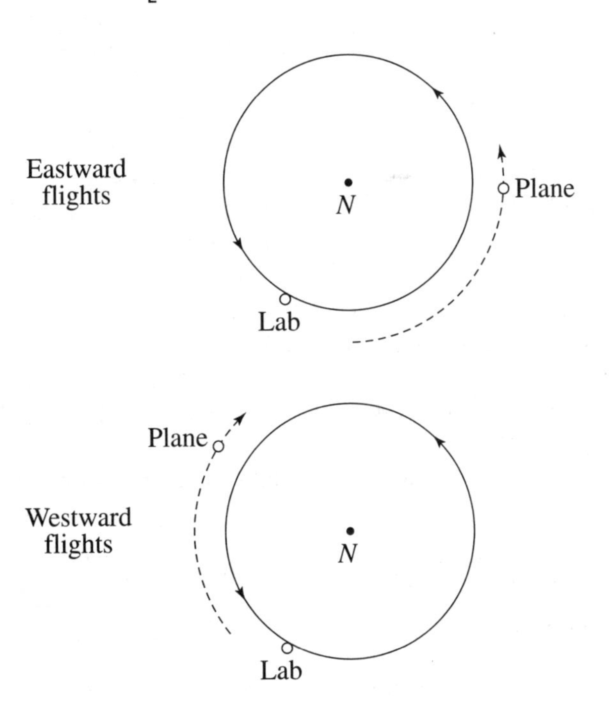

FIGURE 14.4
Eastward and westward airplane flights.

Everything in this equation can be measured at time t (as read by the lab clocks), and then the equation can be summed (that is, integrated) over the duration of the flights to obtain the total time read in the airplane compared with the total time read by ground clocks. This gives the idea of how the theoretical calculations were carried out.[8]

14.4 Satellite Clocks

The equivalence principle shows that clocks at high altitude should run fast compared with clocks at lower altitude by the factor $(1 + gh/c^2)$, where g is the uniform gravitational field acting on the clocks and h is the altitude difference between them. If we tried to compare clock rates on the ground with those in an orbiting satellite we would expect a relatively large altitude effect, because h would then be very large. The problem is that h can be *so* large that the gravitational field is no longer even approximately the same as it is on the ground. In fact, Newton's inverse square law shows that the gravitational field g acting on a clock orbiting at twice Earth's radius would be only 1/4 as large as that on a ground clock. So what should we use for the factor $(1 + gh/c^2)$ in this case?

There is a clue about what we should do: the quantity gh is just the *difference in gravitational potential* $\Delta\phi$ in a uniform field. (In terms of $\Delta\phi$, the potential *energy* mgh of a particle with mass m is $m\Delta\phi$.) The gravitational effect on clocks in a uniform field can therefore be written in the suggestive form

$$\Delta t_{\text{high altitude}} = \Delta t_{\text{low altitude}}(1 + \Delta\phi/c^2) \tag{14.9}$$

where $\Delta\phi$ is positive.

Now what do we do if the gravity is *not* uniform? If we assume that the factor $(1 + \Delta\phi/c^2)$ is still valid for nonuniform gravitational fields, even though $\Delta\phi \neq gh$, we can test the altitude effect also for Earth satellites. In the case of spherically symmetric gravity, for example, the potential is

$$\phi = -\frac{GM_E}{r}, \tag{14.10}$$

as sketched in Fig. 14.5, where r is the distance from Earth's center and M_E is Earth's mass.

Note that although ϕ is negative, it increases with increasing distance from the surface. (Over very small distances, ϕ is approximately *linear* in the distance from point to point, in keeping with the uniform-gravity expression $\phi = gh$.) This suggests that if Δt_∞ is a time interval read by a rest clock at very large distance from Earth,

---

8. See Problem 14–8 and the original articles by Hafele and Keating in *Science* 177 (1972), p. 166 for the predictions and p. 168 for the observations.

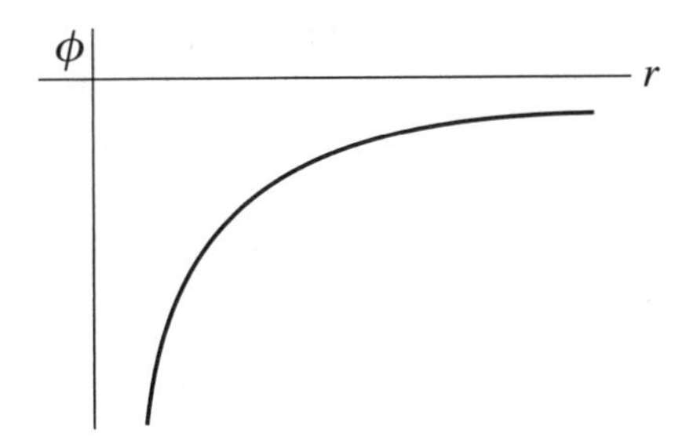

FIGURE 14.5
The gravitational potential around a spherical mass.

then the time interval of a rest clock at distance r from Earth's center would be

$$\Delta t_{\mathrm{r}} = \Delta t_{\infty}\left(1 - \frac{GM_{\mathrm{E}}}{rc^2}\right). \tag{14.11}$$

Note from this expression that the high-altitude clock (with Δt_{∞}) runs fast compared with the clock at r, as expected. Furthermore, we can show that if any two clocks with almost the same altitude are compared, then Eq. 14.4, $\Delta t_{\text{high altitude}} = \Delta t_{\text{low altitude}}(1 + gh/c^2)$, is valid, where g is the strength of gravity at that altitude, in agreement with what we found earlier (see Problem 14–11).

The *difference* in potential between a satellite at radius r_{sat} and the ground at radius r_{E} is

$$\Delta\phi = -\frac{GM_{\mathrm{E}}}{r_{\mathrm{sat}}} + \frac{GM_{\mathrm{E}}}{r_{\mathrm{E}}}. \tag{14.12}$$

Is Eq. 14.9 experimentally correct if we use Eq. 14.12 for $\Delta\phi$? The answer is *yes*; clock rates in satellites have been compared with those on the ground, and the altitude effect is well represented by Eq. 14.9 with Eq. 14.12. These expressions can also be derived from the general theory of relativity in the limit of weak gravity, (that is, if $GM_{\mathrm{E}}/rc^2 \ll 1$), which is appropriate around Earth.

Expression 14.9 is not the whole clock-rate story, however, because it represents the altitude effect alone. The orbiting satellite is moving as well (in fact even clocks on the ground are moving in an inertial frame, because of Earth's rotation). So we have to include the special relativistic time dilation as well to get an accurate prediction of relative clock rates. Both effects have been measured, and are an essential feature of the Global Positioning System (GPS), as described in the next section.

14.5 The Global Positioning System

The Global Positioning System (GPS) was deployed and continues to be maintained by the U.S. military, in part because it allows troops anywhere on Earth to find out

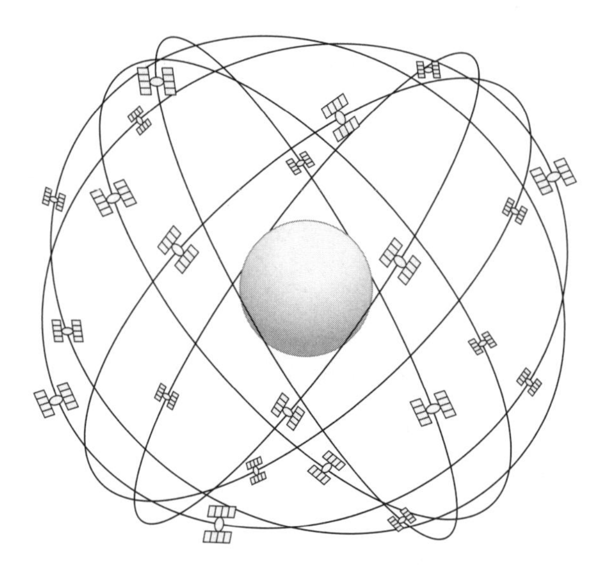

FIGURE 14.6
The six orbital planes of GPS satellites.

where they are at any time. It can also be used to guide missiles and pilotless aircraft. For many years it has been available for civilian use, to locate positions while hiking or boating, for example, or in aircraft so pilots can find their altitude along with their latitude and longitude, and their ground speed as well. The system is based on a set of 24 active satellites (plus a few spares) circling Earth. Each satellite is placed in one of six orbits; the six orbital planes are tilted by 60° with respect to one another, and each is tilted by 55° relative to the equator, as illustrated in Fig. 14.6.

The altitude of all the satellites is fixed at about 20,200 km, because at that altitude they orbit Earth twice each day. The configuration of 24 satellites is chosen so that at least four satellites are "visible" at any time from almost anywhere on Earth. Each satellite contains one or more highly accurate atomic clocks.

Here is the *basic idea* of how the system works.[9] Each satellite continuously broadcasts its position and the time read by its atomic clock. A particular hiker on the ground has strayed off the trail and gotten lost, but carries a hand-held GPS receiver that can pick up the satellite signals. How can the receiver tell her where she is located?

The receiver picks up the signals from (say) four satellites, which we denote by the subscript i ($i = 1, 2, 3, 4$). The four satellites tell the receiver their positions $\mathbf{r}_i = (x_i, y_i, z_i)$, and the times on the satellites t_i. These signals propagate to the receiver at the speed of light, so if $\mathbf{r} = (x, y, z)$ is the position of the receiver and t is the time when the signals are simultaneously received, then the distance between each satellite and the receiver is equal to the speed of light times the time interval between transmission and reception: That is,

$$\left|\mathbf{r} - \mathbf{r}_i\right| = c(t - t_i), \tag{14.13}$$

9. The details are quite complex. See the references at the end of this Chapter.

or, written in terms of the Cartesian components and squared,

$$(x - x_i)^2 + (y - y_i)^2 + (z - z_i)^2 = c^2(t - t_i)^2. \tag{14.14}$$

This is actually a set of four equations, one for each value of i, that is, one equation for each of the four satellites. These amount to four equations in the four unknowns x, y, z, and t. The receiver in effect solves these equations to find both where the receiver is located and what time it is when the signals are received.

This description has been idealized in several ways. One idealization is the assumption that the satellite clocks have been exactly synchronized with one another, so that (for example) t_1 and t_2 both have the same zero point. Otherwise Eqs. 14.14 would contain additional implicit unknowns.

In fact, how accurately *must* the satellite clocks be synchronized with one another? That depends on how accurately one needs to know the receiver's position. Suppose the hiker wants to know where she is located within about 1 meter. Then, roughly speaking, the satellite clocks need to be synchronized within about

$$\Delta t \sim \frac{\Delta x}{c} \sim \frac{1\,\text{m}}{3 \times 10^8\,\text{m/s}} \sim 3 \times 10^{-9}\,\text{s}. \tag{14.15}$$

These satellite clocks, separated from one another by thousands of kilometers, have to be synchronized with one another within a few nanoseconds to provide sufficiently accurate data for the receivers!

How *stable* must the satellite clocks be? That is, how much can a satellite clock be allowed to drift in time if it is to remain synchronized with the others to within a few nanoseconds? In fact, the satellite clocks can be monitored and corrected from a ground station, containing many atomic clocks, about once/orbit, that is, once every 12 hours.[10] So if a clock is to stay synchronized with others within 3 nanoseconds over 12 hours, it must be stable within about

$$\frac{3\,\text{ns}}{12\,\text{h}} = \frac{3 \times 10^{-9}\,\text{s}}{12\,\text{h} \cdot 3600\,\text{s/h}} \sim 10^{-13}\,\text{s/s} \tag{14.16}$$

That is, the atomic clock can deviate only about 10^{-13} seconds in each second. It was not until atomic clocks achieved this precision that anything like the GPS system could be deployed.

The receiver can accurately know where it is located only if the satellite clocks are extremely precise. And that is why the relativistic corrections are essential. Compared with ground clocks, satellite clocks will run slow because of their motion and fast because of their altitude. We need to understand the satellite clock readings within 10^{-13} s/s, so it is important to check whether the clocks need to be corrected for the tiny relativistic effects of time dilation and altitude.

---

10. The primary ground station is the U.S. Naval Observatory in Washington, D.C., which maintains a set of about 50 cesium-beam atomic clocks and 12 hydrogen masers to use as time standards. There are also several other ground stations around the world.

As we have seen in Eq. 14.11, a clock at rest at distance r_{sat} from Earth's center runs slow compared to a rest clock at infinity (that is, very far away) by the fractional amount.

$$\frac{\Delta t}{t} = \frac{GM_E}{r_{sat}c^2}, \tag{14.17}$$

where M_E is Earth's mass and r_{sat} is the satellite's orbital radius.

If the clock is inside a satellite, it is also moving at speed v_{sat} ($v_{sat} \ll c$), and so runs slow due to time dilation by the fractional amount $v_{sat}^2/2c^2$. Now from Newtonian mechanics, the speed of a satellite in circular orbit is

$$v_{sat} = \sqrt{\frac{GM_E}{r_{sat}}}. \tag{14.18}$$

Therefore the fractional time dilation relative to a clock at rest at infinity is

$$\frac{\Delta t}{t} = \frac{v_{sat}^2}{2c^2} = \frac{GM_E}{2r_{sat}c^2}, \tag{14.19}$$

just *half* as large as the altitude effect. Numerically, using the orbital radius $r_{sat} = R_E + h = 6{,}400\text{ km} + 20{,}200\text{ km} = 26{,}800\text{ km}$, we get

$$\text{fractional altitude effect} \sim \frac{GM_E}{r_{sat}c^2} = \frac{\left(6.67 \times 10^{-11}\frac{\text{m}^3}{\text{kg}\cdot\text{s}^2}\right)(5.97 \times 10^{24}\text{ kg})}{(2.68 \times 10^7\text{ m})(3.00 \times 10^8\text{ m/s})^2}$$
$$= 1.65 \times 10^{-10}$$

and the fractional time-dilation effect is half as large. Of course, to compare clocks in the satellite with those on the ground, we have to take account of both the altitude effect and the motion of ground clocks. But it is clear that we *must* take these relativistic effects into account, because they are of order 10^{-10} s/s, a thousand times larger than the maximum permitted drift of 10^{-13} s/s in the satellite clocks!

In summary, relativistic effects that once seemed esoteric and unlikely to play a significant effect in the daily life of very many people turn out to play a critical part in the GPS system, which is now of great importance for many military and civilian uses, affecting the lives of millions.

Physics is a house of cards. What seemed initially in Einstein's postulates to be a statement only about the behavior of light, and not necessarily about the behavior of light in ordinary applications at that, has turned out to revolutionize much of physics and our understanding of the world, and has also led to completely unforeseen applications that have had impacts on the nature of warfare, politics, and the global economy.

References

"Relativity and the Global Positioning System," by Neil Ashby, *Physics Today,* May 2002.
Gravity, by James B. Hartle, Addison-Wesley, San Francisco, 2003.

Sample Problems

1. A driver ties a helium-filled balloon on a string attached to the floor of a car, so the balloon floats up. The driver then accelerates away from a stop sign. While the acceleration remains constant, does the balloon move forward, backward, or remain vertically above the place where it is tied?

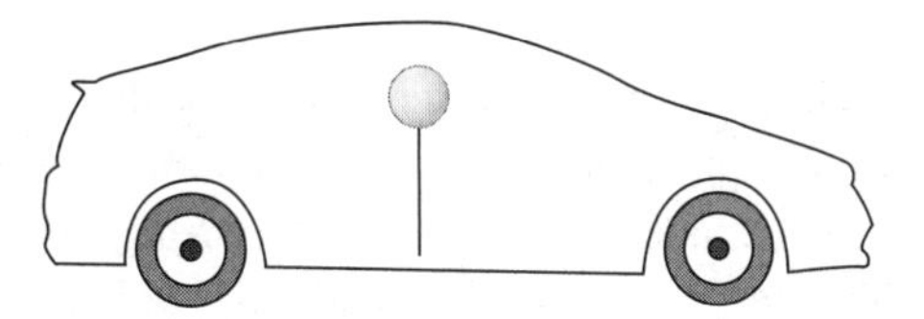

Solution: When the car is at rest, the balloon floats up in a direction opposite to the downward pull of gravity. When the car accelerates, the car defines a reference frame with some acceleration a_0 to the right. From the equivalence principle, a frame accelerating to the *right* is equivalent to a *non*accelerating frame with an artificial gravity g_0 to the *left,* where $g_0 = a_0$. In this nonaccelerating frame, there are now two gravities, the "real" gravity g downward and the artificial gravity to the left. The total effective gravity in the nonaccelerating frame is the vector sum of these two, which is a vector with both downward and leftward components, as shown below. The balloon will float in a direction opposite to this effective gravity; that is, it will rotate as shown, with the balloon moving forward, toward the front of the car. By the equivalence principle, it will do the same thing in the actual, accelerating car: it will move *forward.* (There is another way to see this! A balloon floats because it is under constant bombardment from all sides by rapidly moving air molecules, N_2, O_2, etc., and since the density of air decreases with altitude, slightly fewer molecules strike the top of the balloon than strike the bottom. The net force on the balloon due to striking molecules is therefore *upward*—this is called *buoyancy.* The buoyancy is the same on a ball of iron as

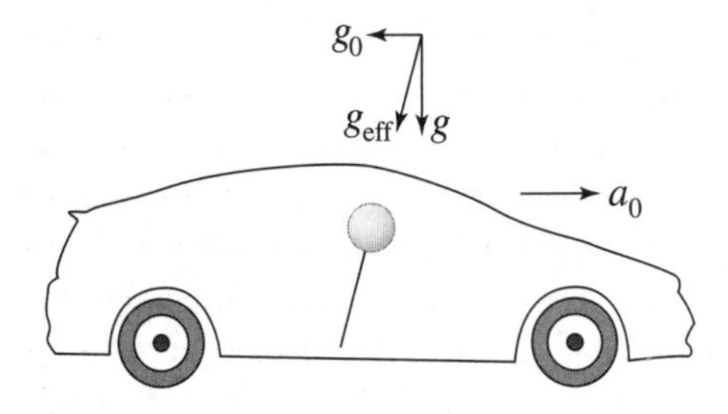

on a balloon of the same size, however. So the reason the balloon rises while the ball of iron falls is that the downward gravitational force on the helium-filled balloon is *less* than the buoyancy, while the gravitational force on the iron ball is *greater* than the buoyancy. The gravitational force on a sphere of air is often equal to its buoyancy, in which case the sphere of air is in equilibrium, and so neither rises nor falls. When the car accelerates forward, air molecules flow somewhat toward the rear of the car, so the density of air is greater at the rear than at the front. The balloon is therefore pushed forward by the greater number of molecules bombarding the rear of the balloon than the front of the balloon.)

2. In Eq. 14.11 it was argued that the relation between the rates of rest clocks at an arbitrary radius r outside a spherical mass and at infinity is

$$\Delta t_{\mathrm{r}} = \Delta t_{\infty}(1 - (GM_{\mathrm{E}}/rc^2)),$$

so that the clock at r runs slow compared with the clock at infinity, since $\Delta t_{\mathrm{r}} < \Delta t_{\infty}$. This expression turns out to be an approximation, valid only if $GM_{\mathrm{E}}/rc^2 \ll 1$. (a) Evaluate GM_{E}/rc^2 at Earth's surface, and show that it is indeed very small there, and so is even smaller at higher altitudes. (b) The general theory of relativity shows that the exact relationship for the rates of clocks at rest outside of a spherical mass M is $\Delta t_{\mathrm{r}} = \Delta t_{\infty}\sqrt{1 - (2GM_{\mathrm{E}}/rc^2)}$. Show that this exact expression is consistent with Eq. 14.11 if $GM_{\mathrm{E}}/rc^2 \ll 1$. (c) Find the radius, according to the exact expression, at which a rest-clock rate goes to *zero*.

Solution: (a) The quantity

$$\frac{GM_{\mathrm{E}}}{rc^2} = \frac{(6.67 \times 10^{-11}\,\mathrm{m^3/kg\,s^2})(6.0 \times 10^{24}\,\mathrm{kg})}{(6400\,\mathrm{km})(3.0 \times 10^8\,\mathrm{m/s})^2} = 6.9 \times 10^{-10},$$

which is indeed $\ll 1$.

(b) The exact relationship is

$$\Delta t_{\mathrm{r}} = \Delta t_{\infty}\sqrt{\left(1 - \frac{2GM}{rc^2}\right)} = \Delta t_{\infty}\left(1 - \frac{2GM}{rc^2}\right)^{1/2}$$

$$= \Delta t_{\infty}\left(1 + \frac{1}{2}\left(-\frac{2GM}{rc^2}\right) + \cdots\right) = \Delta t_{\infty}\left(1 - \frac{GM}{rc^2} + \cdots\right)$$

where we have used the binomial expansion of Appendix A. This *does* reduce to Eq. 14.11 if $GM/rc^2 \ll 1$, since we can neglect the subsequent even smaller terms in the series.

(b) Note from $\Delta t_{\mathrm{r}} = \Delta t_{\infty}\sqrt{1 - (2GM/rc^2)}$ that $\Delta t_{\mathrm{r}} \to 0$ if $r \to 2GM/c^2$. If a spherical mass M is entirely contained within a radius $r = 2GM/c^2$, then a *black hole* has been formed. The spherical surface of radius $r = 2GM/c^2$ is called the *event horizon* of the black hole, because observers outside this radius cannot see what is inside—things inside are "beyond their horizon." This is because light cannot escape out through the event horizon, so whatever is inside is invisible to people outside. It is possible in principle for a clock to be at rest (with constant r) anywhere *outside* the event horizon, but not at or within the event horizon. So in particular there can be no clock at rest at $r = 2GM/c^2$. A clock *can* be at rest just outside this radius, and from the equation $\Delta t_r = \Delta t_\infty \sqrt{1 - (2GM/rc^2)}$ we know that it will run very slow compared with clocks at rest far away.

How much would Earth have to be compressed to turn it into a black hole? The answer is that, keeping its mass constant, Earth's radius would have to shrink from 6400 km to

$$r = \frac{2GM}{c^2} = \frac{2(6.67 \times 10^{-11}\,\text{m}^3/\text{kg s}^2)(6.0 \times 10^{24}\,\text{kg})}{(3.0 \times 10^8\,\text{m/s})^2} = 8.9 \times 10^{-3}\,\text{m},$$

slightly less than 1 centimeter!

Problems

14–1. (a) A balloon filled with helium is released at rest in the middle of a drifting spaceship in gravity-free space. The ship is otherwise filled with air. What does the balloon do? (b) Now the spaceship accelerates forward. Does the balloon drift toward the front or back of the rocket?

14–2. A satellite orbiting Earth in a circle of radius r with constant speed v is not moving in a straight line at constant speed, so is accelerating. In fact, it is accelerating toward Earth's center with $a = v^2/r$, independent of its mass. (a) Find the period (time to orbit once around) of a hypothetical satellite with orbital radius $r = r_E$. (b) Find the altitude $h = r - r_E$ of a "geosynchronous" satellite, a satellite that stays above a particular point on the equator and therefore has orbital period 24 hours. Express the result in kilometers and as a multiple of Earth's radius.

14–3. (a) You are holding a cup of hot chocolate in a coasting car. The driver steps on the brakes. Which way should you tip the cup so its contents will not spill out? (b) A car is at rest with one of its doors open, as seen from above in the figure below. Using the principle of equivalence, describe how to close the door by accelerating the car.

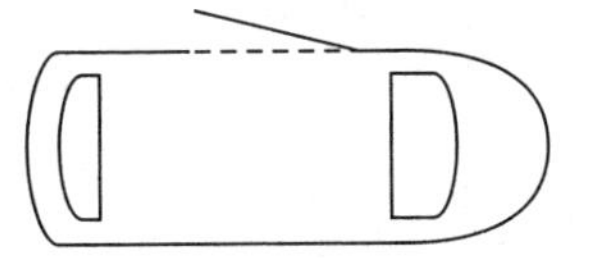

14–4. On Einstein's 76th (and last) birthday, March 14, 1955, his neighbor, Prof. Eric Rogers, gave him a toy constructed from a heavy brass ball, a spring, and other components, as illustrated below. The ball was attached to a string that hung outside a metal cup into which the ball could fit snugly. The string passed through a hole in the cup and down through a pipe, where it was tied to a spring. The assembly was mounted on a curtain rod so that one could easily hold onto the whole contraption. Finally, the cup and ball assembly was enclosed in a transparent glass sphere.

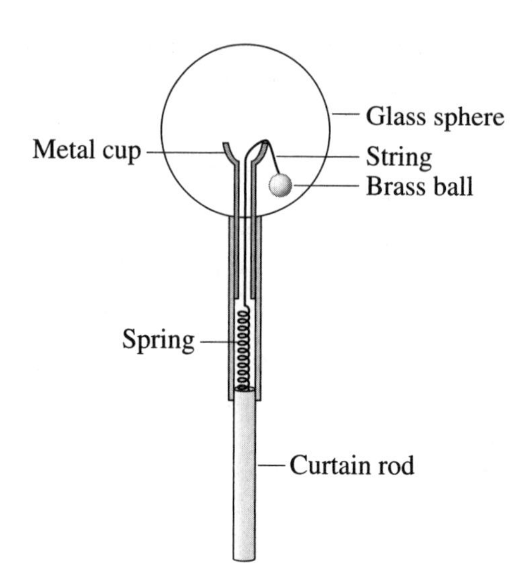

If the spring had been strong enough it could have pulled the ball into the cup; however, it was too weak to counteract gravity, so the ball hung limply outside the cup. By shaking the curtain rod, it would be possible to pop the ball into the cup. However, this turned out to be very difficult. The challenge was to find a way to pop the ball into the cup every time.

Einstein was delighted. He recognized immediately that the necessary trick hinged on a physical principle he himself had thought up half a century earlier. He was pleased that his friend took the trouble to remind him of what he had described as one of the happiest moments in his life. (See *An Old Man's Toy,* by A. Zee, Collier Books, Macmillan Publishing Company, New York, 1989.) *What was the trick?*

14–5. A laser is aimed horizontally in a room on Earth, a distance y_0 above the floor, and a pulse of light is emitted. (a) How much will the pulse fall by the time it reaches the opposite wall, a distance L away? The gravitational field is g. (b) What must be the value of L if $g = 9.8\ \text{m/s}^2$ and the pulse falls by 1 Å? (1 "Ångstrom," where $1\ \text{Å} = 10^{-10}$ m, is roughly the diameter of a hydrogen atom.)

14–6. Identical twins are separated at birth. One lives at sea level and the other lives on top of a mountain at altitude h. Pretending that everything else is equal between them, (a) which twin lives longer? (b) If the sea-level twin dies at age T_0, how much sooner or later will the mountain twin die, in terms of any or all of g, h, c, and T_0?

14–7. In the Hafele–Keating experiment, the special-relativistic time-dilation effect on clocks carried eastward around the world made them run slow compared with ground clocks, while it made clocks carried westward run fast compared with ground clocks. Explain why.

14–8. In the Hafele–Keating experiment, the several planes took off and landed, continuously, changing their ground speeds, altitudes, and latitudes. Suppose instead that the experiment could have been done using a single plane each way, one eastward and one westward, flying directly over the equator. Suppose also that each plane stayed at the constant altitude of 10 km (32,800 ft), and that each flew at constant speed, returning in 48 hours to the same place it began above the ground. What theoretical results would one get for the altitude and motion effects in this case? In other words, what numbers would we place in Table 14.2 for these flights?

14–9. Derive Eq. 14.8 from previous equations, carefully justifying each step.

14–10. (a) Explain why clocks in a satellite in a low-altitude circular Earth orbit run slow compared with ground clocks, and why clocks in a satellite in a very high-altitude Earth orbit run fast compared with ground clocks. (b) Find the altitude (expressed as a fraction of one Earth radius) of a satellite whose clock runs at the *same rate* as ground clocks. (Pretend that Earth is not itself rotating.)

14–11. Show that $\Delta\phi = -GM/r_2 + GM/r_1$, the difference in gravitational potential between points r_1 and $r_2 = r_1 + h$ (with $h \ll r_1$) around a spherical mass M, is approximately $\Delta\phi = gh$, where $g = GM/r_1^2$. *Hint:* use the binomial approximation.

14–12. Using a spectrograph attached to a telescope, astronomers can measure the frequencies of photons emitted from atoms on the Sun and compare these frequencies with those emitted by the same kind of atoms in the laboratory. (a) Calculate the fractional frequency redshift $\Delta\nu/\nu$ they should observe in viewing the Sun, due to the gravitational effect of light traveling from the Sun to Earth, to two significant figures. [*Hint:* There are gravitational potentials due to the Sun, both when the light is emitted from the Sun and when it reaches Earth; there are also gravitational potentials due to Earth, both when the light is emitted from the Sun and when it reaches Earth. Show that of these four potentials, one is so much larger than the others that all but that one may be neglected to two significant figures. (The masses of Sun and Earth are 1.99×10^{30} kg and 5.98×10^{24} kg, respectively; the radii of the Sun and Earth are 6.96×10^5 km and 6.37×10^3 km; and the distance from Sun to Earth is 1.49×10^8 km.)] (b) White dwarf stars have masses comparable with that of the Sun, and radii comparable to that of Earth. Estimate the fractional frequency redshift expected when measuring spectral lines emitted by white dwarfs. (Gravitational redshifts have long been observed from the Sun and other stars, and white dwarfs in particular.)

14–13. Suppose that someday GPS satellites are set in orbit around the planet Mars, orbiting Mars twice each Mars day. (a) What are the orbital radii of these satellites? (b) What are their orbital speeds? (c) If explorers on Mars want to know their positions to about 1 meter accuracy, how closely (in seconds) must the orbiting clocks be synchronized? (d) How stable must the orbiting clocks be (in seconds/second), assuming they can be corrected from the ground once each orbit? (e) Estimate the fractional effects on clocks due to time dilation and the altitude effect on Mars. *Given:* The mass, radius, and rotational period of Mars are $M = 6.4 \times 10^{23}$ kg, $R = 3.4 \times 10^3$ km, and $P = 24.62$ hours.

14–14. One of the signals (called $L1$) from GPS satellites has frequency 1575.42 MHz $\equiv 1575.42 \times 10^6$ cycles/s. (a) Find the wavelength of the signal. Then estimate the frequency change (in Hz) in this signal as seen from the ground on Earth due (b) to the altitude effect and (c) to time dilation.

APPENDIX A

The Binomial Approximation

THE SO-CALLED BINOMIAL APPROXIMATION, based on the binomial series, is one of the most useful formulas in physics. We use it often in this book.

The binomial *series* is the following: If a quantity x is such that its absolute value is less than unity, that is, $|x| < 1$, then if n is any power,

$$(1+x)^n = 1 + nx + \frac{n(n-1)}{2!}x^2 + \frac{n(n-1)(n-2)}{3!}x^3 + \cdots. \tag{A.1}$$

The binomial series is the special case of the well-known Taylor series

$$f(x) = f(x_0) + f'(x_0)(x - x_0) + \frac{1}{2!}f''(x_0)(x - x_0)^2 + \frac{1}{3!}f'''(x_0)(x - x_0)^3 + \cdots \tag{A.2}$$

in which we choose $x_0 = 0$ and $f(x) = (1+x)^n$. (Here $f'(x_0)$ is the first derivative of $f(x)$ evaluated at $x = x_0$, and so on.)

As an example of the binomial series, suppose that $x = 0.1$ and $n = 3$. Then

$$\begin{aligned}(1.1)^3 &= 1 + 3(0.1) + \frac{3(2)}{2!}(0.1)^2 + \frac{3(2)(1)}{3!}(0.1)^3 + \cdots \\ &= 1 + 0.3 + 0.03 + 0.001 + \cdots = 1.331 + \cdots.\end{aligned} \tag{A.3}$$

Now it is easy to show by multiplication that $(1.1)^3$ is *exactly* 1.331. That is consistent with the binomial series, because the next term in Eq. A.3 is $[n(n-1)(n-2)(n-3)/4!]x^4 = 0$ (since $n = 3$), and all subsequent terms vanish as well. The series is *bound* to truncate like this sooner or later if n is a positive integer. However, the binomial series is also valid if n is a fractional power or a negative integer. In those cases we obtain an infinite series.

Suppose, for example, that $x = 0.1$ and $n = 1/2$. Then according to the series,

$$\begin{aligned}\sqrt{1.1} \equiv (1.1)^{1/2} &= 1 + (1/2)(0.1) + \frac{(1/2)(-1/2)}{2}(0.1)^2 \\ &\quad + \frac{(1/2)(-1/2)(-3/2)}{6}(0.1)^3 + \cdots \\ &= 1 + \frac{1}{2}(0.1) - \frac{1}{8}(0.01) + \frac{1}{16}(0.001) + \cdots \\ &= 1 + 0.05 - 0.00125 + 0.0000625 + \cdots \\ &= 1.0488125 + \cdots . \end{aligned} \tag{A.4}$$

Note that the terms in the series keep getting smaller (quite a lot smaller), so the number 1.0488125 should be a pretty good approximation to $\sqrt{1.1}$. How good is it? We can estimate the error in the approximation by computing the *next* term in the series, $[(1/2)(-1/2)(-3/2)(-5/2)/24](0.1)^4 = -0.00000390625$, which starts to enter only at the sixth decimal place. Therefore we should be confident about our answer at least to the fourth decimal place, that is, $\sqrt{1.1} = 1.0488$. (The correct answer through the sixth decimal place is 1.048809, according to a hand calculator.)

So much for the binomial *series*. The *first-order binomial approximation* consists in keeping only a single term in the series after the initial "1"; that is,

$$\textit{1st-order binomial approximation:} \quad (1 + x)^n \cong 1 + nx. \tag{A.5}$$

The second-order approximation keeps the second-order term,

$$\textit{2nd-order binomial approximation:} \quad (1 + x)^n \cong 1 + nx + \frac{n(n-1)}{2!}x^2, \tag{A.6}$$

and so on. The first-order (or second-order) approximation often provides a quick way to get answers to sufficient accuracy.

You might well ask: Can't a good hand calculator do better than the binomial approximation, with a lot less work? The answer is *sometimes*. It could find $\sqrt{1.1}$ much more quickly than our calculations above, for example. Consider, however, the quantity $\sqrt{1 + 10^{-32}}$. According to the second-order binomial approximation,

$$\sqrt{1 + 10^{-32}} \cong 1 + \frac{1}{2}(10^{-32}) + \frac{(1/2)(-1/2)}{2}(10^{-32})^2, \tag{A.7}$$

which equals $1 + (10^{-32})/2$ to about 64 significant figures, since the second-order term doesn't kick in until then! Can your hand calculator tell you that?

The binomial approximation is also often used for *algebraic* calculations. In relativity, for example, we know that a moving clock runs slow by the factor $\sqrt{1 - v^2/c^2}$. Suppose that the clock is not moving very fast, so that $v/c \ll 1$. The quantity

$\sqrt{1 - v^2/c^2}$ has the form $(1 + x)^n$ where $x = -v^2/c^2$ and $n = 1/2$, so this slow-moving clock runs slow by the factor

$$\sqrt{1 - v^2/c^2} \cong 1 - \frac{1}{2}\left(\frac{v}{c}\right)^2 \tag{A.8}$$

to first order. At some time t according to a stationary clock, the time difference between the stationary clock and the slow-moving clock is

$$\Delta t = t - t\sqrt{1 - v^2/c^2} \cong t\left[1 - \left(1 - \frac{1}{2}\left(\frac{v}{c}\right)^2\right)\right] = \frac{1}{2}\left(\frac{v}{c}\right)^2 t. \tag{A.9}$$

This simple result is an excellent approximation as long as v/c is small. How small? That depends on how much accuracy you need. The error in the result can be estimated by the next term in the binomial series. To a better approximation,

$$\sqrt{1 - v^2/c^2} \cong 1 - \frac{1}{2}\left(\frac{v}{c}\right)^2 + \frac{(1/2)(-1/2)}{2!}\left[-\left(\frac{v}{c}\right)^2\right]^2, \tag{A.10}$$

so the error in using the result of Eq. A.9 is about $[(v/c)^4/8]t$. As a fraction of the result of Eq. A.9, this correction is only $(v/c)^2/4$, which represents an error of less than 1% as long as $v < 0.2c$.

As another example (shown also in Chapter 11), we can approximate the exact kinetic energy

$$KE = mc^2\left[\frac{1}{\sqrt{1 - v^2/c^2}} - 1\right] \tag{A.11}$$

of a particle, assuming $v \ll c$. Then the first-order binomial approximation is

$$\frac{1}{\sqrt{1 - v^2/c^2}} = (1 - v^2/c^2)^{-1/2} = 1 + (-1/2)(-v^2/c^2), \tag{A.12}$$

so the kinetic energy is

$$KE = mc^2\left[\frac{1}{\sqrt{1 - v^2/c^2}} - 1\right] \cong mc^2\left[1 + \left(-\frac{1}{2}\right)\left(-\frac{v^2}{c^2}\right) - 1\right] = \frac{1}{2}mv^2, \tag{A.13}$$

which is the well-known expression for the kinetic energy of a nonrelativistic particle.

Problems

A–1. Evaluate $(1.01)^{3/2}$ to four significant figures, using (a) the binomial approximation, (b) a hand calculator. (c) Also find $(0.99)^{3/2}$ by both methods.

A–2. Evaluate $\sqrt{401}$ using the first binomial approximation. *Hint:* First write $(401)^{1/2} = (400 + 1)^{1/2} = (400)^{1/2}(1 + 1/400)^{1/2}$.

A–3. (a) Evaluate $(9.00036)^{1/2}$ using the first binomial approximation. (b) Then find $(9 + 3.6 \times 10^{-16})^{1/2}$.

A–4. How fast would a meterstick have to move so it would be Lorentz contracted by $1\,\text{nm} \equiv 10^{-9}$ m? Express the answer in the form v/c.

A–5. The momentum of a particle moving in one dimension is $p = \gamma m v$ where $\gamma = (1 - v^2/c^2)^{-1/2}$, so it becomes simply $p = mv$ in the nonrelativistic limit. (a) Using the binomial approximation, what is the first correction term beyond $p = mv$? (b) How fast would the particle have to move (expressed in the form v/c) to make the first correction term as large as 1% of the nonrelativistic momentum?

A–6. Derive the binomial series from the Taylor series.

APPENDIX B

The "Paradox" of Light Spheres

AT THE END OF CHAPTER 3, an apparently paradoxical conclusion was drawn from the fact that light moves at the same speed in all inertial frames. If a bomb explodes at the origins of two frames just as they pass, the resulting light rushes out in a spherical wave front in each frame, and furthermore a set of observers in each frame finds that the sphere's center is at the origin of *their own frame.* This seems impossible, since after the origins pass, they no longer coincide!

This result seems paradoxical because we are used to assuming implicitly that simultaneity is an *absolute* rather than a *relative* concept. Using the results of the discussion in Chapter 6 on the readings of moving clocks, we can resolve this paradox, showing why observers in one frame measure the sphere center to be at their own origin, but also understand why observers in the *other* frame measure the center to be at *their* origin.

How can observers in a frame set up an experiment capable of discerning the spherical nature of a wave front? One way would be to station a number of clocks at equal distances from the origin, synchronize them, and then measure the time read by each clock as the light passes by. If they all read the same time, the observers can conclude that the light expands as a sphere centered about their origin.

We will suppose that observers in each frame choose to use this method and that in particular they station their clocks at a distance D from their own origin. Besides measuring this distance, observers in each frame properly synchronize their clocks according to the prescription of Chapter 6. We'll view events from the unprimed frame, verifying that the wave spreads out symmetrically from our origin, and also we'll see why primed observers think the wave is centered at *their* origin.

Let the primed frame be moving to our right at speed V. Figure B.1 shows the situation as the frame origins coincide and the bomb explodes. Four clocks are shown in each frame, with all unprimed clocks synchronized at $t = 0$. The primed clocks on the x' axis are unsynchronized from our point of view: The one on the right reads $t' = -VD/c^2$, and the one on the left reads $t' = +VD/c^2$, according to the rules of Chapter 6, where D is the rest distance of each clock from the origin of its frame. The

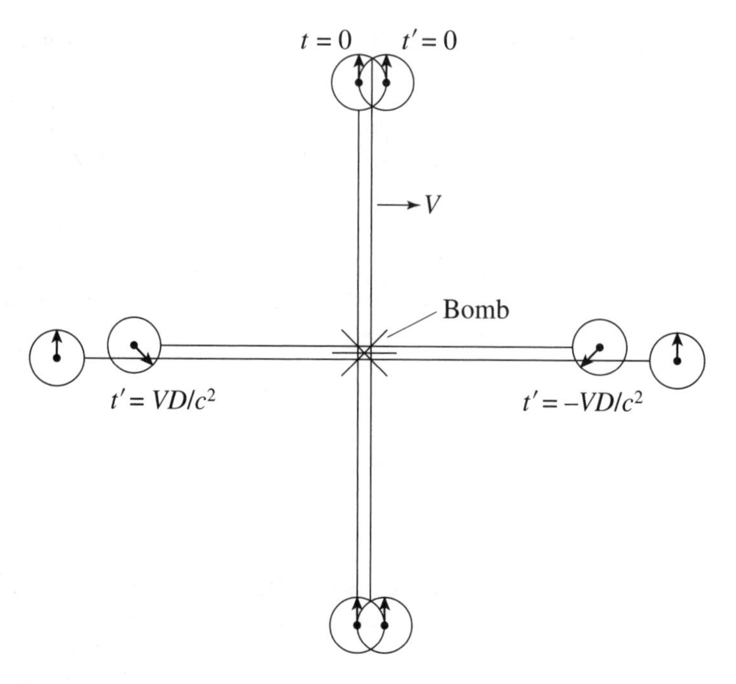

FIGURE B.1
A primed frame moves to the right at speed V relative to an unprimed frame, just as their origins coincide and a bomb goes off at the origins. Eight clocks are shown, four at rest in each frame. Each is a distance D from the origin of its frame. In our unprimed frame, the moving clocks at the right and left are closer to the origin because of the Lorentz contraction; they also lag and lead our clocks in time by the leading-clocks-lag rule.

distance of each of these clocks from our origin is only $D\sqrt{1-V^2/c^2}$ in our frame, owing to the Lorentz contraction.

As the light spreads out with the same speed in all directions in our frame, the first clock it meets is the left-hand primed clock, since that clock is moving toward our origin. This clock was originally a distance $D\sqrt{1-V^2/c^2}$ from our origin, so when this clock meets the expanding light wave, the distance traveled by the clock plus the distance traveled by the light wave must be

$$V\Delta t + c\Delta t = (V+c)\Delta t = D\sqrt{1-v^2/c^2}. \tag{B.1}$$

Therefore, in our frame this clock meets the light wave when our clocks have advanced by

$$\Delta t = D\frac{\sqrt{1-V^2/c^2}}{c+V} = \frac{D}{c}\sqrt{\frac{1-V/c}{1+V/c}}. \tag{B.2}$$

This moving clock runs slow by the factor $\sqrt{1 - V^2/c^2}$, however, so will advance by only

$$\Delta t' = \left(\frac{D}{c}\sqrt{\frac{1 - V/c}{1 + V/c}}\right)\sqrt{1 - V^2/c^2} = \frac{D}{c}\left(1 - \frac{V}{c}\right) \tag{B.3}$$

since $1 - V^2/c^2 = (1 + V/c)(1 - V/c)$. This clock began (as in Fig. B.1) with the reading VD/c^2, so when it intercepts the light sphere from the explosion it reads

$$t' = \frac{VD}{c^2} + \frac{D}{c}\left(1 - \frac{V}{c}\right) = \frac{D}{c}. \tag{B.4}$$

The situation is then as shown in Fig. B.2.

The next important event occurs when the wave front intercepts all four of our unprimed clocks, as shown in Fig. B.3. They are all a distance D from the explosion, so this event happens at time $t = D/c$. The moving clocks have all advanced by $(D/c)\sqrt{1 - V^2/c^2}$ from their starting values, as shown.

The next important event takes place when the light reaches the two primed clocks on the y' axis, as shown in Fig. B.4. By the law of Pythagoras the distance of these

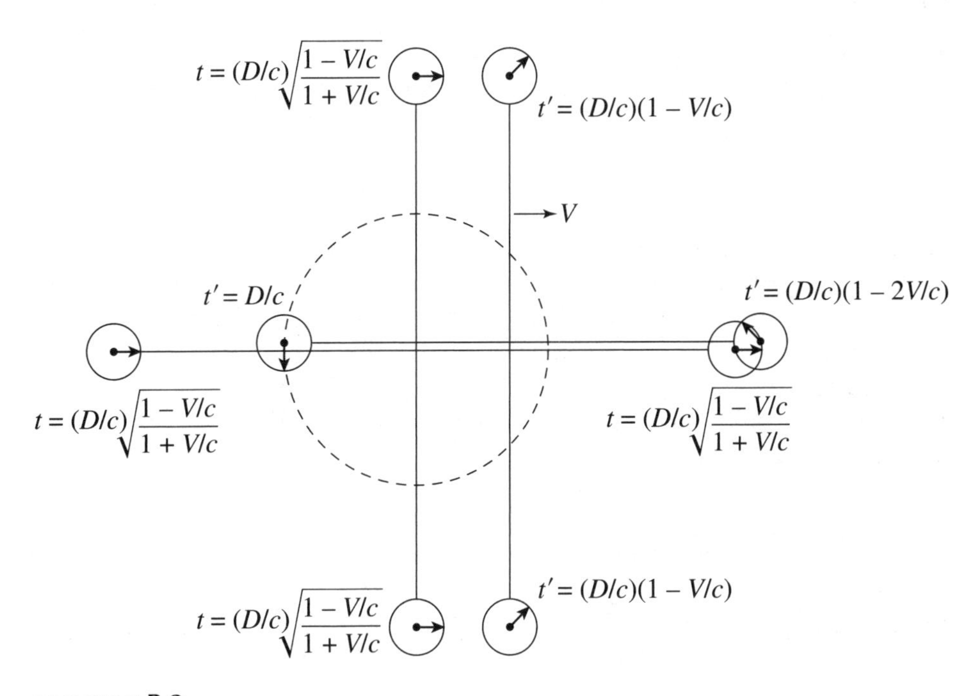

FIGURE B.2
The situation when the light sphere meets the left-hand moving clock. The stationary clocks have all advanced by time $(D/c)\sqrt{(1 - V/c)/(1 + V/c)}$. The moving clocks have all advanced by $(D/c)(1 - V/c)$, and so in particular the left-hand moving clock reads $t' = D/c$.

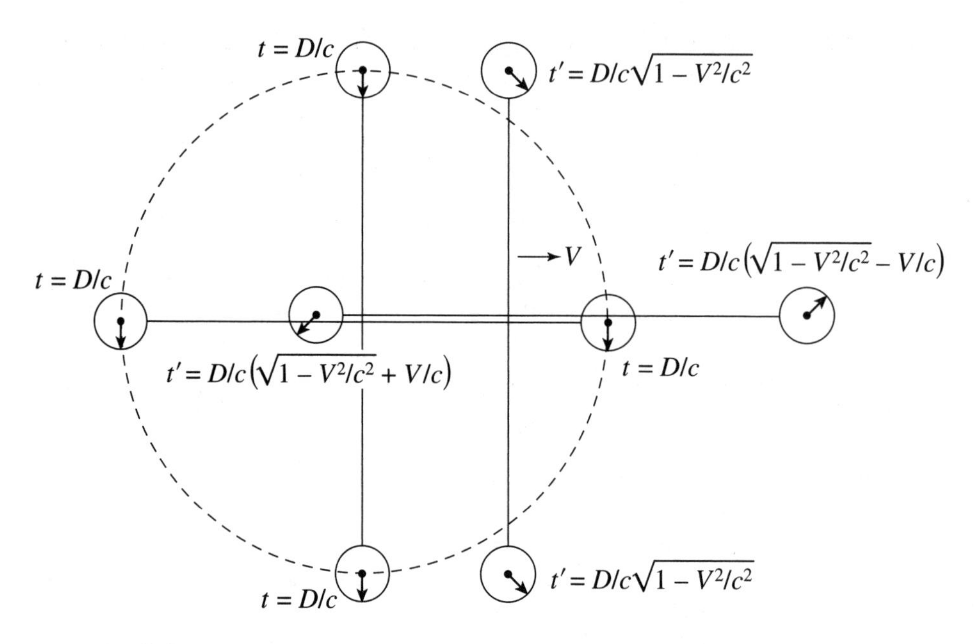

FIGURE B.3
The wave front reaches all four unprimed clocks when they read D/c.

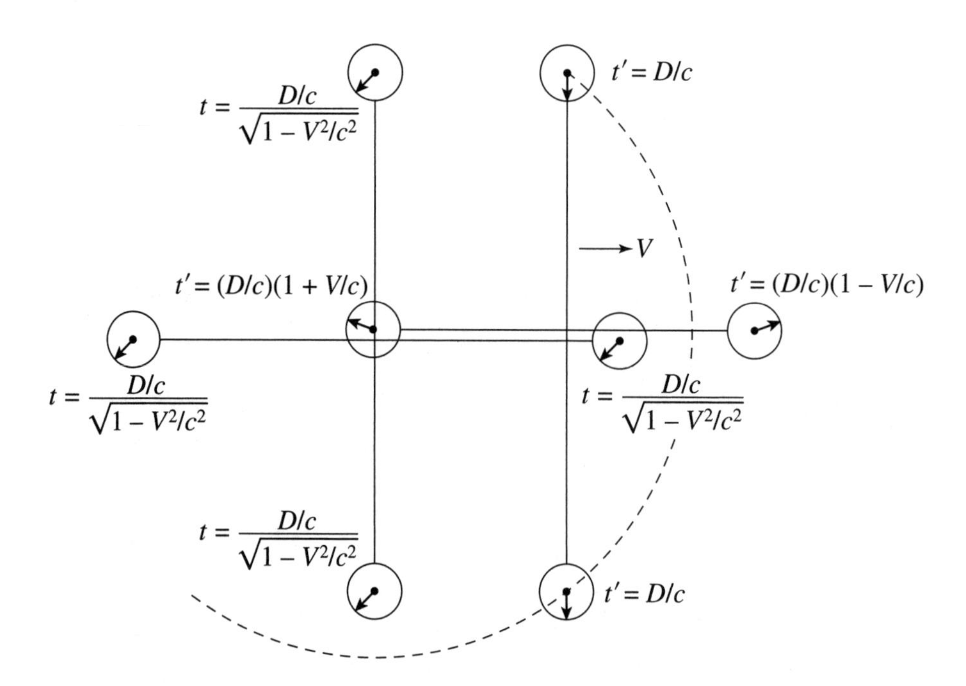

FIGURE B.4
The light reaches the two primed clocks on the y' axis. At this time these two clocks read D/c.

clocks from our origin is $\sqrt{D^2 + V^2t^2}$, where t is the time read by our unprimed clocks when the event takes place. That is because D and Vt are the y and x coordinates of the *upper* primed clock, and $-D$ and Vt are the y and x coordinates of the *lower* primed clock. This distance is also ct, because light left the origin and arrived at time t, so $c^2t^2 = D^2 + V^2t^2$. Solving for t, we get $t = (D/c)/\sqrt{1 - V^2/c^2}$.

The primed clocks have run slow during this interval, so they have advanced by only $t\sqrt{1 - V^2/c^2} = D/c$ since the beginning. The primed clocks on the axis started at zero, so they will actually read D/c, as shown in the figure. The other clocks will read as shown.

Finally, light from the explosion catches up with the right-hand primed clock, as shown in Fig. B.5. If this time is t (in our unprimed frame), the clock is a distance $D\sqrt{1 - V^2/c^2} + Vt$ away from our origin in this figure, since at time $t = 0$ the clock was already a distance $D\sqrt{1 - V^2/c^2}$ away. This total distance is also covered by the light sphere in time t, so

$$ct = D\sqrt{1 - V^2/c^2} + Vt. \tag{B.5}$$

Solving for t, we get

$$t = D\frac{\sqrt{1 - V^2/c^2}}{c - V} = \frac{D}{c}\sqrt{\frac{1 + V/c}{1 - V/c}}, \tag{B.6}$$

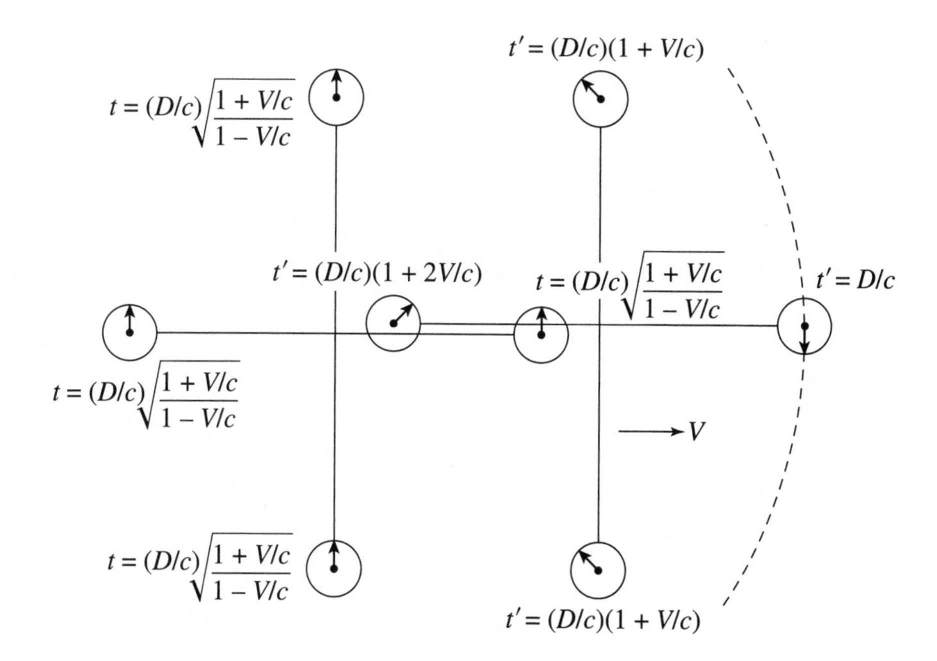

FIGURE B.5
Light finally catches up with the right-hand moving clock. That clock reads D/c when it does so.

which is the amount by which the unprimed clocks have advanced since the beginning. The moving clocks all advance by the lesser amount

$$\Delta t' = \frac{D}{c}\sqrt{\frac{1+V/c}{1-V/c}}\sqrt{1-V^2/c^2} = \frac{D}{c}\left(1+\frac{V}{c}\right). \tag{B.7}$$

The right-hand primed clock began in Fig. B.1 with time $-VD/c^2$, however, so in this final picture it reads $-VD/c^2 + (D/c)(1+V/c) = D/c$.

Notice that from our point of view, all of our unprimed clocks read D/c when the light reaches them, as shown in Fig. B.3. Therefore we see that the light is spreading out equally in all directions, as though it came from our origin, which it did. In the other figures we have seen that from our point of view the light reaches the primed clocks at *different* times. Nevertheless, all the primed clocks also read D/c when the light reaches them. Therefore the primed observers have an equal right to claim that the light sphere is centered at their own origin: all their clocks receive the light at the same time from their point of view. There is complete symmetry between the two frames—neither frame is more fundamental than the other in describing this experiment. There is no paradox; it is possible for both sets of observers to say the light spreads out equally fast in all directions as it emerges from the origin of their frame.

APPENDIX C

The Appearance of Moving Objects

C.1 An Approaching Spaceship

SOMETIME IN THE FUTURE, with an enormously powerful telescope, we see a planet orbiting a distant star. Not only that–we see an alien race moving about on the surface! They have built a spaceship and are preparing to pay us a visit. One day we observe them blast off, headed in our direction. We know that their planet is 10 light-years away, however, so we figure that we have at least 10 years to prepare for the visit, appoint welcoming committees, write speeches, and so on.

So if they arrive only a little over one month later, should we be stunned? Should we throw away relativity books and proclaim that the aliens have somehow invented faster-than-light travel?

Not at all. As shown in Fig. C.1, they could in principle achieve this trick without ever moving faster than light. Suppose for example their ship travels at $0.99\,c$. They

FIGURE C.1
(a) A distant spaceship blasts off from a planet $10\,c \cdot$ years away. (b) We *observe* the distant spaceship blast off, 10 years later. (c) The spaceship arrives 0.1 years after that. It *appears* to have moved at speed $100\,c$, but it *actually* moved at only $0.99\,c$.

blast off in Fig. C.1(a). By the time we see them blast off, which is 10 years later, as shown in Fig. C.1(b), they are already 99% of the way to Earth, since the distance they have moved is $vt = 0.99\,c \cdot 10\text{ yrs} = 9.9\,c \cdot \text{yrs}$. With only $0.1\,c \cdot$ yrs to go, they will arrive about 36 days after we see them depart, at apparent speed $10\,c \cdot \text{yrs}/0.1\,\text{yrs} = 100\,c$, as shown in Fig. C.1(c).

C.2 Quasar Jets

Needless to say, we have never observed an alien ship coming toward us, at least so far. But a very similar phenomenon *has* been observed in distant quasar jets. Quasars are very distant objects that are nonetheless bright. They look like stars at first glance, so they are also called "quasi-stellar objects," or "QSOs." They were first observed at radio frequencies, but hundreds have also been observed at optical frequencies. They are thought to be located at the centers of distant galaxies, connected perhaps with the formation of giant black holes in the galactic nuclei. For reasons still not entirely understood, distant quasars occasionally eject a pair of jets, both consisting of fast-moving material that emerges from the quasar in opposite directions. Sometimes we see only one of the jets, since the other may be obscured behind the quasar.

An example is shown in Fig. C.2, a series of radio images obtained of the quasar 3C 273. The bright object at the upper left is the quasar, and the "blob" at the lower right is a jet.

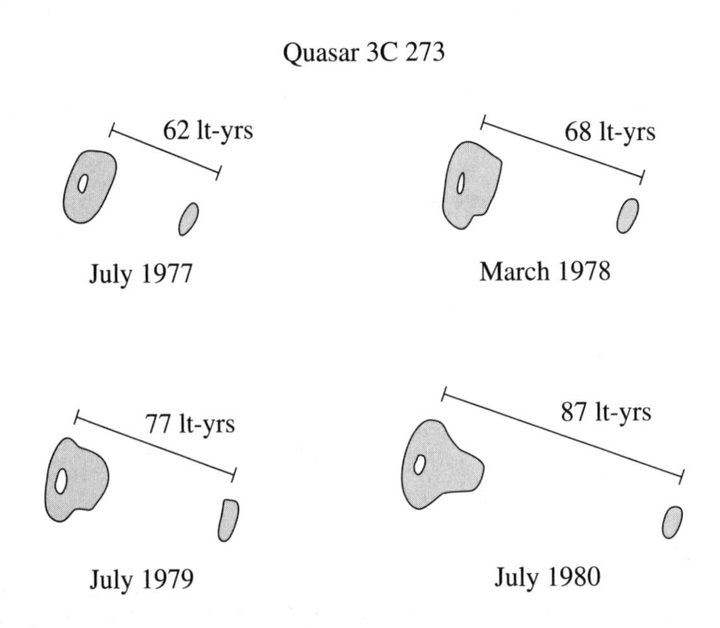

FIGURE C.2
Four radio images of quasar 3C 273, taken at various times months apart.

Note from the pictures that the blob *seems* to be moving faster than the speed of light! Between July 1977 and July 1980 it moved a distance $(87 - 62)\,c \cdot \text{yrs} = 25\,c \cdot \text{yrs}$, so it *appeared* to be departing from the central object at the approximate speed $25\,c \cdot \text{yrs}/3\,\text{yrs} = 8.3\,c$. Is the blob really moving faster than light? Do these observations show that special relativity is wrong?

Suppose we are a very large distance d from a quasar, and that at time $t = 0$ a blob leaves the quasar at speed v, headed at angle θ from a line between us and the quasar, as shown in Fig. C.3(a). (We neglect the speed of the quasar itself in the following.) At time $t = \tau$, the blob has moved a distance $v\tau$. During this time it has gotten closer to us by the distance $v\tau \cos\theta$, and it has moved sideways in the sky a distance $v\tau \sin\theta$, as shown in Fig. C.3(b).

At the much later time $t = d/c$, light that was emitted at $t = 0$ finally reaches us. Through our telescope we see that the blob has just left the quasar, as shown in Fig. C.3(c). A bit later still, we see the light that left the blob at time $t = \tau$. When does this light reach us? It left at time $t = \tau$ and had to travel a distance of about $d - v\tau \cos\theta$ toward us at the speed of light, so we see it at time $t = \tau + (d - v\tau \cos\theta)/c$, as shown in Fig. C.3(d).

Now we can figure out how fast the blob appears to move in the sky. We cannot see that it has gotten closer to us; the distance is negligibly small compared with the distance between the quasar and Earth. We *can* see the so-called "proper" motion, the motion perpendicular to our line or sight. From our point of view the blob moves

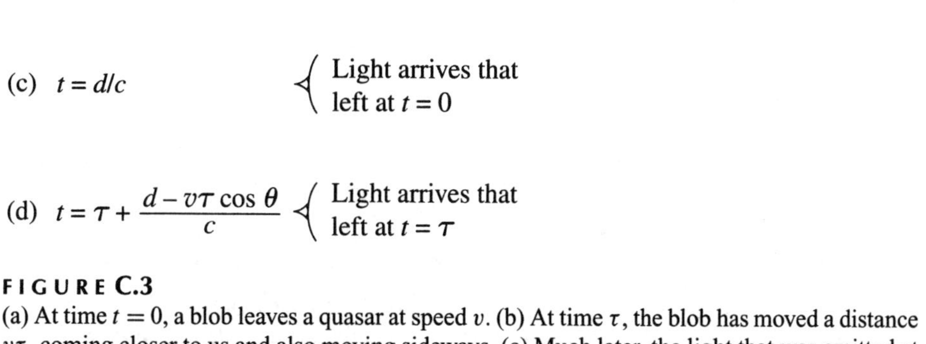

(a) $t = 0$ — d, θ, v

(b) $t = \tau$ — $v\tau \sin\theta$, $v\tau$, θ, $v\tau \cos\theta$

(c) $t = d/c$ — Light arrives that left at $t = 0$

(d) $t = \tau + \dfrac{d - v\tau \cos\theta}{c}$ — Light arrives that left at $t = \tau$

FIGURE C.3
(a) At time $t = 0$, a blob leaves a quasar at speed v. (b) At time τ, the blob has moved a distance $v\tau$, coming closer to us and also moving sideways. (c) Much later, the light that was emitted at $t = 0$ from the quasar has finally reached us. We see the blob just starting to leave the quasar. (d) A bit later still, we see the blob as it was at time τ.

sideways in the sky a distance $v\tau \sin\theta$ during the time interval Δt (from our point of view), where

$$\Delta t = \left(\tau + \frac{d - v\tau\cos\theta}{c}\right) - \frac{d}{c} = \tau\left(1 - \frac{v\cos\theta}{c}\right). \tag{C.1}$$

The apparent sideways velocity in the sky is therefore

$$v_{\text{apparent}} = \frac{\text{distance}}{\text{time}} = \frac{v\tau\sin\theta}{\tau\left(1 - \frac{v\cos\theta}{c}\right)} = \frac{v\sin\theta}{\left(1 - \frac{v\cos\theta}{c}\right)}, \tag{C.2}$$

a very interesting result! If the blob were not approaching us as time goes on, the apparent sideways velocity would be just the *numerator* of Eq. C.2. The denominator plays a critical role, however. Note that if v is close to the speed of light and $\cos\theta$ is nearly unity (meaning that θ is close to zero, since $\cos 0 = 1$), the denominator can get very small, raising the apparent velocity in the sky.

Take the special case $v/c = 0.995$ (the blob is moving at a highly relativistic speed) and $\theta = 10^\circ$ (the blob is coming nearly straight toward us.) Then $\sin\theta = 0.174$ and $\cos\theta = 0.985$, so

$$v_{\text{apparent}} = \frac{(0.995\,c)(0.174)}{(1 - (0.995)(0.985))} = \frac{.173\,c}{.020} = (8.7)\,c. \tag{C.3}$$

Even though the blob is moving at less than the speed of light, it appears to be moving nearly nine times the speed of light sideways in the sky!

Apparent velocities are not necessarily real velocities. The "optical illusion" of faster-than-light velocities is due to the approach of the blob. Of course if the blob came directly at us we would see no motion in the sky at all. Faster-than-light apparent velocities require that the object come *nearly* toward us, at highly relativistic speeds. Special relativity is not overturned by these observations.

C.3 The "Terrell Twist"

Having demonstrated in Section C.2 a real *observed* optical illusion of sorts, we turn to a so-far *un*observed optical illusion predicted by special relativity. The Lorentz–Fitzgerald contraction was discussed in Chapter 5, and it was concluded that it is essential in measuring the length of a moving body to measure the position of the two ends *simultaneously* in our frame. Any other kind of measurement would fail to correspond to our usual definition of length. Obviously, measuring the position of one end at one time, and the other end at a different time, would be a poor definition.

Yet it is just this kind of measurement that is being made when a single observer watches a moving object, or when a snapshot is taken. A snapshot of a moving train won't necessarily record its "true" length, since photons that arrive at the camera when the picture is taken will generally have left different parts of the train at different times.

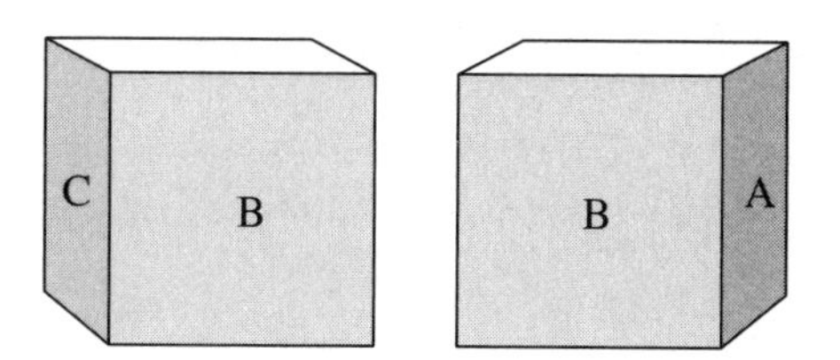

FIGURE C.4
Three sides of a cubical block.

Thus a snapshot made of a train traveling on a track running past us will make the train appear longer during its approach and shorter during its departure. (On its approach, for example, photons from the rear of the train would have to leave the train earlier than those from the front in order to reach our camera simultaneously. The train is moving, however, so during the time the rear photons are moving toward us the front of the train has gotten closer to us when its photons are emitted, so the train appears to the camera to be longer than it really is.) This effect is just a consequence of the finite speed of light, and should not be blamed on the theory of relativity.

A detailed study of how a moving three-dimensional object would appear to a single observer (or camera) was carried out by J. L. Terrell.[1] We have thought of a train as a one-dimensional object having an apparent length depending on its speed and position, but the visual appearance of three-dimensional objects is more subtle. The result is somewhat surprising. It turns out that if an observer watches a distant moving object, that object will appear to be simply *rotated*!

A convenient three-dimensional object to look at is a cubical block of wood with lettered sides for purposes of identification, as in Fig. C.4. Two views of the block are shown. We suppose that this block has unit edge length, and that it moves past us from left to right at high speed, so that side B is parallel to the track and faces us just as the block goes by. If we don't think very carefully, we might suppose that the process would look as shown in Fig. C.5. (The figure shows five views of the block as it approaches, passes, and recedes.) The block is shown stretched out as it approaches, Lorentz-contracted as it passes by, and compressed as it leaves.

In fact, the block would *actually* look as shown in Fig. C.6! It does *not* appear to be distorted, but rather it looks like an ordinary cubical block that has been rotated about a vertical axis. As it approaches, it suffers an apparent partial twist, so that somewhere between the first and second views we see only side B. Then in the second view we have entirely lost side A, which is supposed to be *toward* us, and are beginning to see side C, which is supposed to be *away* from us. This twist persists as the block passes by, but becomes less pronounced as the block departs.

1. J. L. Terrell, *Physical Review* **116**, 1041 (1959). See also the discussion of V. F. Weisskopf in *Physics Today* **13** (1960).

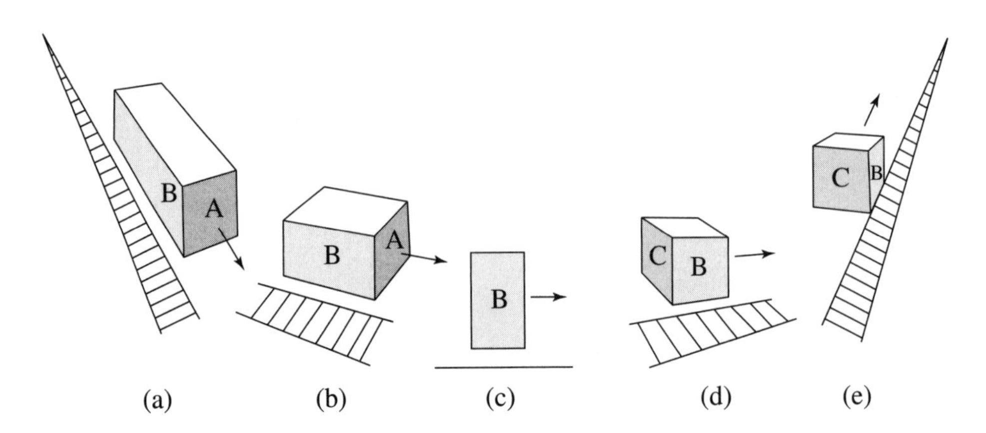

FIGURE C.5
The way we might *think* a fast-moving block would look as it passes by, accounting for the finite speed of light.

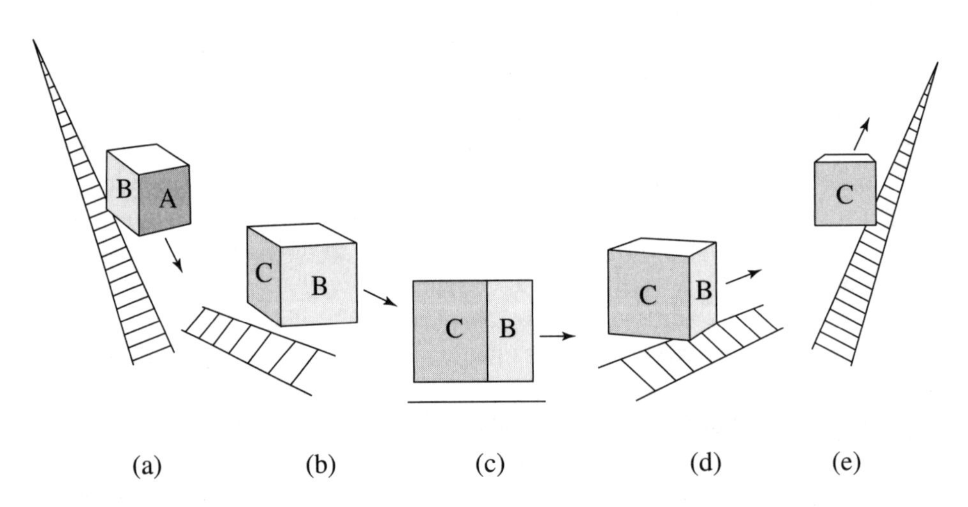

FIGURE C.6
As a fast-moving block passes by, it will look rotated rather than distorted.

Why does the block *appear* to be twisted? The basic idea is that if the block is moving very fast, the block can get out of the way of the light from side C as long as the light moves at a sufficiently large angle with respect to the block's direction of motion. Similarly, the block tends to "run into" the light from side A and so that light never escapes from the block by the time the block reaches the second position in Fig. C.6.

The reason why the visual appearance of a moving block is identical to that of a rotated block is most easily understood when the block is at its closest position to us. Normally in this position we would expect to see only side B, contracted by the factor

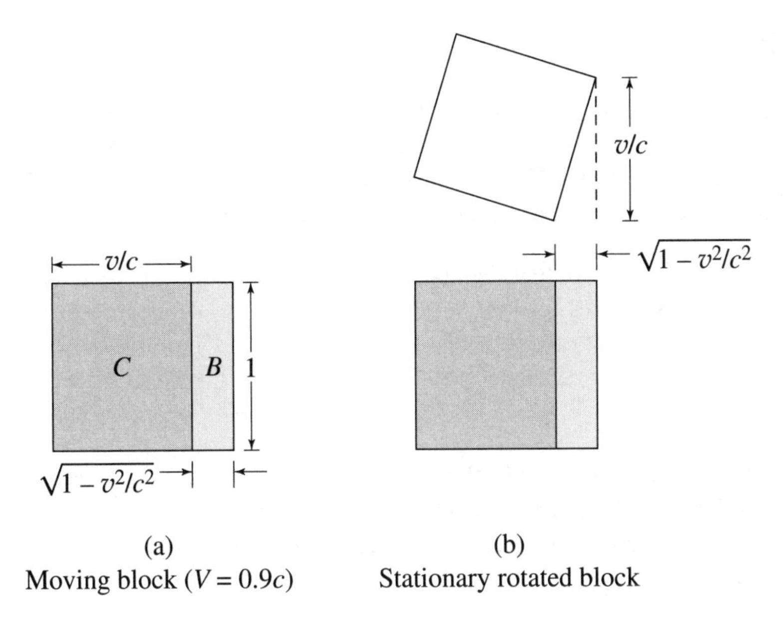

FIGURE C.7
Snapshots of a fast-moving nonrotated block and a stationary rotated block. They appear to be the same.

$\sqrt{1 - v^2/c^2}$. In fact, side C can also be seen in this position because the block moves out of the way of the light being emitted from that side, allowing us to see it.

The light from the *back* edge of side C had to be emitted earlier than the light from the front edge in order for it to reach our eye at the same time. Figure C.7 shows that the apparent width of side C is v/c. Therefore the block will appear as in Fig. C.7(a). Figure C.7(b) shows a similar block, stationary but physically rotated. A snapshot of this object is the same as that of the fast-moving but unrotated block.

Other objects also look rotated, so that for example a sphere will always look spherical, and will not appear to be squashed in its direction of motion. To be precise, we should say that the exactly rotated appearance holds true only when the object subtends a small angle at our eye. Large objects will be somewhat distorted, and the analysis is more complicated.

It is interesting that since the Sun and Moon move with respect to us, the twist effect makes them appear very slightly rotated from their actual orientations. It is straightforward to show that the rotation angle $\Delta\theta$ satisfies

$$\cos \Delta\theta = \sqrt{1 - v^2/c^2} \tag{C.4}$$

where v is the relative sideways velocity of the object. This rotation angle can be used to find out how much of the object is hidden and how much is uncovered due to the twist effect.

Problems

C–1. (a) An alien spaceship leaves a distant planet, headed for Earth at speed v. In terms of v, what is its *apparent* speed toward us, in the sense of Section C.1? (b) One of our spaceships leaves Earth at speed v, headed for the distant planet. What is its *apparent* speed away from us, in the sense of Section C.1?

C–2. Equation C.2 shows that the apparent speed of an object perpendicular to our line of sight is

$$v_{\text{apparent}} = \frac{v \sin\theta}{1 - (v/c)\cos\theta},$$

where v is the object's actual velocity and θ is the angle between its actual velocity and our line of sight. (a) For fixed v, for what angle θ will v_{apparent} be maximized? (b) What is the maximum v_{apparent} we could observe if (i) $v/c = 0.9$, (ii) $v/c = 0.99$, and (iii) $v/c = 0.999$?

C–3. *The Doppler effect revisited.* (a) Clock B moves directly away from us at speed v while we watch it through a telescope. How fast does it *appear* to tick, compared with our own clock A? *Hint:* suppose that when our clock A reads $t = 0$, the moving clock B is some distance d away and also happens to read $t' = 0$, as shown in the first picture below. At this instant a light ray leaves B in our direction. A bit later, when A reads t_1, as shown in the second picture, B reads $t_1\sqrt{1 - v^2/c^2}$ in our frame because of time dilation, and a second light ray leaves B in our direction. Draw two additional pictures: The first new picture should show when the first ray reaches us, when we see B reading zero; the second new picture should show when the second ray reaches us, when we see B reading $t_1\sqrt{1 - v^2/c^2}$. Then find the ratio between the apparent time interval $\Delta t'$ on B and the actual interval Δt on A between the receptions of the two signals. By what factor does B appear to run slow?

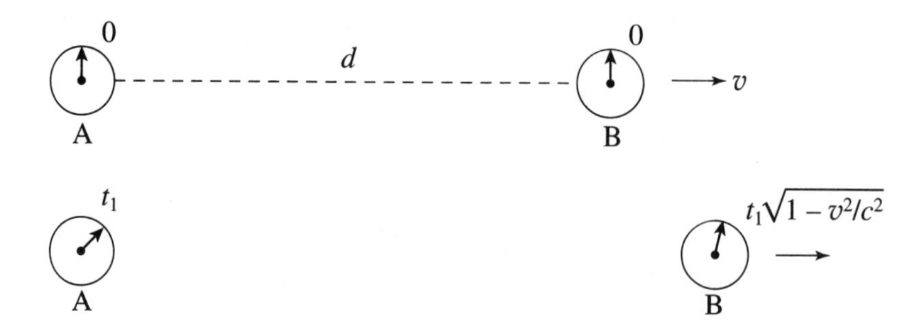

(b) Use this result to derive the Doppler frequency shift ratio

$$\frac{\nu'}{\nu_0} \equiv \frac{\nu_{\text{observed}}}{\nu_{\text{emitted}}}$$

for a source moving directly away from us with speed v. *Hint:* Suppose the second hand of clock B is coated with a material that emits monochromatic light of frequency ν_0 cycles/second (that is, the number of wavelengths emitted per second is ν_0.) Therefore B emits light of frequency ν_0 in its frame, and if B's second hand advances by some time interval $\Delta t'$, it emits $\nu_0 \Delta t'$ wavelengths of light. We receive all of these wavelengths in time interval Δt, so the frequency (number of wavelengths per second) we observe must be

$$\nu_{\text{observed}} = \frac{\nu_0 \Delta t'}{\Delta t}.$$

Thus the ratio of the frequency we receive to the frequency emitted is

$$\frac{\nu_{\text{observed}}}{\nu_{\text{emitted}}} = \frac{\nu_0 \Delta t'/\Delta t}{\nu_0} = \frac{\Delta t'}{\Delta t}.$$

(c) Redo parts (a) and (b) for a clock *approaching* us at speed v.

C–4. Two views of a cube are shown below. In the first view we see only a single face, and in the second view we see equal amounts of two faces. Now suppose the cube is moving by us (some distance away) at speed v. At its point of closest approach, (a) suppose the block actually has one face exactly toward us as in the first picture but it *looks* like the second picture, due to the Terrell twist. What is v/c for this block? (b) Suppose instead the block is actually turned so one edge is toward us, as in the second picture, but it *looks* like the first picture. What is v/c for this block?

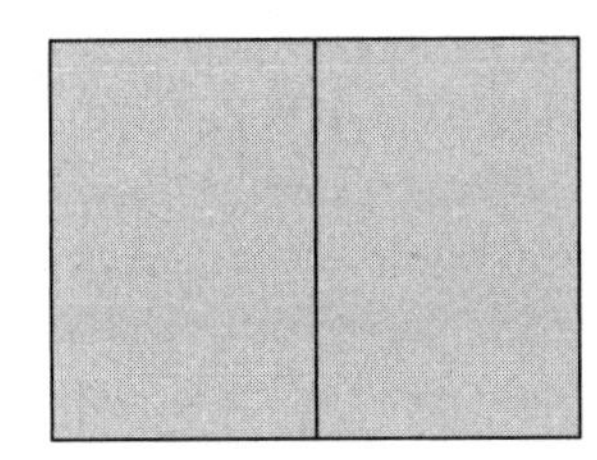

APPENDIX D

The Twin Paradox Revisited

AS DISCUSSED IN CHAPTER 7, the time-dilation effect of special relativity can lead to apparently paradoxical conclusions. For each of two relatively moving inertial observers to claim that the other observer's clock runs slow seems contradictory and nonsensical. Nevertheless, as illustrated in Section 7.1, there is actually no contradiction. When account is taken of the Lorentz contraction and the fact that two clocks in one frame are not synchronized when viewed from another frame, the "paradox" is resolved. In Section 7.1, only inertial clocks and inertial observers were considered. No clocks or observers were accelerated during any portion of the experiment. Therefore that discussion does not help us understand the story of twins Al and Bertha, in which Bertha departs for Alpha Centauri and returns, necessarily accelerating during at least part of the trip. To understand this situation from both points of view, it is necessary to know how to deal with *accelerating* clocks and *accelerating* observers.

As mentioned several times already, special relativity deals only with observations made by *inertial* observers. Therefore the twin who accelerates can't use the Lorentz transformation in any given inertial frame or any deductions made from it, such as time dilation or length contraction. This should *not* be construed as a claim that special relativity can't be used to analyze accelerating *objects,* including rods and clocks. As long as the acceleration, momentum, energy, and other properties of a particle are measured by inertial *observers,* the usual rules of special relativity can be applied.

So our first project here is to discuss accelerating clocks, so that the twin who stays at home can calculate the aging of the traveling twin while that twin is accelerating. Secondly, in order to analyze events from the viewpoint of the traveler, we have to know how clocks behave when viewed from accelerating frames of reference.

With regard to accelerating *clocks,* we'll assume all our clocks are "ideal." Such a clock is not affected by acceleration per se, but simply runs slow by the time-dilation factor $\sqrt{1 - v^2/c^2}$ appropriate for the velocity it has at that moment. No real clock is completely ideal, since acceleration would have some effect on any clock we might devise. A watch that decelerates as it hits the floor may have its rate dramatically altered. Similarly, the traveling twin would not survive an overly sudden acceleration when turning around to come back home. An atomic or nuclear clock, whose rate is measured by the frequency of emitted radiation, is a nearly ideal clock for reasonable accelerations. Certainly for the accelerations likely to occur in spaceships, such a clock could be considered ideal. Of course any other kind of clock could be used if it were near enough to ideal, or if it could be corrected for the effects of acceleration.

Assuming that we have such a clock, we would like to know what it reads as compared with a similar clock at rest in an inertial frame. During an infinitesimal time interval dt (as measured by the inertial clock), the accelerating clock will have some velocity v, so will record a time interval $d\tau = dt\sqrt{1 - v^2/c^2}$ due to time dilation. The velocity v depends on time, so this expression has to be integrated for finite time intervals. If we choose $\tau = 0$ when $t = 0$, then the time read by the accelerating clock is

$$\tau = \int_0^t dt\sqrt{1 - v(t)^2/c^2}, \tag{D.1}$$

where $v(t)$ is the time-dependent speed of the clock. The time τ is called the "integrated proper time."

Now return to Bertha's excellent adventure as she leaves the solar system bound for Alpha Centauri (α-C) at speed $(4/5)c$, as shown in Fig. D.1(a), leaving her twin brother Al at home. Both of their clocks read *zero* at her departure. We assume that α-C is in the same inertial reference frame as the Sun, and that α-C's clock also reads zero in this picture. The distance to α-C is $4\,c \cdot$ yrs in Al's inertial frame, so Bertha arrives at a time $t = D/v = 4\,\text{yrs}/(4/5\,c) = 5$ yrs later, as shown in Fig. D.1(b). Al's clock and α-C's clock both read 5 yrs, but her clock reads only $5\,\text{yrs}\sqrt{1 - v^2/c^2} = 5\,\text{yrs}(3/5) = 3$ yrs, since her clocks have been running slow in Al's inertial frame. Bertha then decelerates, turns around and accelerates again, heading back home at $(4/5)c$, as shown in Fig. D.1(c). We suppose this turnaround is essentially instantaneous on the scale of years, so that her clock still reads 3 yrs and the α-C clock still reads 5 yrs by the time she is headed back home. ("Bertha" may have to be an atomic clock in this case, because the acceleration would be large!) Finally, she arrives back home, 5 yrs later on Al's clock and 3 yrs later on her clock, as shown in Fig. D.1(d). Altogether, he has aged 10 years while she has aged only 6 years.

The pictures and calculations presented in Fig. D.1 were all made in Al's inertial reference frame. Now let's see what the trip looks like in *Bertha's* frame. There are actually *two* inertial frames needed for Bertha: her rest frame while she is approaching α-C, and her rest frame while she is returning home. In addition, there is the *non*inertial *accelerating* frame during her turnaround. She cannot use the simple rules of special relativity during that relatively brief period! Figure D.2 shows the sequence of events in both of Bertha's inertial frames. The solar system departs from her at speed $(4/5)c$ in Fig. D.2(a), when both the Sun's clock and her clock read zero. In her frame the Sun is *leading* α-C, so (from the leading-clocks-lag rule) the clock on the Sun *lags* the clock on α-C by $vD/c^2 = (4/5)4\,c \cdot \text{yrs}/c = 3.2$ yrs. Since the Sun clock reads zero in this picture, the clock on α-C must read 3.2 yrs in Fig. D.2(a). Also, the distance between the Sun and α-C in her initial inertial frame is only $4\,c \cdot \text{yrs}\sqrt{1 - v^2/c^2} = 4c \cdot \text{yrs}\,(3/5) = 2.4\,c \cdot$ yrs, due to Lorentz contraction.

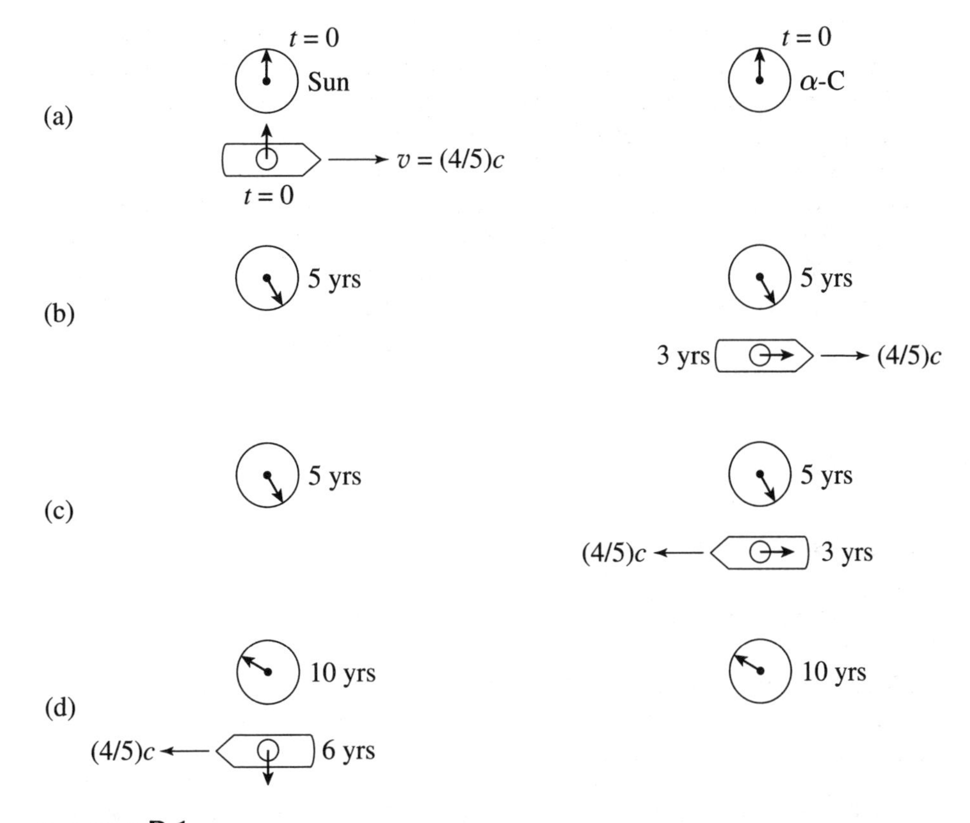

FIGURE D.1
(a) Bertha leaves brother Al at speed $(4/5)c$. (b) B arrives at Alpha Centauri 5 yrs later (on Al's clock and α-C's clock), but only 3 yrs later on her clock. (c) B quickly turns around and heads home. (d) She arrives home when Al's clock reads 10 yrs and her clock reads 6 yrs. She is 4 yrs younger than Al upon her return.

Alpha Centauri finally reaches Bertha in Fig. D.2(b). It has moved the distance $2.4\,c \cdot$ yrs in her frame at speed $(4/5)c$, so her clock has advanced by $t = d/v = 2.4\,c \cdot \text{yrs}/(4/5\,c) = 3\,\text{yrs}$, as we knew must happen. During this period of time the clock on α-C has advanced by only $3\,\text{yrs}\sqrt{1 - v^2/c^2} = 3\,\text{yrs} \cdot 3/5 = 1.8\,\text{yrs}$, because its clock has been running slow in Bertha's frame. It already read 3.2 yrs in Fig. D.2(a), however, so in Fig. D.2(b) it will read $3.2\,\text{yrs} + 1.8\,\text{yrs} = 5\,\text{yrs}$, again as expected. *Just as we found in Al's frame, Bertha's clock will read* 3 yrs *and the clock on α-C will read* 5 yrs *when they meet.* Meanwhile, the Sun's clock has also advanced by 1.8 yrs, so (since it read $t = 0$ in the first picture) it will read 1.8 yrs in the second picture, when Bertha arrives at α-C.

Now Bertha briefly enters an accelerating frame, turning around at α-C to go back home. When she emerges from the acceleration, she is essentially still right next to α-C, so her clock still reads 3 yrs and the α-C clock still reads 5 yrs. She now finds herself in an inertial frame in which the Sun and α-C are moving to the *right,* so that now α-C is *leading* the Sun, so lags the Sun in time, by 3.2 yrs. That is, the clock on

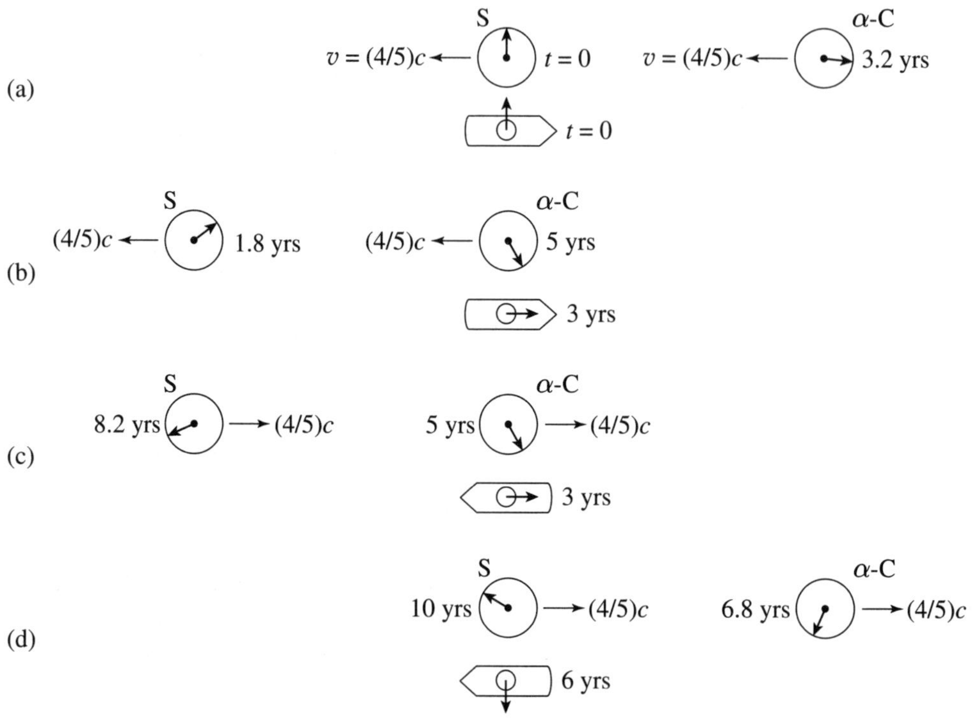

(a) S $v = (4/5)c$ $t = 0$ α-C $v = (4/5)c$ 3.2 yrs
$t = 0$

(b) S $(4/5)c$ 1.8 yrs α-C $(4/5)c$ 5 yrs
3 yrs

(c) S 8.2 yrs $(4/5)c$ α-C 5 yrs $(4/5)c$
3 yrs

(d) S 10 yrs $(4/5)c$ α-C 6.8 yrs $(4/5)c$
6 yrs

FIGURE D.2
(a) In Bertha's original inertial frame, the Sun and α-C both move at $(4/5)c$ to the left, and are a distance $2.4\, c \cdot$ yrs apart. Bertha's clock and the Sun's clock both read $t = 0$, but the α-C clock reads 3.2 yrs. (b) α-C arrives at Bertha; her clock reads 3 yrs and the α-C clock reads 5 yrs, as expected. The Sun clock reads 1.8 yrs, since it must lag the α-C clock by 3.2 yrs. (c) After a quick acceleration, Bertha is still at α-C, but is now in a different inertial frame in which the Sun and α-C are both moving to the *right* at $(4/5)c$. Her clock and the α-C clock still read 3 yrs and 5 yrs, respectively, but in this new inertial frame, α-C now *leads* the Sun clock, so must lag the Sun clock by 3.2 yrs. Therefore in this picture the Sun clock reads 5 yrs + 3.2 yrs = 8.2 yrs. It has quickly advanced from 1.8 yrs to 8.2 yrs (that is, by 6.4 yrs) during Bertha's brief period of acceleration! (d) The Sun now reaches Bertha; her clock advances by another 3 yrs and the Sun clock advances by the time-dilated amount of 3 yrs $\cdot$ 3/5 = 1.8 yrs. So her clock now reads 6 yrs and the Sun clock reads 8.2 yrs + 1.8 yrs = 10 yrs. This is the same result found in Al's frame of reference—he is 10 years older and she is 6 years older than when they parted.

the Sun in Fig. D.2(c) will read the later time 5 yrs + 3.2 yrs = 8.2 yrs. In her frame(s) of reference, during her comparatively brief period of acceleration, the Sun clock has quickly advanced from 1.8 yrs to 8.2 yrs, a time interval of 6.4 yrs!

Now the Sun (and α-C) move to the right at $(4/5)c$, until the Sun reaches Bertha, as shown in Fig. D.2(d). This requires a time $t = d/v = 2.4\, c \cdot \text{yrs}/(4/5\, c) = 3$ yrs in her frame, so she is another 3 years older. The Sun clock runs slow, so advances by only 3 yrs $\cdot$ 3/5 = 1.8 yrs. Therefore in the final picture the Sun clock will read

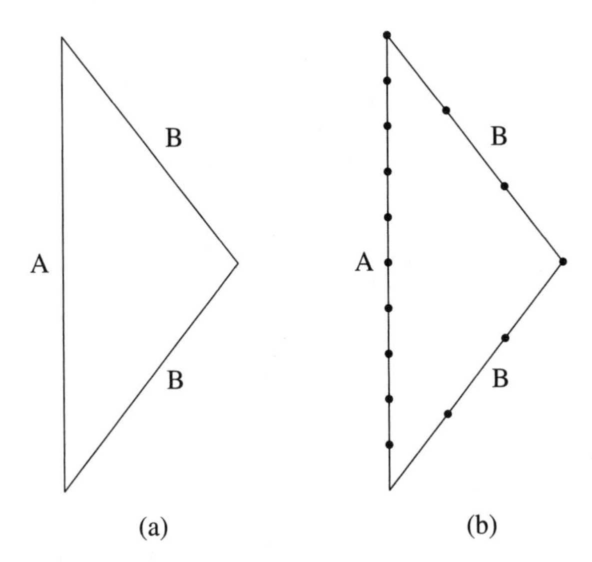

FIGURE D.3
(a) Spacetime diagram of the paths of Al (A) and Bertha (B). (b) The same spacetime diagram, with dots placed at birthdays. Note that Al has 10 birthdays and Bertha has 6 after the beginning of the trip.

8.2 yrs + 1.8 yrs = 10 yrs. This is exactly the same as found from Al's point of view. During the entire trip she has aged 6 years while he has aged 10 years. In her frame(s) of reference, it is true that he ages more slowly than she does while their relative velocity stays constant, but during her brief period of acceleration he ages so much that by the time they get back together, he is 4 years older than she is! There is no paradox here; the results are consistent.

The adventures of Al and Bertha are neatly summarized by the spacetime diagram of Fig. D.3(a), with time plotted on the vertical axis and distance on the horizontal axis. Al stays in one place, so advances only in time; Bertha first travels to the right, and then travels back to the left. Constant-velocity (that is, *inertial*) motion corresponds to those portions of a path that have constant slope. The only part of a path with non-constant slope is where Bertha turns around halfway into her trip. Figure D.3(b) reproduces the diagram, with dots placed on each path corresponding to the birthdays of each twin, assuming the original departure corresponded to one of their mutual birthdays.

The diagrams will help us answer a very interesting question: Suppose Al watches Bertha through a powerful telescope as she travels to α-C and back. How will she seem to age throughout her trip? Likewise, suppose Bertha watches Al through a telescope on board her ship. How will he seem to age as she watches him? We know that in her frame(s) of reference, he ages more slowly than she does during the two constant-velocity portions of her trip, but that he ages very quickly during the time she turns around, so quickly in fact that his total age at the end is greater than hers. Will she actually see this rapid aging as she watches him through a telescope?

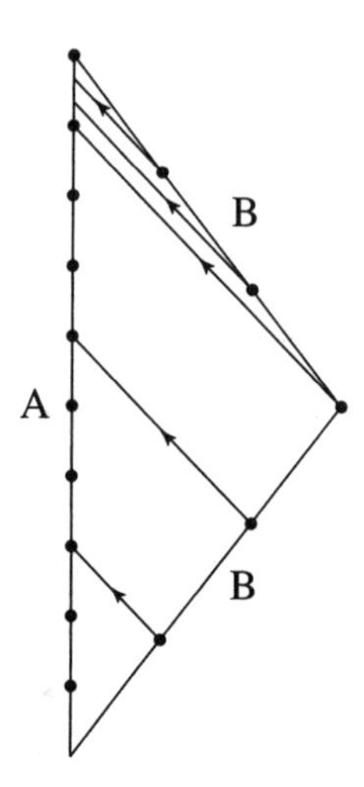

FIGURE D.4
One light ray from each of Bertha's birthdays travels to Al. He sees these rays through a telescope and finds that she *appears* to age one-third as fast as he does on her outbound voyage, and three times as fast as he does while she is returning.

We start by finding how Bertha looks to Al as he watches her. This can be illustrated by taking light paths from Bertha to Al, one on each of her birthdays, as shown in the spacetime diagram of Fig. D.4. Each of these light paths is tilted at 45°. Note that the light ray from her *first* birthday after departure intersects Al on his *third* birthday after departure, and the next ray intersects Al on his sixth birthday after departure. On Al's *ninth* birthday he finally sees that Bertha has arrived at Alpha Centauri, when she is having her *third* birthday after departure. So Bertha has aged only one-third as fast as Al while she is outbound, as seen by Al through his telescope. Then between his ninth and tenth birthdays, Al sees Bertha age by three additional years while she is returning home (her fourth, fifth, and sixth birthdays after departure.) That is, she *appears* to age three times as fast as Al on her way home, as he watches her through his telescope.

Why the factors of 1/3 and 3? The easiest way to get them is to use the Doppler-effect formulas derived in Chapter 13. As we watch an object receding from us at speed $v = (4/5)c$, the frequency of light (or the frequency of signals) *observed* ($\nu_{\rm ob}$) in terms of the frequency *emitted* ($\nu_{\rm em}$) is given by

$$\nu_{\rm ob} = \nu_{\rm em}\sqrt{\frac{1 - v/c}{1 + v/c}} = \nu_{\rm em}\sqrt{\frac{1 - 4/5}{1 + 4/5}} = \frac{1}{3}\nu_{\rm em}. \tag{D.2}$$

Similarly, if we watch an object approach us, its frequency is increased to

$$\nu_{\rm ob} = \nu_{\rm em}\sqrt{\frac{1 + v/c}{1 - v/c}} = \nu_{\rm em}\sqrt{\frac{1 + 4/5}{1 - 4/5}} = 3\nu_{\rm em}. \tag{D.3}$$

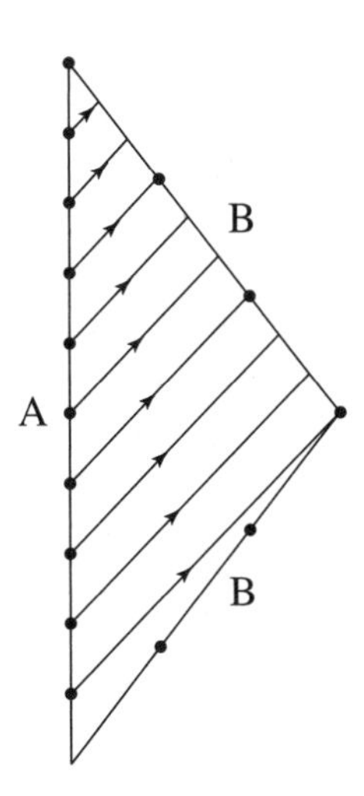

FIGURE D.5
Light rays from Al to Bertha, one on each of his birthdays while she is *en route.* As Bertha watches through her telescope, he appears to age slowly during the outbound part of the trip, and quickly while she is returning home.

We can also find how Al looks to Bertha as she watches him through her on-board telescope. Light paths from Al to Bertha, one on each of his birthdays, are shown in Fig. D.5. The light ray from his *first* birthday after departure intersects her on her *third* birthday after departure, when she has just arrived at Alpha Centauri. During this time he seems to have aged only one-third as fast as she. Then on the return trip, he appears to age three times as fast as she. (The factors of 1/3 and 3 again follow from the Doppler shift formulas.) Between her third and fourth birthdays, for example, she sees him age from his first to his fourth birthday after departure. Note that when viewed through her telescope, she does *not* observe his very rapid aging during her brief acceleration period. Instead, he *appears* to age faster by only a factor of three during her return trip, but this somewhat faster aging appears to last for the entire 3 years of her return. The end result is exactly the same. When they get back together, he is 10 years older and she is only 6 years older. Everyone agrees with that. The twin "paradox" only *seems* paradoxical.

APPENDIX E

The "Cosmic Speed Limit"

IT IS OFTEN SAID THAT "nothing can go faster than the speed of light." We are naturally suspicious! People *used* to say that no piloted aircraft could ever travel faster than the speed of sound. That limitation was hard to overcome, but it *was* overcome. Who says we can't travel faster than light?

E.1 Some Difficulties

Needless to say, there are difficulties. In Appendix F it is shown that if a constant force is applied to an object, it keeps speeding up, but its acceleration decreases so that it never quite reaches the speed of light, no matter how long the force is applied. Suppose, however, there were some more exotic way to give an object a velocity $v > c$. Then the quantity $1 - v^2/c^2$ would become negative, so the famous factor $\sqrt{1 - v^2/c^2}$ would become imaginary. If the "superluminal" (that is, faster-than-light) object happened to be a clock, for example, it would read imaginary time according to the time dilation formula and have an imaginary length according to the Lorentz-contraction formula. Any moving particle of mass m would also have imaginary momentum and imaginary energy. None of this makes sense.

Nevertheless, people have speculated about the possible existence of *single particles* called *tachyons,* which *always* travel faster than light. These hypothetical particles are pointlike and without any kind of clock, so time dilation and length contraction don't matter. Also, tachyons have *real* momentum $\mathbf{p} = \gamma m\mathbf{v}$ and *real* energy $E = \gamma mc^2$, because they are assigned an imaginary mass m, so that the imaginary number $i \equiv \sqrt{-1}$ in m cancels out with the i in $\gamma = 1/\sqrt{1 - v^2/c^2} = 1/i\sqrt{v^2/c^2 - 1}$. If tachyons exist, and if they interact with ordinary matter (including massless photons or particles with real mass like electrons and protons), there are ways to infer their presence.[1] High-energy physicists have carried out experiments to look for them in these indirect ways, but so far none has been found.

If tachyons exist, then the world of particles is divided into three parts: (i) ordinary particles with real mass and velocities $v < c$, (ii) photons (and possibly other massless particles) with velocity $v = c$, and (iii) tachyons, with imaginary mass, that always travel with $v > c$. A pretty picture in some ways, but up to now there is no evidence that the third category exists.[2]

What about *signals* of any kind? Might we someday be able to send some *signal* faster than the speed of light? As we have seen, the agent carrying the message cannot be an ordinary material object and cannot be light itself. Nevertheless, suppose *somehow* that a message could be sent faster than light, using tachyons, "wormholes," quantum entanglement, mental telepathy, or any other hypothetical means. Would such a signal cause any problems?

E.2 Causality Paradoxes

It turns out that if we could send signals faster than light we could contruct a paradox that is completely different than those described in Chapter 7. Those "paradoxes" were only *apparent* paradoxes; they were not self-contradictions, but only counterintuitive. We could explain all of them just by being careful about using special relativity consistently.

As an example of what can happen with superluminal signals, consider the following story about the twins Al and Bertha. One day Bertha departs from Earth, moving at constant speed $(3/5)c$. Thirty-two days after she leaves, Al decides to send her a signal that moves at speed $v = 7c$. A spacetime diagram of the events is shown in Fig. E.1.

How far away is Bertha when she receives the message, and what does her clock read at that instant? At time t in Al's frame, she has moved away a distance $(3/5\, c)t$. This is also the distance the signal moves at speed $7c$ during the time interval $\Delta t = t - 32$ days; that is,

$$x = \frac{3}{5}ct = 7c(t - 32\text{ d}). \tag{E.1}$$

Solving this equation for t gives $t = 35$ d. That is, in Al's frame she receives the signal 3 days after it was sent; at that time she is a distance $x = 7c \cdot 3\text{ d} = 21c \cdot \text{d}$ away. In Al's frame her clock has been running slow, so although *his* clock reads 35 days, hers reads only $35\text{ d} \cdot \sqrt{1 - (3/5)^2} = 28$ d when she receives the signal.

1. See Problem 12–18.

2. By the way, we call c the speed of light for historical reasons—light was the first thing discovered that moved at that speed. At heart, c is the speed of *any* massless particle, which may include not only photons but other particles like the graviton, which has not been discovered but is thought to exist and to carry the gravitational force.

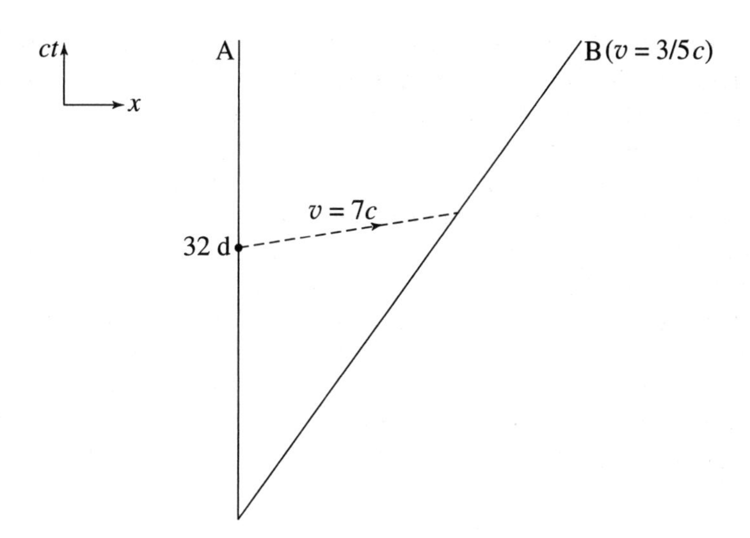

FIGURE E.1
A spacetime diagram in A's frame of reference. B is moving to the right at speed $(3/5)c$.

There is nothing particularly strange in this; the communication is very fast, covering $21c \cdot \mathrm{d}$ in only 3 days, but it does not seem in any way paradoxical.

Now in Fig. E.2 we draw the same events from *Bertha's* point of view. To her, Al moves off to the *left* at $(3/5)c$, as shown. When his clock reads 32 days he sends the signal. However, since his clock has been running slow from her point of view, her clock reads $32\ \mathrm{d}/\sqrt{1-(3/5)^2} = 40$ days when he sends the signal.

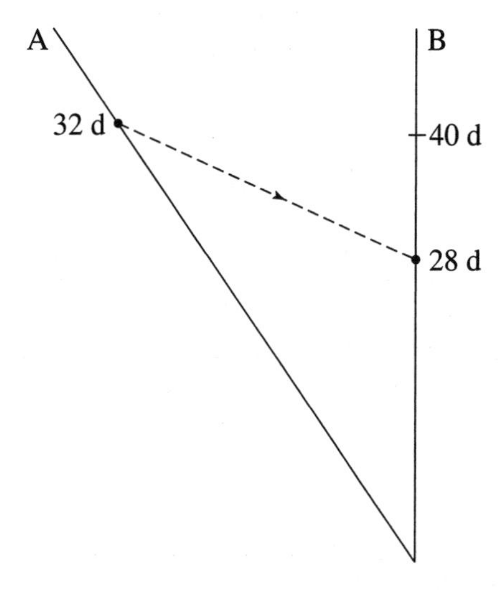

FIGURE E.2
The spacetime diagram in Bertha's rest frame.

And when the signal arrives, her clock reads 28 days (this fact must be the same in all frames, because the arrival of the signal and the instant when her clock reads 28 days are at the same point), so in her frame the signal *left* at 40 days and *arrived* at 28 days. It traveled backward in time!

This result is weird. It is a special consequence of analyzing a faster-than-light signal. It cannot happen with signals that go at speed c or less. However, strange as it is to have a signal travel backward in time, it is still not obvious that this represents a *paradox.*

However, let's continue the story. Suppose that upon receiving the signal when her clock reads 28 days, Bertha decides she doesn't *like* the message. In fact, 4 days later, when her clock reads 32 days, she sends a return message to Al, which travels at speed $v = 7c$ from *her* point of view. Her return message asks him not to send the first message! What is the effect of this return message? Note by symmetry that it will arrive back at Al when her clock reads 35 days and his reads 28 days (his clock has been running slow from her point of view). That is, the return message arrives at Al 4 days before he sent the original message! The spacetime diagram in her rest frame is shown in Fig. E.3.

The same complete scenario is shown in Fig. E.4 from the point of view of Al's rest frame.

This is a paradox of a very different sort than we have met before. To make it even more clear, suppose that Al has a combined transmitter/receiver that is programmed as follows: it *will* send a message to Bertha when its clock reads 32 days *unless* it has received a prior message that indicates it should *not* send the message. Bertha's transmitter/receiver, on the other hand, is programmed to send a message back to Al when its clock reads 32 d, *only* if it receives a prior message from Al.

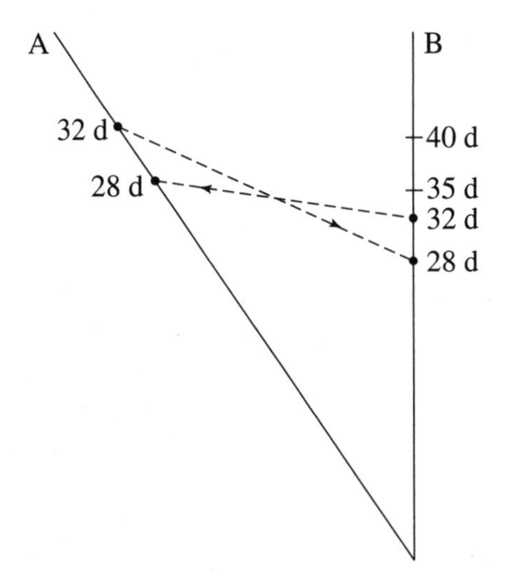

FIGURE E.3
The original message and return message in Bertha's frame.

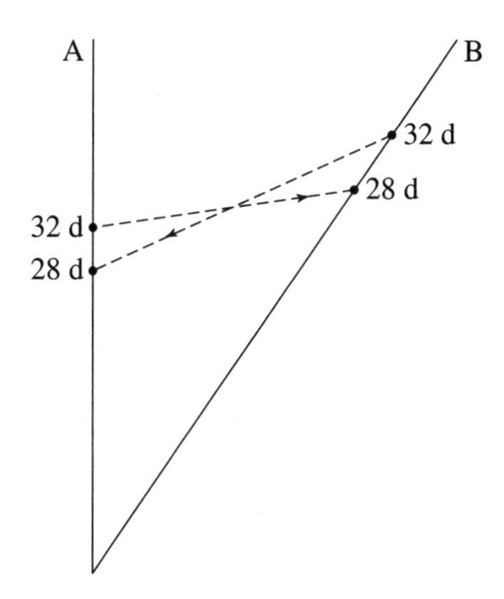

FIGURE E.4
The two messages from the point of view of Al's rest frame.

So if Al *does* send the original message, it must be that his transmitter/receiver has *not* received a prior message from Bertha. But in receiving his message, Bertha *will* send the return message. So Al should have received the prior message from Bertha. Contrariwise, if Al does *not* send the original message, his transmitter/receiver must have received a message from Bertha. But Bertha does not send a message unless she receives one from Al. Both scenarios appear to be contradictory.

Such a paradox is called a *causality paradox*. Causality paradoxes can arise if messages travel faster than light. Some people have suggested ways to avoid a true paradox; all of them call for dramatic changes in how we usually think about things. The reader is invited to try for herself/himself to suggest a way out!

E.3 "Things" That Go Faster Than Light

In spite of all our discussions so far, there certainly *are* phenomena that travel faster than light, or could in principle travel faster than light. Suppose for example we aim a laser at the Moon and observe through a telescope the bright dot where the laser beam is partially reflected from the Moon's surface. If we turn the laser from left to right, we observe the reflected dot on the Moon move from left to right. Is it possible to turn the laser so quickly that the dot will move across the Moon's surface faster than the speed of light?

Picture the laser as a hose sending out a stream of photons toward the Moon. Figure E.5 shows the photon stream at discrete times. The dot stays at the left edge of the Moon until Fig. E.5(e), when it starts to move to the right, and finally reaches the right side of the moon in Fig. E.5(g). Once it starts to move, the dot actually *can*

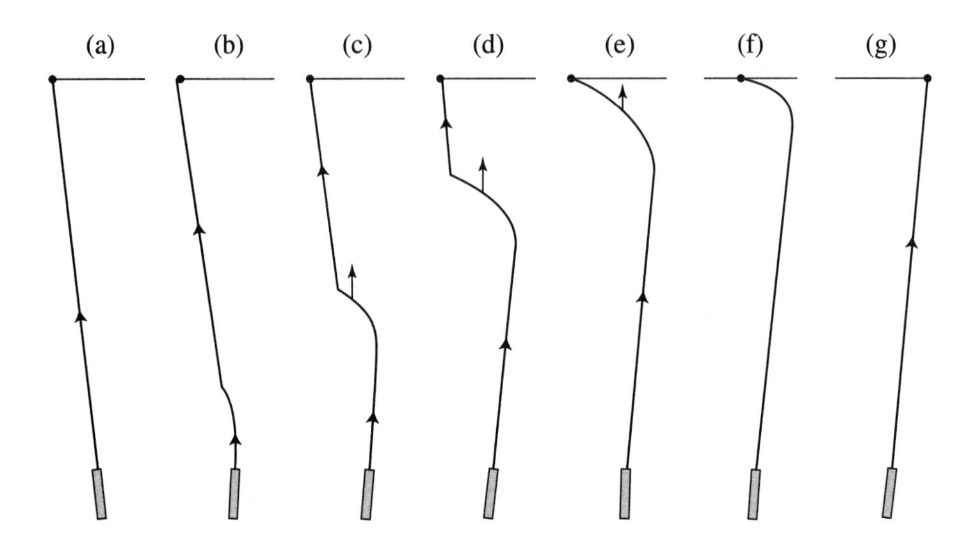

FIGURE E.5
The stream of photons from a laser aimed at the Moon. (a) Before the laser has been turned, the beam has already been directed for some time at the left edge of the Moon. (b) Then the laser has been turned a bit, so some photons are now directed further to the right, while photons emitted some time before are still directed to the left. (c) The laser is now aimed at the right edge of the Moon, and the laser will not be moved from now on. (d) Still later. (e) Photons emitted from the laser at the time it started to turn finally reach the Moon at its left edge. (f) The dot has moved across half of the Moon. (g) The dot has now reached the right edge of the Moon.

travel across the Moon's surface faster than the speed of light, as shown in Problem E–2. Note that no single photon travels with the dot, and note also that no one living on the left edge of the Moon can send a signal to someone living on the right edge—all potential signals are controlled by the person turning the laser on Earth. Therefore such a faster-than-light dot violates no principle of relativity.

Another "thing" that could travel faster than light is the point of intersection of the two blades of a very long pair of scissors as the scissor blades close. A sequence of pictures is shown in Fig. E.6. While the atoms in each blade move toward one another at speeds less than c, the abstract point of intersection can move faster than c! It is essential that the blades close so that each of them remains straight. This cannot be done by simply squeezing the handles together in the usual way, however, which is related to the fact that no superluminal *message* can be sent using the pair of scissors. See Problem E–3 for details.

From time to time other real or imaginary superluminal phenomena are reported. These do not threaten relativity unless it is claimed they can be used to send messages at speeds greater than c. If the tachyons discussed in Section E.1 are ever discovered, they will cause some sleepless nights to relativity theorists unless they can show that these tachyons cannot be used to send superluminal signals.[3]

---

3. Many relativity theorists work at night anyway; they would not be affected.

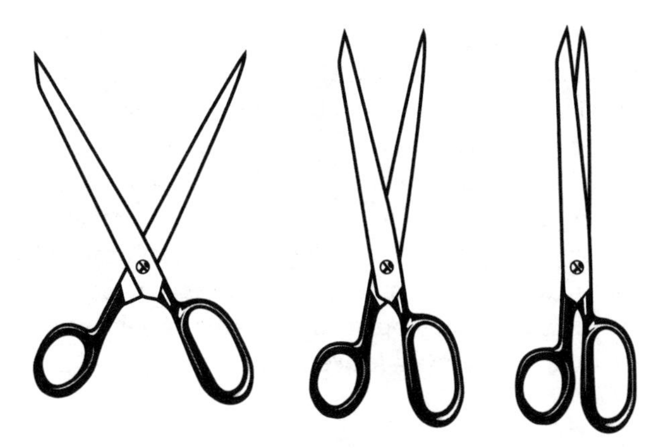

FIGURE E.6
A long-bladed pair of scissors is closed. The point of intersection of the blades can in principle move faster than light, even though none of the atoms in the blades achieves the speed of light. Also, one can show that no superluminal signal can be sent using scissors.

Problems

E–1. In A's reference frame, a tachyon moves to the right with velocity $v = \alpha c$, where $\alpha > 1$. B's reference frame moves to the right relative to A's frame with velocity $V = \beta c$, where $\beta < 1$. (a) Find β (in terms of α) such that in B's frame the tachyon moves at *infinite* velocity. (b) Show that if β is larger than this value (although β must remain < 1), then the tachyon can be thought of either as moving to the *left* in B's frame or as moving to the *right* but backward in time. (c) In particular, suppose that in A's frame the tachyon moves at speed $2c$, and that B is to the right of A. What is the minimum velocity at which B must travel relative to A so that such a tachyon emitted by A will be received at B before it was sent, according to observers in B's frame?

E–2. A laser beam is aimed at the left edge of the Moon, as described in Section E.3. The laser is then turned in a short time Δt to aim at the *right* edge of the Moon. (a) Show that the velocity of the bright dot on the Moon travels from left to right at speed $v = D/\Delta t$, where D is the diameter of the Moon. (b) The Moon's diameter is $D = 3476$ km. What is the value of Δt below which the dot will move across the Moon's surface at superluminal speed? (c) Does this short time interval seem feasible? (It is adequate here to model the Moon as a flat circle of diameter 3476 km perpendicular to our line of sight.)

E–3. Suppose that each of the very long blades of a pair of scissors is perfectly straight and of width $2D$, as shown below. The two blades are pivoted about a point P in the middle of each blade, as shown. (a) Show that the distance of the point of intersection from the pivot point P is given by $y = D/\sin\theta$, where θ is the angle of the centerline of each blade relative to the overall centerline, as shown in the figure.

(b) The blades are then closed with constant angular velocity $\omega_0 = -d\theta/dt$ where ω_0 is measured in radians per second. (Note that ω_0 here is positive, since $d\theta/dt$ is negative.) Assume that neither D nor ω_0 is extremely large, so that $D\omega_0/c \ll 1$. Show then that the point of intersection reaches the speed c when the blades are at angle $\theta_0 = \sqrt{D\omega_0/c}$ and the distance from the pivot point is $y_0 = \sqrt{cD/\omega_0}$. (For smaller θ or larger y, the point of intersection moves faster than c.) (c) If we could make continual slight variations in the angular velocity of the blades as they are closing, it seems that we could send a coded message up the point of intersection at speeds greater than c! What property of the blades would prevent us from doing that? *Hint:* Are rigid bodies possible in relativity?

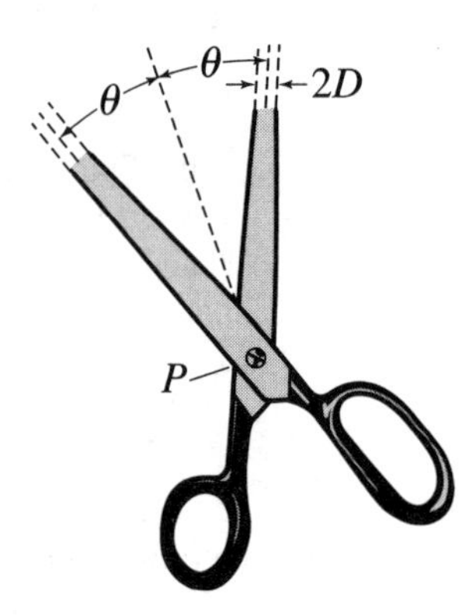

APPENDIX F

"Relativistic Mass" and Relativistic Forces

SPECIAL RELATIVITY makes us take a second look at physical quantities we know well from classical physics, because radical revisions may be needed. What about mass and force, for example? Does Einstein's theory require us to change our thinking about either of them? Also, we need to know whether Newton's second law of motion, one of the central pillars of classical mechanics, survives in relativistic mechanics.

F.1 "Relativistic Mass"

In this book we consistently use the symbol "m" to represent the mass of an object, taken to be constant, independent of the object's velocity. It is often called the "rest mass" of the object, the mass measured when the object is at rest. In contrast, some people speak (or write) about relativity using a so-called "relativistic mass," m_R, defined by[1]

$$m_R \equiv \gamma m = \frac{m}{\sqrt{1 - v^2/c^2}}. \tag{F.1}$$

This concept leads to statements like "mass increases with velocity" and "nothing can reach the speed of light because then its mass would be infinite." Such conceptions are common and have some intuitive appeal, so we need to discuss their meaning. How did the idea of "relativistic mass" arise? What does it mean physically? What are its advantages and disadvantages?

1. Confusingly, the symbol "m" is often used for relativistic mass instead of "m_R," and the symbol "m_0" is used for rest mass rather than "m."

As shown in Chapter 10, the momentum of a particle is

$$\mathbf{p} = \gamma m \mathbf{v} = \frac{m\mathbf{v}}{\sqrt{1 - v^2/c^2}}. \tag{F.2}$$

The idea of "relativistic mass" arose mainly because people could then retain the classical definition momentum = mass × velocity even for relativistic particles, as long as the constant mass m is replaced by the "relativistic mass" m_{R}.

Some of the *advantages* of using relativistic mass are

(i) There is a simple form for the momentum, $\mathbf{p} = m_{\mathrm{R}}\mathbf{v}$.

(ii) There is a simple form for the total energy, $E = m_{\mathrm{R}}c^2$.

(iii) It gives a simple intuitive reason for the increased inertia, or "sluggishness," of objects at high speed: they are more "massive," so it is harder to make them move faster. *However, this is deceptive! The resistance to acceleration of a relativistic object is not necessarily m_{R}, but depends upon which way you are pushing on it!*

(iv) Although it is beyond the scope of this book, it turns out that if you weigh a large bunch of particles in a box when they are at rest, they weigh still *more* if they are moving around inside. For example, if the box contains air, it will get slightly heavier if you heat up the air. However, gravity is more complicated than simply replacing all masses by m_{R}. A thorough understanding comes only from Einstein's *general* theory of relativity.

Some of the *disadvantages* of using relativistic mass are

(i) By hiding $\sqrt{1 - v^2/c^2}$ in the mass, we may forget it is there. It is safer to exhibit all such factors explicitly in the formulas, by writing, for example,

$$\mathbf{p} = \frac{m\mathbf{v}}{\sqrt{1 - v^2/c^2}} \quad \text{and} \quad E = \frac{mc^2}{\sqrt{1 - v^2/c^2}}.$$

(ii) One may get the mistaken impression that to go from classical to relativistic mechanics it is only necessary to replace *all* masses by m_{R}. This certainly works for the momentum $\mathbf{p} = m\mathbf{v}$, but it does not work, for example, for kinetic energy: The relativistic kinetic energy is *not* $(1/2)m_{\mathrm{R}}v^2$. And it works in Newton's second law $\mathbf{F} = m\mathbf{a}$ only in the very special case where the force exerted on a particle is perpendicular to its velocity. In all other cases, $\mathbf{F} \neq m_{\mathrm{R}}\mathbf{a}$!

(iii) Relativity fundamentally serves to correct our notions about time and space. That is, it is really the dynamical equations dealing with motion, like energy and momentum, that ought to be changed, and not the properties of individual particles, like mass.[2]

(iv) When relativity is cast in four-dimensional spacetime form, as in Chapters 10 and 11, the idea of relativistic mass is out of place and clumsy.

In summary, although there are some intuitive and therefore pedagogical reasons for introducing a velocity-dependent relativistic mass, these advantages are greatly outweighed by the disadvantages. Therefore we do not use relativistic mass elsewhere in this book.

F.2 Forces and Newton's Second Law

Newton's second law of motion is central to classical mechanics. One version ($\mathbf{F} = m\mathbf{a}$) gives a particle's acceleration in terms of the net force on it; another version ($\mathbf{F} = d\mathbf{p}/dt$) gives the rate of change of the particle's momentum in terms of the force. The two versions are equivalent in nonrelativistic mechanics, but they *aren't* equivalent for a relativistic particle with momentum $\mathbf{p} = \gamma m\mathbf{v}$, because then $d\mathbf{p}/dt \neq m\mathbf{a}$. At least one version of Newton's second law has to be wrong for a relativistic particle!

The correct law of motion in relativity depends partly on what we accept as a relativistic force. A natural choice to use is the *electromagnetic* force.[3] One reason for this is that electromagnetic forces are commonly used to accelerate charged particles to relativistic speeds, as in high-energy accelerators. Another reason is that relativity is greatly indebted to electromagnetism, because Einstein had the insight that electromagnetism is best understood if special relativity is correct. In fact, both electromagnetic theory and experiment show that if $\mathbf{F}$ is the electromagnetic force on a particle and $\mathbf{p}$ its momentum, then

$$\mathbf{F} = \frac{d\mathbf{p}}{dt} \tag{F.3}$$

is the correct law of motion, while the well-known classical equation $\mathbf{F} = m\mathbf{a}$ is *not correct* for a relativistic particle!

---

2. Albert Einstein himself, in a letter to L. Barnett (as quoted by L. B. Okun, in his article "The Concept of Mass" in *Physics Today* 42, 31, June 1989), said: "It is not good to introduce the concept of the mass $M = m/\sqrt{1 - v^2/c^2}$ of a moving body for which no clear definition can be given. It is better to introduce no other mass concept than the 'rest mass' m. Instead of introducing M it is better to mention the expression for the momentum and energy of a body in motion."

3. If a particle has an electrical charge q, it can be acted upon by the *electric* force $\mathbf{F_E} = q\mathbf{E}$ due to an electric field $\mathbf{E}$, and the *magnetic* force $\mathbf{F_B} = q\mathbf{v} \times \mathbf{B}$ due to a magnetic field $\mathbf{B}$. The magnetic force depends upon the velocity $\mathbf{v}$ of the particle as well as its charge q and the strength of the magnetic field, and involves the "cross product" $\mathbf{v} \times \mathbf{B}$ of the vectors $\mathbf{v}$ and $\mathbf{B}$. The total electromagnetic force on a nonrelativistic particle is therefore $\mathbf{F} = q(\mathbf{E} + \mathbf{v} \times \mathbf{B})$. The cross product $\mathbf{v} \times \mathbf{B}$ is a *vector.* The *magnitude* of $\mathbf{v} \times \mathbf{B}$ is $vB \sin\theta$, where θ is the smaller angle between $\mathbf{v}$ and $\mathbf{B}$, while the *direction* of $\mathbf{v} \times \mathbf{B}$ is perpendicular to both $\mathbf{v}$ and $\mathbf{B}$, as given by the "right-hand rule." The rule is to extend the fingers of the right hand in the $\mathbf{v}$ direction, and then curl them toward the vector $\mathbf{B}$. Your thumb will then point in the $\mathbf{v} \times \mathbf{B}$ direction. See any college-level textbook on electromagnetism for a more thorough explanation.

F.3 Constant-Force Motion

A nice feature of Eq. F.3 is what it predicts if we keep applying a force that is constant in both magnitude and direction. Suppose that the particle is initially at rest. The entire motion is then one-dimensional, so we can remove the vector signs in Eq. F.3 and integrate over time to give

$$\int_0^t F\,dt = \int_0^p dp = p. \tag{F.4}$$

We have assumed that F is constant, so it can be brought outside the integral. The momentum of the particle therefore increases at a steady rate,

$$p = Ft, \tag{F.5}$$

as shown in Fig. F.1.

Even though the particle's momentum keeps increasing at a steady rate, its velocity never reaches the speed of light! That is because, using the definition of momentum,

$$p \equiv \frac{mv}{\sqrt{1 - v^2/c^2}} = Ft. \tag{F.6}$$

As the particle's velocity approaches the speed of light, the numerator mv stops growing very much, but the denominator keeps getting smaller and smaller, increasing the momentum a lot without increasing the velocity much at all. (Note that if $F = ma$ were correct, a constant force could eventually boost a particle to the speed of light and beyond, which is not possible!)

We can find the particle's velocity as a function of time by multiplying Eq. F.6 by $\sqrt{1 - v^2/c^2}$ and squaring the result; this gives

$$m^2v^2 \equiv m^2c^2(v^2/c^2) = (1 - v^2/c^2)(Ft)^2.$$

Solving for v/c then gives

$$\frac{v}{c} = \frac{Ft/mc}{\sqrt{1 + (Ft/mc)^2}}. \tag{F.7}$$

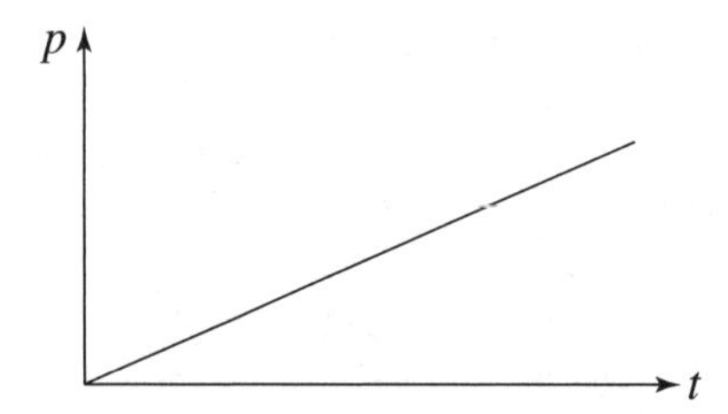

FIGURE F.1
Momentum increase for constant force.

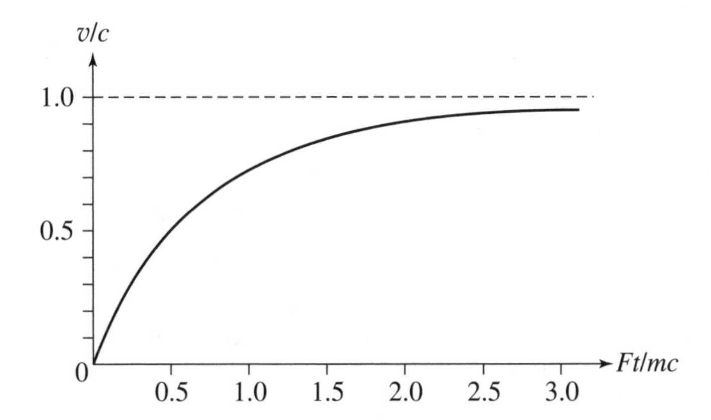

FIGURE F.2
The velocity v/c as a function of time for constant force.

Note that $v = 0$ at $t = 0$, consistent with starting the object from rest. For very small times t, the quantity $(Ft/mc)^2 \ll 1$, so the equation simplifies to $v/c \cong Ft/mc$. This linear increase in velocity with time is expected in nonrelativistic physics if the acceleration $a = F/m =$ constant. For small times (and therefore small velocities), we retrieve the usual nonrelativistic behavior.

For very *large* times, the "one" in the denominator of Eq. F.7 is more and more overwhelmed by the term $(Ft/mc)^2$, so $v/c \to 1$ as $t \to \infty$. The particle gets closer and closer to the speed of light as time goes on, but it never quite reaches it. A graph of v/c as a function of time is shown in Fig. F.2.

Figures F.1 and F.2 illustrate the difference between momentum and velocity in relativity. For small velocities, they both increase linearly with time, as we expect nonrelativistically, but they differ markedly for relativistic velocities.

F.4 General One-Dimensional Motion

Suppose now that the force on a particle is still entirely one-dimensional but not necessarily constant in magnitude. In that case Eq. F.3 gives

$$F = \frac{dp}{dt} = \frac{d}{dt}\left(\frac{mv}{\sqrt{1 - v^2/c^2}}\right) = \frac{ma}{(1 - v^2/c^2)^{3/2}}, \tag{F.8}$$

where the acceleration $a = dv/dt$. (See Problem F–5.)[4] The acceleration of the particle is therefore

$$a = (F/m)(1 - v^2/c^2)^{3/2}, \tag{F.9}$$

4. Note from Eq. F.8 that for one-dimensional motion we could *not* have simply replaced m by the "relativistic mass" $m_R \equiv m/\sqrt{1 - v^2/c^2}$ and then used $F = m_R a$.

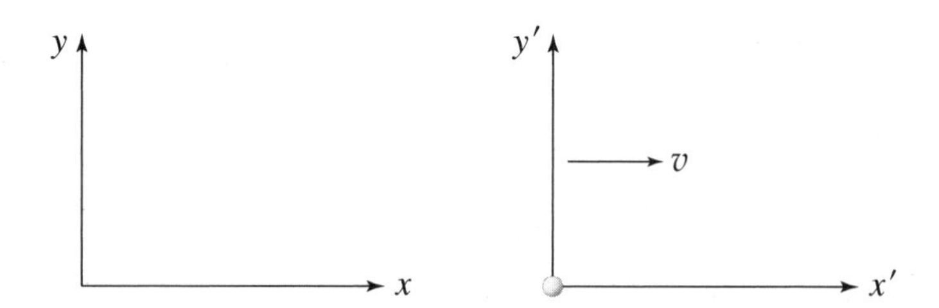

FIGURE F.3
Two inertial frames; the unprimed frame is the original rest frame of the particle, and the primed frame is the frame in which the particle is instantaneously at rest at some later time. Note that the particle is moving at velocity v relative to the unprimed frame.

reducing to the usual nonrelativistic result $a = F/m$ for small velocities but becoming smaller and smaller as the velocity increases.

The acceleration a we have been describing is measured in some fixed inertial frame, say the (*unprimed*) frame in which the particle started from rest. Now compare a with the acceleration a' measured in a *primed* inertial frame in which the particle is instantaneously at rest at some later time. That is, relative to the initial rest frame, we choose the primed frame to move at velocity $V = v$, as shown in Fig. F.3. The relationship between these two accelerations is (see Problem F–6)

$$a = a'(1 - v^2/c^2)^{3/2}, \tag{F.10}$$

so that the acceleration of the particle as measured in its original rest frame is *less* than it is in the particular inertial frame that happens to be moving along with the particle at some later instant of time.

In the primed frame we can still use $F' = ma'$, since the particle is instantaneously at rest in that frame and so is obviously nonrelativistic. Combining Eqs. F.8 and F.10, we obtain

$$F = \frac{ma}{(1 - v^2/c^2)^{3/2}} = \frac{ma'(1 - v^2/c^2)^{3/2}}{(1 - v^2/c^2)^{3/2}} = ma' = F', \tag{F.11}$$

so the force acting on the particle in this case is the *same,* whether the force is measured in the original rest frame or in any frame in which the particle later finds itself momentarily at rest![5]

Problems

F–1. A constant force $F_0 = 10^5$ newtons is exerted on a particle of mass $m = 1\,\text{g} = 10^{-3}$ kg. If it starts from rest, how long does it take it to reach velocity $(4/5)c$?

---

5. In deriving Eq. F.11, we have assumed that the motion is one-dimensional and that the relative velocity of the primed and unprimed frames is in this same direction. In more general circumstances, the force acting on a particle is *different* in different frames. See Problem F–10.

F–2. With what constant force F_0 must we push on a particle of mass m, initially at rest, if it is to gain kinetic energy $2mc^2$ in a given time t_0?

F–3. An electron in a particular linear accelerator is subject to a force $F_0 = qE_0$, where q is its electric charge and E_0 is a *constant* electric field acting upon it. Starting from rest, how long does it take the electron to achieve the kinetic energy $10\,mc^2$? (b) What is this time if the electric field is $E_0 = 500$ newtons/coulomb? (The mass and charge of an electron are $m = 9.1 \times 10^{-31}$ kg and $q = 1.60 \times 10^{-19}$ coulombs (C). Note that 1 newton (N) = 1 kg m/s^2.)

F–4. A 1.0 kg space probe packed with electronics is ejected from the Moon by a powerful laser that pushes on the probe with the constant force 1000 N. How long, according to Moon clocks, does it take the probe to reach speed $c/2$?

F–5. Take the derivative in Eq. F.8 to show that for one-dimensional motion, $F = ma/(1 - v^2/c^2)^{3/2}$.

F–6. (a) Show, by differentiating the inverse velocity transformation of Section 8.3 with respect to time t, that the acceleration $a = dv/dt$ of a particle in an unprimed frame, in terms of its acceleration $a' = dv'/dt'$ in a primed frame, is

$$a = \frac{(1 - V^2/c^2)^{3/2}}{(1 + v'V/c^2)^3} a'$$

for one-dimensional motion, where V is the relative velocity between the two frames and v' is the particle's velocity in the primed frame. (b) Suppose the particle happens to be at rest relative to the primed frame at some instant, so $v' = 0$ at that instant. Show in that case that $a = a'(1 - v^2/c^2)^{3/2}$.

F–7. (a) Show that the acceleration of a particle of mass m that moves in one dimension under a constant force F is

$$a = \frac{F/m}{[1 + (Ft/mc)^2]^{3/2}}$$

where t is the time measured in its initial rest frame since it began to accelerate. (b) Plot $a(t)$. What are the limiting values of a for small t and large t?

F–8. In classical mechanics, the equation $\mathbf{F} = m\mathbf{a}$ ensures that the force $\mathbf{F}$ acting upon a particle is always parallel to the particle's acceleration $\mathbf{a}$. (a) Show that in *relativistic* mechanics, the force is *not necessarily* parallel to the acceleration. (b) Show that $\mathbf{F}$ *is* parallel to $\mathbf{a}$ in the two special cases: (*i*) $\mathbf{v} \parallel \mathbf{F}$, and (*ii*) $\mathbf{v} \perp \mathbf{F}$, where $\mathbf{v}$ is the velocity of the particle. *Hint:* In (*ii*), it is helpful to take the vector dot products $\mathbf{v} \cdot \mathbf{F}$ and $\mathbf{v} \cdot d\mathbf{p}/dt$.

F–9. (a) Show that if the force $\mathbf{F}$ on a particle always acts in a direction *perpendicular* to the particle's velocity vector $\mathbf{v}$, then $\mathbf{F}$ is related to the particle's acceleration vector $\mathbf{a} = d\mathbf{v}/dt$ by $\mathbf{F} = \gamma m\mathbf{a}$. (b) Then show that if in addition $\mathbf{F}$ has the *constant* magnitude F, the particle moves in a circular path of radius $r = \gamma mv^2/F$, where

v is its speed. (This sort of motion is executed by an electrically charged particle moving in a uniform magnetic field.) *Hint:* The momentum is $\mathbf{p} = \gamma m\mathbf{v}$, so $\mathbf{p}$ is in the same direction as $\mathbf{v}$. So if $\mathbf{F}$ is perpendicular to $\mathbf{v}$, then it is perpendicular to $\mathbf{p}$ as well. However, Newton's second law $\mathbf{F} = d\mathbf{p}/dt$ tells us that for short time intervals Δt, the *change* in momentum is $\Delta\mathbf{p} = \mathbf{F}\Delta t$, so even though $\mathbf{p}$ is perpendicular to $\mathbf{F}$, $\Delta\mathbf{p}$ is parallel to $\mathbf{F}$, and so $\mathbf{p}$ and $\Delta\mathbf{p}$ are perpendicular to one another, as shown below. If $\mathbf{p}(t)$ is the momentum at time t, then the momentum a time Δt later is $\mathbf{p}(t + \Delta t) = \mathbf{p}(t) + \Delta\mathbf{p}$, a vector that points in a slightly different direction but which has the same magnitude. Likewise, the velocity vector points in a slightly different direction but retains the same magnitude.

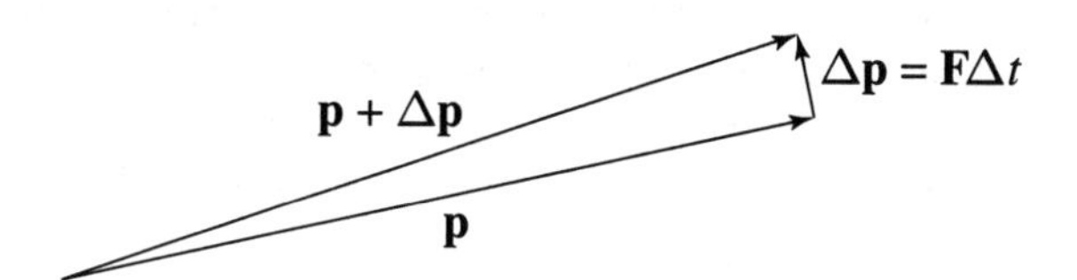

F–10. (a) Show that the components of the net force acting on a particle in the usual primed and unprimed frames are related by

$$F'_x = \frac{F_x - V/c^2(\mathbf{v}\cdot\mathbf{F})}{1 - v_x V/c^2}, \quad F'_y = \frac{F_y\sqrt{1 - V^2/c^2}}{1 - v_x V/c^2}, \quad F'_z = \frac{F_z\sqrt{1 - V^2/c^2}}{1 - v_x V/c^2}$$

where $\mathbf{v}\cdot\mathbf{F} \equiv v_x F_x + v_y F_y + v_z F_z$. *Hint:* By definition, $F_x = dp_x/dt$, $F'_x = dp'_x/dt'$, and so on, where, from the Lorentz transformation, $t' = \gamma(t - Vx/c^2)$ and $p'_x = \gamma(p_x - VE/c^2)$, $p'_y = p_y$, $p'_z = p_z$. Also useful is Eq. 11.16, $E^2 = (p_x^2 + p_y^2 + p_z^2)c^2 + m^2c^4$, together with $E = mc^2/\sqrt{1 - v^2/c^2}$ and $\mathbf{p} = m\mathbf{v}/\sqrt{1 - v^2/c^2}$. (b) Then show that if the motion is entirely along the x direction, it follows that $F'_x = F_x$, confirming the result derived in Eq. F.11 in an entirely different way.

F–11. The *Minkowski Force* K_μ acting on a particle has (by definition) the four components $K_\mu = dp_\mu/d\tau$ ($\mu = 0, 1, 2, 3$), where p_μ is the momentum–energy four-vector and τ is the proper time—the time read by a hypothetical clock carried along with the particle. (The usual force components $F_x = dp_x/dt$, $F_y = dp_y/dt$, $F_z = dp_z/dt$ are not themselves the spatial components of a four-vector, because the time t read by a clock in some inertial frame is not a scalar in spacetime, but a component of the position four-vector x_μ.) Show that the components of K_μ can be written

$$K_\mu = \left(\gamma\frac{\mathbf{v}\cdot\mathbf{F}}{c}, \gamma F_x, \gamma F_y, \gamma F_z\right),$$

where $\gamma = (1 - v^2/c^2)^{-1/2}$ is the gamma factor for the particle and v its speed, and where $\mathbf{v}\cdot\mathbf{F} \equiv v_x F_x + v_y F_y + v_z F_z$. (Four-vectors transform with the Lorentz transformation, which provides an alternative way to derive the ordinary force tranformation equations given in Problem F–10.)

APPENDIX G

The Ultimate Relativistic Spaceflight

"YOU MAY NOW remove your seatbelts, stand up, and walk around. Gravity will feel just the same as on Earth. You are embarked on a relativistic voyage that will take several years to complete. You should feel quite comfortable, because the on-board gravity will be the same as you are used to."

FIGURE G.1
Artist's conception of a Bussard interstellar ramjet (NASA).

We are far from Earth's gravity, but because of our ship's acceleration, we will feel an effective gravity, according to Einstein's principle of equivalence.[1] The effective gravity g_{eff}, which is numerically equal to the ship's acceleration a' in an inertial frame in which it happens to be instantaneously at rest, seems to "pull" us toward the rear of the ship, just as we are "pulled" toward the seat-back of a car when we accelerate forward.

Suppose that this effective gravity $g_{\text{eff}} = a'$ is *constant,* and suppose also that our ship has *constant mass*. According to Appendix F (see Eq. F.11), this can be accomplished by applying a constant force to the ship (constant in the initial rest frame or in any subsequent instantaneous rest frame of the ship—they are the same, so it doesn't matter).[2]

If the force produces an acceleration $g_{\text{eff}} = 9.8\ \text{m/s}^2$, for example, we can walk around feeling an effective gravity just like Earth's, directed toward the rear of the ship.

Setting $F = ma' = mg_{\text{eff}}$, where m is the mass of the ship, Eq. F.7 of Appendix F gives the velocity of the ship to be

$$\frac{v}{c} = \frac{g_{\text{eff}}t/c}{\sqrt{1 + (g_{\text{eff}}t/c)^2}} \tag{G.1}$$

in terms of the time since it started to accelerate. We will now rewrite this equation using the *hyperbolic functions* sinh, cosh, and tanh (often pronounced "sinch," "cawsh," and "tanch"), which are analogs of the more familiar *circular functions* sine, cosine, and tangent. We do this because it makes several subsequent results remarkably concise.

Recall that the circular functions *sine* and *cosine* can be defined in terms of the unit circle $x^2 + y^2 = 1$ shown in Fig. G.2. If a right triangle is drawn inside this unit circle, with horizontal and vertical legs and a hypotenuse along a radius of the circle,

---

1. See Chapter 14.

2. How might a constant force be exerted on a constant-mass ship? One possibility is the push from the beam of an extremely powerful laser based on the far side of the moon. Photons from the beam would bounce off a reflector on the rear of the ship (without vaporizing it!), transferring momentum to the ship. However, this might work at best for only a relatively short time. Three effects would tend to weaken the force with time: As the ship increases in speed, (*i*) fewer photons would strike it per unit time; (*ii*) reflected photons would have lower frequency and lower momentum, so each bouncing photon would transfer less momentum to the ship. (*iii*) As the ship gets farther away, a smaller and smaller fraction of the spreading laser beam would strike the ship's reflector. A different possibility is to design a ship that could pick up deuterons (using electromagnetic fields) from the hydrogen in space, and "burn" the deuterons onboard by controlled fusion reactions, as described in Appendix H (Section H.2) emitting the products as exhaust from the ship's rear. This is an "interstellar ramjet" or "Bussard ramjet," proposed in 1960 by the physicist Robert Bussard. (Bussard, R. W., "Galactic Matter and Interstellar Flight," *Astronautica Acta,* 6, 1960.) It is problematical whether enough deuterons could be swept up to make this work, and there are numerous other difficulties as well. (See, for example, Heppenheimer, T. A. "On the Infeasibility of Interstellar Ramjets," *Journal of the British Interplanetary Society* 31, 1978.) So perhaps there will never be a practical way to accelerate a constant-mass ship for very long, but it is interesting to think about anyway.

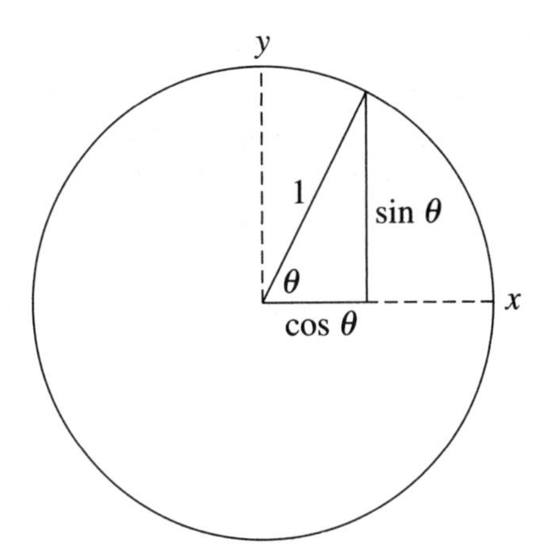

FIGURE G.2
Definitions of $\sin\theta$ and $\cos\theta$.

then $\cos\theta$ is the length of the horizontal leg, $\sin\theta$ is the length of the vertical leg, and $\tan\theta$ is the ratio $\tan\theta = \sin\theta/\cos\theta$, where θ is the angle between the horizontal leg and the hypotenuse, as shown.

The identity $\sin^2\theta + \cos^2\theta = 1$ is then an immediate consequence of the Pythagorean theorem. The derivatives of $\sin\theta$ and $\cos\theta$ are

$$\frac{d}{d\theta}(\sin\theta) = \cos\theta \quad \frac{d}{d\theta}(\cos\theta) = -\sin\theta, \tag{G.2}$$

and the corresponding indefinite integrals are $\int \sin\theta\, d\theta = -\cos\theta$ and $\int \cos\theta\, d\theta = \sin\theta$.

The *hyperbolic* functions, on the other hand, can be defined in terms of the unit *hyperbola* $x^2 - y^2 = 1$, whose right-hand branch is shown in Fig. G.3. Draw a right triangle whose hypotenuse extends from the origin to an arbitrary point on the hyperbola, and set the horizontal leg equal to $\cosh u$ (short for $\cosh(u)$, just as $\cos\theta$ is short for $\cos(\theta)$), the vertical leg to $\sinh u$, and define $\tanh u = \sinh u/\cosh u$, where u is a parameter. The identity $\cosh^2 u - \sinh^2 u = 1$ follows from the fact that $x^2 - y^2 = 1$ for points on the hyperbola.

The derivatives of $\sinh u$ and $\cosh u$ are

$$\frac{d}{du}(\sinh u) = \cosh u \quad \frac{d}{du}(\cosh u) = \sinh u, \tag{G.3}$$

and the corresponding indefinite integrals are $\int \sinh u\, du = \cosh u$ and $\int \cosh u\, du = \sinh u$.

The hyperbolic functions $\sinh u$, $\cosh u$, and $\tanh u$ are drawn as a function of u in Fig. G.4. The cosh function has the shape of a hanging chain, called the *catenary*. Note that $\sinh u \to \cosh u$ and $\tanh u \to 1$ as $u \to \infty$.

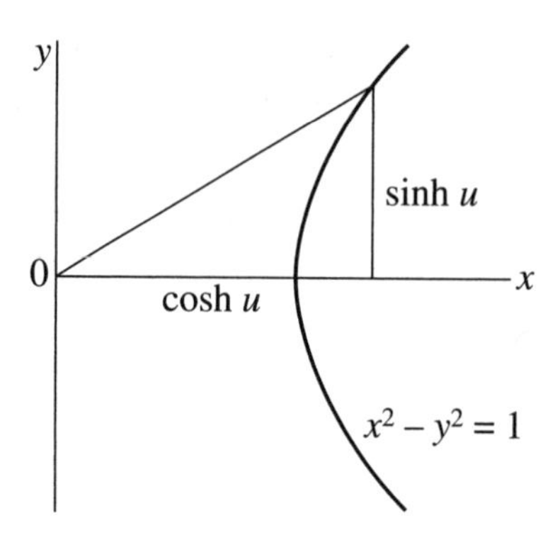

FIGURE G.3
Definitions of sinh u and cosh u.

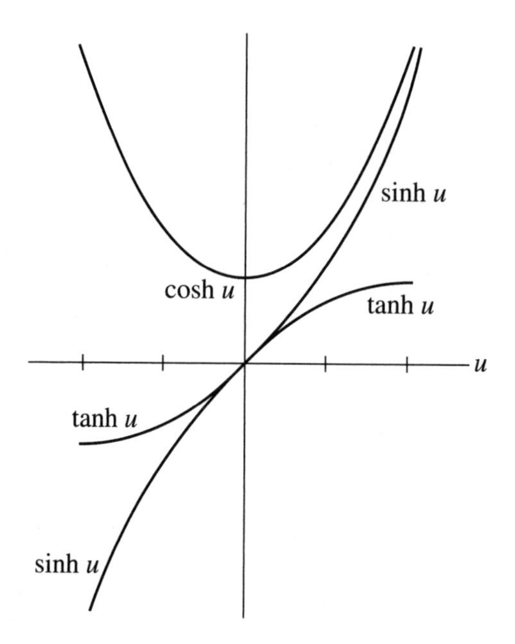

FIGURE G.4
The hyperbolic functions sinh u, cosh u, and tanh u.

The hyperbolic functions can be expressed in terms of exponentials,

$$\sinh u = \frac{e^u - e^{-u}}{2}, \qquad \cosh u = \frac{e^u + e^{-u}}{2}, \qquad \tanh u = \frac{\sinh u}{\cosh u} = \frac{e^u - e^{-u}}{e^u + e^{-u}}. \tag{G.4}$$

These identities can be used to verify the derivatives (G.3), for example.

Now we can use the hyperbolic functions to describe the motion of our constant effective-gravity spaceship. First, let $g_{\text{eff}}t/c = \sinh u$. (We will soon figure out what

u means physically; in the meanwhile, note that $u = 0$ when the ship began its voyage at $t = 0$, and that $u \to \infty$ as $t \to \infty$.) Then Eq. G.1 becomes simply

$$\frac{v}{c} = \frac{g_{\text{eff}}t/c}{\sqrt{1 + (g_{\text{eff}}t/c)^2}} = \frac{\sinh u}{\sqrt{1 + \sinh^2 u}} = \frac{\sinh u}{\cosh u} = \tanh u, \tag{G.5}$$

using the identity $\cosh^2 u - \sinh^2 u = 1$. Next we can find a simple expression for the ship's gamma factor,

$$\gamma \equiv \frac{1}{\sqrt{1 - v^2/c^2}} = \frac{1}{\sqrt{1 - \tanh^2 u}} = \cosh u, \tag{G.6}$$

which follows from the identity

$$1 - \tanh^2 u = 1 - \frac{\sinh^2 u}{\cosh^2 u} = \frac{\cosh^2 u - \sinh^2 u}{\cosh^2 u} = \frac{1}{\cosh^2 u}. \tag{G.7}$$

Next, we can learn how ship passengers age relative to the aging of people left behind. Using integrated proper time,[3] ship passengers age by

$$\begin{aligned}
\tau &= \int_0^\tau d\tau = \int_0^t dt\sqrt{1 - v^2/c^2} = \int_0^t dt\sqrt{1 - \tanh^2 u} \\
&= \int_0^u d\left(\frac{c}{g_{\text{eff}}} \sinh u\right) \times \left(\frac{1}{\cosh u}\right) \\
&= \frac{c}{g_{\text{eff}}} \int_0^u du \cosh u \left(\frac{1}{\cosh u}\right) = \frac{c}{g_{\text{eff}}} \int_0^u du = \frac{c}{g_{\text{eff}}} u,
\end{aligned} \tag{G.8}$$

so now we know the physical meaning of the parameter "u"! It is simply

$$u = \left(\frac{g_{\text{eff}}}{c}\right) \tau, \tag{G.9}$$

a constant multiple of the age increase τ of the passengers.

3. See Appendix D.

Finally, we can find the distance traveled by the spaceship,

$$x = \int_0^t v\,dt = \int_0^u c \tanh u\, d\left(\frac{c}{g_{\text{eff}}} \sinh u\right) = \frac{c^2}{g_{\text{eff}}} \int_0^u \tanh u \cosh u\, du$$

$$= \frac{c^2}{g_{\text{eff}}} \int_0^u \sinh u\, du = \frac{c^2}{g_{\text{eff}}} \cosh u \Big|_0^u = \frac{c^2}{g_{\text{eff}}}(\cosh u - 1). \quad \text{(G.10)}$$

We now rewrite all of these results in terms of the physical quantity τ. First of all, the relation between the advances in onboard time τ and original-frame time t becomes[4]

$$g_{\text{eff}} t/c = \sinh(g_{\text{eff}}\tau/c), \quad \text{(G.11)}$$

and the velocity of the ship as a function of τ is

$$\frac{v}{c} = \tanh(g_{\text{eff}}\tau/c), \quad \text{(G.12)}$$

both as shown in Fig. G.5.

The gamma factor of the ship is $\gamma = \cosh(g_{\text{eff}}\tau/c)$, so the *energy* of the ship, expressed as a fraction of its mass energy, is

$$E/mc^2 = \gamma = \cosh(g_{\text{eff}}\tau/c), \quad \text{(G.13)}$$

as shown in Fig. G.6. Finally, the distance the ship has moved in the original inertial frame, in terms of the onboard aging, is

$$x = \frac{c^2}{g_{\text{eff}}}(\cosh(g_{\text{eff}}\tau/c) - 1), \quad \text{(G.14)}$$

also as shown in Fig. G.6.

Table G.1 gives examples of the shipboard time τ and Earth time t, the ship velocity v, its gamma factor γ and energy ratio E/mc^2, and the distance x traveled. The table assumes an effective gravity $g_{\text{eff}} = 9.8$ m/s^2 and that the ship keeps accelerating all the way to the destination, without slowing down.[5]

---

4. This equation, and all subsequent equations, can be greatly simplified for large travel times, that is, for trips with $g_{\text{eff}}\tau/c \gg 1$. See Problem G–1.

5. In Table G.1, the astute reader may rightly protest that we have carried the trip too far! We have assumed that the universe is static, and that we can use a single special-relativistic inertial frame out to 10^{10} light-years. Neither is true! In fact, the universe is expanding, faster and faster as time goes on. Also, it is necessary to use *general* relativity, not special relativity, in understanding the universe on such large scales. When done "right," a one-gee spaceship can still move out to enormous distances in one human lifetime, but the details depend on the choice of cosmological model. In any case, such accelerating observers would be able to see parts of the universe we can never see from Earth.

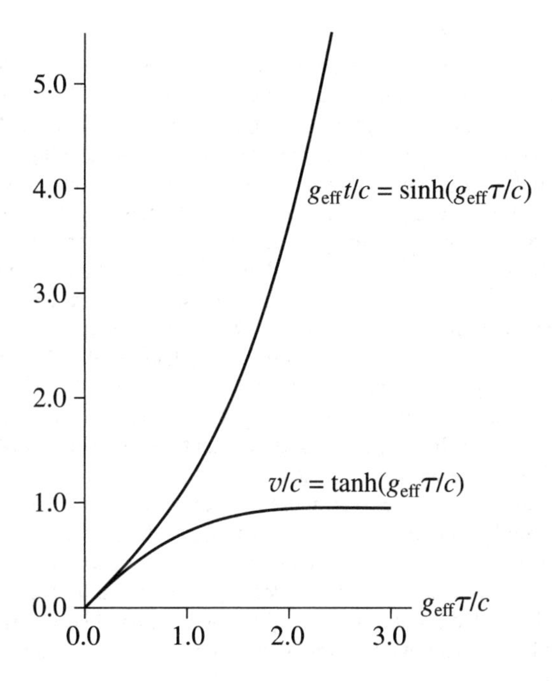

FIGURE G.5
Aging of people who stayed home and the velocity of the ship, in terms of the aging of ship passengers. Note that ship passengers age slowly, and that $v/c \to 1$ as $\tau \to \infty$.

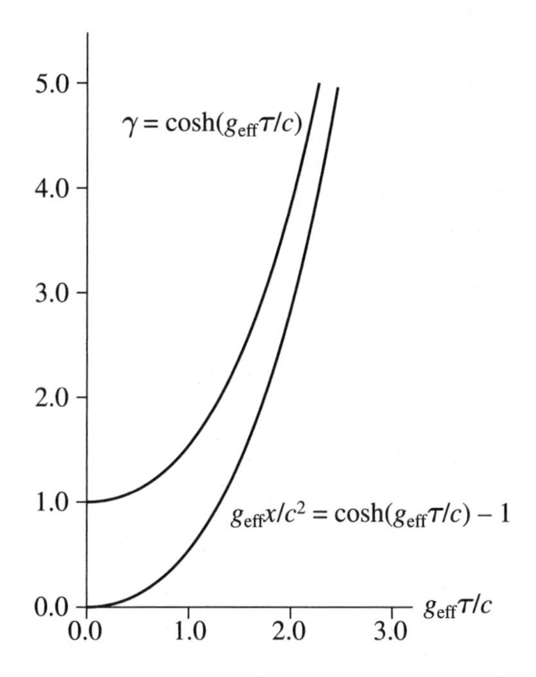

FIGURE G.6
The energy ratio $\gamma = E/mc^2$ and the distance the ship has moved, both in terms of the aging of its passengers.

TABLE G.1
A sample of times, velocities, energies, and distances for a constant $g_{\text{eff}} = 9.8\ \text{m/s}^2$ spaceflight. At the end of 1.18 years to the people left at home, ship passengers have aged 1 year and the ship is moving at 77% the speed of light and has traveled 0.56 light-years. At the end of 3.75 years on Earth, ship passengers have aged 2 years and the ship is traveling at 97% the speed of light and has traveled 2.9 light-years, not yet as far as the nearest star, Proxima Centauri, 4.0 light-years away. The ship's energy is nearly four times its mass energy. After 10.8 years on Earth, ship passengers have aged 3 years and the ship is moving at 99.95% the speed of light and has traveled 9.78 light-years, farther away than many local stars. When ship passengers have aged 10 years, Earth is 15,000 years older and the ship has traveled about 15,000 light-years, an appreciable fraction of the width of our Milky Way galaxy. After 20 years to the passengers, Earth has aged 440 million years and the ship has traveled 440 million light-years, having departed our galaxy and neighboring galaxies long before. When the passengers are about 3 years older still, they will have traveled roughly as far as we can currently see out in the universe, on the order of 10 billion light-years. After that time, who knows what they will meet up with! (Note that beginning with $\tau = 5$ years, v/c is listed simply as 1^-, meaning that v/c is only slightly less than unity.) If the passengers want to slow down and stop when they reach a destination, they should accelerate for half the distance to the destination, and decelerate for the rest of the trip. If, for example, they accelerate for 2 years according to their clocks, they will have traveled 2.90 light-years; if they then decelerate at the same rate, they will have aged a total of 4 years and traveled a total of 5.80 light-years. If they then invert the trip to return home, then they will have been gone 8 years by the time they return, and people on Earth will have aged 4 × 3.75 years = 15 years.

τ (yrs)	$g_{\text{eff}}\tau/c$	$g_{\text{eff}}t/c$	t (yrs)	v/c	$\gamma = E/mc^2$	$x(c \cdot \text{yrs})$
1.00	1.03	1.22	1.18	0.77	1.58	0.56
2.00	2.06	3.86	3.75	0.97	3.99	2.90
3.00	3.10	11.1	10.8	0.996	11.1	9.78
4.00	4.13	31.1	30.2	0.9995	31.1	29.2
5.00	5.16	87	84	1^-	87	83
10.0	10.3	1.5×10^4	1.5×10^4	1^-	1.5×10^4	1.5×10^4
20.0	20.6	4.5×10^8	4.4×10^8	1^-	4.5×10^8	4.4×10^8
23.0	23.7	1.03×10^{10}	10^{10}	1^-	1.03×10^{10}	10^{10}

Problems

G–1. Show that if $g_{\text{eff}}\tau/c \gg 1$, then Eqs. G.11, G.13, and G.14 reduce to

$$\text{(G.11A):}\quad g_{\text{eff}}t/c \cong \frac{1}{2}e^{g_{\text{eff}}\tau/c}$$

$$\text{(G.13A):}\quad E = \gamma mc^2 \cong mc^2\frac{1}{2}e^{g_{\text{eff}}\tau/c}$$

$$\text{(G.14A):}\quad x \cong \frac{c^2}{g_{\text{eff}}}\left(\frac{1}{2}e^{g_{\text{eff}}\tau/c} - 1\right).$$

G–2. How large must the quantity $g_{\text{eff}}\tau/c$ be, to ensure that the approximate equations given in Problem G–1 are valid to three significant figures?

G–3. A major expedition is being planned to send a constant effective-gravity Bussard ramjet to the nucleus of our Milky Way galaxy, 30,000 light-years away. If the effective gravity on board is to be kept at $g_{\text{eff}} = 10\,\text{m/s}^2$ for the convenience of the human crew, (a) how long will it take the ship to get there, according to clocks on Earth? (b) How much older will the crew be by the time they arrive? (c) After flying by the galactic nucleus, carefully avoiding the massive black hole that lurks there, the ship will slow down, still with $g_{\text{eff}} = 10\,\text{m/s}^2$, reaching a point 30,000 light-years on the other side of the galaxy, and then accelerate back again with $g_{\text{eff}} = 10\,\text{m/s}^2$, pass by the galactic nucleus once more at high speed, and finally slow down again at the same rate on its way back to Earth. By the end of the round trip, how much time will have passed on Earth, and how much older will the crew be than when they originally departed? *Hint:* See the equations in Problem G–1.

G–4. (a) Use Eq. G.11, Eq. G.14, and the identity $1 + \sinh^2 u = \cosh^2 u$, to show that the Earth time t required for a constant-g_{eff} spaceship to travel a distance x is given by

$$t = \frac{c}{g_{\text{eff}}}\sqrt{\frac{g_{\text{eff}}x}{c^2}\left(\frac{g_{\text{eff}}x}{c^2} + 2\right)}.$$

(b) Explorers seek to travel to the star Alpha Centauri, 4 light-years from Earth. Starting from rest, they board a spaceship in which the constant effective onboard gravity is $g_{\text{eff}} = 10\,\text{m/s}^2$. How long does it take them to fly by Alpha Centauri according to Earth clocks?

G–5. (a) Show in general that the ship time τ in terms of distance x, the distance traveled, for a ship starting from rest can be written in terms of a natural logarithm as

$$\tau = \frac{c}{g_{\text{eff}}}\ln\left(\frac{g_{\text{eff}}x}{c^2} + 1 + \sqrt{\frac{g_{\text{eff}}x}{c^2}\left(\frac{g_{\text{eff}}x}{c^2} + 2\right)}\right).$$

Hint: Use the formula for $\cosh u$ given in Eq. G.4. (b) Explorers board a $g_{\text{eff}} = 10\,\text{m/s}^2$ spaceship bound for Alpha Centauri, 4 light-years from Earth. When they fly by Alpha Centauri, how much older are they than when they left?

G–6. Prove the identity $\sinh^2(u/2) = \frac{1}{2}(\cosh u - 1)$, and then show that the distance x traveled by a constant-g_{eff} spaceship starting from rest, in terms of the proper time τ on shipboard, is given by

$$x = \left(\frac{2c^2}{g_{\text{eff}}}\right)\sinh^2\left(\frac{g_{\text{eff}}\tau}{2c}\right).$$

G–7. Travelers on Earth board the shuttle to Barnard's star, 6.0 light-years from Earth. The shuttle accelerates with $g_{\text{eff}} = 10\,\text{m/s}^2$ to the halfway point, and then decelerates with the same g_{eff} until it comes to rest near Barnard's star. How much older are the travelers than when they left? *Hint:* Use the equation for $\tau(x)$ quoted in Problem G–5.

G–8. Explorers start on a constant $g_{\text{eff}} = 10\text{ m/s}^2$ journey to Sirius, the brightest star in Earth's sky, 8.6 light-years away. Halfway there they turn the ship around and slow down at the same rate, so they come to rest at Sirius. (a) How long does it take them to arrive, according to Earth clocks? (b) How much older are they upon arrival? (c) How fast are they traveling at the middle of the journey? *Hint:* You can use the formulas quoted in Problems G–4 and G–5.

G–9. Explorers embark on a constant $g_{\text{eff}} = 10\text{ m/s}^2$ journey to study planets forming around the very young star *T Tauri,* which is 460 light-years away. Halfway there they turn the ship around and slow down at the same rate, so they come to rest in the *T Tauri* system. (a) How long does it take them to arrive, according to Earth clocks? (b) How much older are they upon arrival? (c) How fast are they traveling at the middle of the journey? *Hint:* You can use the formulas quoted in Problems G–4 and G–5.

G–10. Explorers depart to fly past the Andromeda galaxy (M 31), 2 million light-years distant, in a spaceship with constant effective onboard gravity g_{eff}. What must be the value of g_{eff} if the explorers want to age only 20 years during the trip?

G–11. *Dimensionless variables for spaceships with constant* g_{eff}. Four of the interesting variables for a spaceship that starts at rest in some stationary frame and then accelerates with constant onboard effective gravity g_{eff} are v, t, τ, and x. Here v is the velocity of the ship in the stationary frame, t and τ are the times after departure in the stationary frame and ship frame, respectively, and x is the distance the ship has traveled in the stationary frame. Formulas can be generated for any of these variables in terms of any *one* of the other three, together with g_{eff} and c. These formulas are particularly simple if the four variables are reexpressed in terms of the dimensionless quantities $V \equiv v/c$, $T \equiv g_{\text{eff}}t/c$, $T_{\text{ship}} \equiv g_{\text{eff}}\tau/c$, and $X \equiv g_{\text{eff}}x/c^2$. (a) Show that Eqs. G.11, G.12, and G.14 then assume the simple forms $T = \sinh T_{\text{ship}}$, $V = \tanh T_{\text{ship}}$, and $X = \cosh T_{\text{ship}} - 1$. (b) There are altogether 12 equations (including those given in part (a)) relating each of the four variables to each of the other three. *Find all 12 of these equations.* For example, the equation for T_{ship} in terms of T can be found from

$$T = \sinh T_{\text{ship}} = \frac{e^{T_{\text{ship}}} - e^{-T_{\text{ship}}}}{2} \equiv \frac{z + 1/z}{2}$$

where $z = e^{T_{\text{ship}}}$. This gives the equation $z^2 - 2Tz - 1 = 0$, which can be solved, using the quadratic formula, to give $z = T \pm \sqrt{T^2 + 1}$. We have to choose the plus sign, because $z = e^{T_{\text{ship}}} > 0$, so finally

$$T_{\text{ship}} = \ln\left(T + \sqrt{T^2 + 1}\right).$$

(One could alternatively write simply $T_{\text{ship}} = \sinh^{-1} T$, in terms of the *inverse hyperbolic sine,* but natural logarithms are sometimes easier to calculate or look up than inverse hyperbolic functions. So avoid inverse hyperbolic functions in this problem.)

APPENDIX H

Nuclear Decays, Fission, and Fusion

WE INTRODUCED THE TOPIC of nuclear binding energies in Chapter 12, along with the relationship between binding energy, mass energy, and kinetic energy of nuclear reactions. Here we illustrate these concepts using nuclear decays, nuclear fusion, and nuclear fission.

H.1 Nuclear Decays

Some nuclei are stable—they never decay into anything else. For example, the nuclei of the two lightest isotopes of hydrogen are stable: neither the proton ^{1}H nor the deuteron ^{2}H has ever been observed to decay. Many nuclei are *un*stable, however. For example, the nucleus of the third isotope of hydrogen, the triton ^{3}H, is unstable, with a half-life $t_{1/2} = 12.3$ years, meaning that half of a large sample of ^{3}H decays by the end of 12.3 years, and half of the *remaining* ^{3}H decays by the end of another 12.3 years, and so on.[1] *All* uranium nuclei are unstable, including ^{238}U, the nucleus of the most common isotope. The only reason we find uranium on Earth is that ^{238}U and some other uranium nuclei have very long half-lives. In fact, ^{238}U has a half-life of 4.5 billion years ($= 4.5 \times 10^9$ yrs $= 4.5$ gigayears $= 4.5$ Gyr), which by coincidence is pretty close to the age of Earth, so that rocks on Earth contain about half the ^{238}U they had when Earth was formed. Whatever the decay, there is always a partial conversion of mass energy into kinetic energy.

1. The *half-life* $t_{1/2}$ of a particle or radioactive nucleus is different from its mean (that is, average) lifetime. If the number of unstable particles at time $t = 0$ is N_0, then the number remaining at later time t is $N = N_0 e^{-t/\tau}$, where τ is the mean lifetime. The half-life $t_{1/2}$ is the time required for half the particles to decay, that is, when $N = N_0/2$. This corresponds to $1/2 = e^{-t_{1/2}/\tau}$, so $t_{1/2} = \ln(2)\tau = (0.693\ldots)\tau$. The mean lifetime of the triton is therefore 12.3 yrs/(.693) = 17.7 yrs.

$\leftarrow\bullet\,^{234}$Th $\bullet\,^{238}$U $\alpha\,\bullet\rightarrow$

FIGURE H.1
An alpha decay.

Early in the study of radioactivity, physicists and chemists classified nuclear decays into three types: *alpha* (α), *beta* (β), and *gamma* (γ). The decay of ^{238}U is an example of alpha decay, written

$$^{238}\text{U} \rightarrow {}^{234}\text{Th} + \alpha, \tag{H.1}$$

where the decay products include a thorium nucleus ^{234}Th and an α particle, as illustrated in Fig. H.1. All thorium nuclei contain 90 protons, and ^{234}Th in particular contains 234 nucleons = 90 protons + 144 neutrons. Because the number of protons and neutrons is conserved in α decays, the α particle must contain two protons and two neutrons. It is therefore identical to ^{4}He, the nucleus in the most common isotope of helium. In the rest frame of the initial ^{238}U nucleus there was no kinetic energy, so all the energy was contained in the mass energy of the ^{238}U nucleus. After the decay, the ^{234}Th nucleus and α particle each have kinetic energy, so by energy conservation we know that the sum of the mass energies of the ^{234}Th nucleus and α particle must be *less* than the mass energy of ^{238}U. We can also say that the total binding energy per nucleon must be larger in the decay products than it was in the original nucleus. The larger binding energy after the decay means that kinetic energy can be released.

The heavy-hydrogen nucleus ^{3}H, the triton, is an example of a particle that undergoes *beta decay.* The reaction in this case is

$$^{3}\text{H} \rightarrow {}^{3}\text{He} + e^{-} + \bar{\nu}, \tag{H.2}$$

in which the triton decays into a ^{3}He nucleus, an electron e^{-}, and an *antineutrino* $\bar{\nu}$, as shown in Fig. H.2.[2] As with α decays, the final particles in β decays must together have less mass than the initial particle, to provide the final kinetic energy.

A particularly famous beta decay is the radioactive carbon decay

$$^{14}\text{C} \rightarrow {}^{14}\text{N} + e^{-} + \bar{\nu} \tag{H.3}$$

with a half-life of 5730 years; this decay is used to date ancient manuscripts, clothing, bones, and other organic material that is in the approximate range 100–60,000 years old. Free neutrons also undergo beta decay,

$$n \rightarrow p + e^{-} + \bar{\nu}, \tag{H.4}$$

with a half-life of 614 s (10.2 minutes).

2. More specifically, this particle is an *electron*-type antineutrino. There are two other neutrino and antineutrino "flavors": *muon*-type and *tauon*-type.

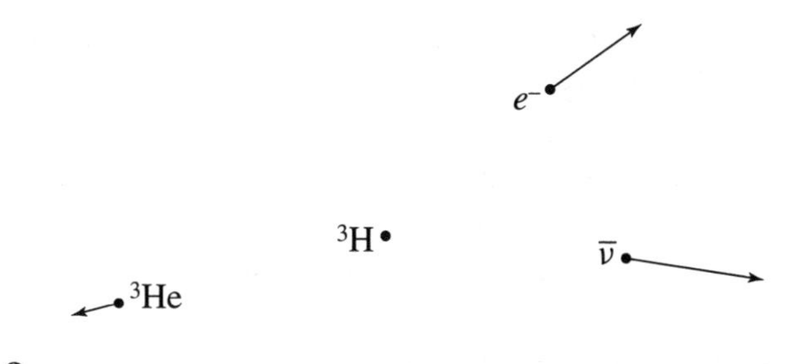

FIGURE H.2
A beta decay.

Finally, *gamma decays* are simply decays of any nucleus that happens to be in an excited energy state into the *same* nucleus in a lower energy state, with the emission of a photon. If that nucleus happens to be in the most common isotope of oxygen, for example, the reaction is

$$^{16}\text{O}^* \rightarrow {}^{16}\text{O} + \gamma \tag{H.5}$$

where the star means that the initial nucleus is in an excited state. The gamma (γ) here is a *gamma-ray* photon, distinguished from other photons only in that it typically has very high energies, of perhaps tens of kilo-electron volts ($\sim 10\,\text{keV}$) up to a few million electron volts ($\sim 1\,\text{MeV}$) or even higher. In gamma decays the gamma-ray photon has a lot of kinetic energy, and the final nucleus has some too, so the initial excited nucleus must have greater mass than the final nucleus. In terms of binding energy, the final nucleus has more binding energy than the initial excited nucleus, and this greater binding energy means that some of the initial mass energy is converted to kinetic energy of the final particles.

H.2 Nuclear Fusion

There is yet another way to convert mass energy partly into kinetic energy. From the curve of binding energy of Fig. 12.2, it is clear that most of the lightest nuclei tend to have smaller binding energies per nucleon than somewhat heavier nuclei. So if we can somehow merge lighter nuclei together to form heavier ones, the resulting increase in binding energy means that there will be energy left over for kinetic energy. That is, fusing lighter nuclei into heavier ones is a potential source of useful energy.

As a matter of fact, most of the energy we use on Earth originates from sunlight, including fossil fuels like coal, oil, and natural gas. And sunlight in turn ultimately derives from nuclear fusion reactions in the Sun's core. These reactions typically begin with

$$p + p \rightarrow d + e^+ + \nu, \tag{H.6}$$

in which two protons convert into a deuteron (^{2}H), a positron, and a neutrino. The initial protons have mass energy

$$2m_pc^2 = 2(938.3\,\text{MeV}) = 1876.6\,\text{MeV},$$

and the deuteron, consisting of a proton and neutron bound together, has mass energy

$$m_dc^2 = (m_p + m_n)c^2 - BE = (938.3 + 939.6)\,\text{MeV} - 2.2\,\text{MeV} = 1875.7\,\text{MeV}.$$

The positron has mass energy 0.5 MeV, and the neutrino has negligible mass energy on this scale, so the total mass energy after the reaction is 1876.2 MeV, which is 0.4 MeV *less* than the initial mass energy of the two protons. This net loss of mass energy implies a corresponding gain in kinetic energy of the final particles over the initial particles. Even more importantly, a deuteron is created!

The deuteron just created can be struck by another proton in the Sun to cause a second fusion reaction

$$p + d \rightarrow {}^3\text{He} + \gamma, \tag{H.7}$$

with a release this time of 5.5 MeV of kinetic energy. Eventually, a sufficient number of ^{3}He nuclei are created so that two of them can find one another and fuse in the reaction

$${}^3\text{He} + {}^3\text{He} \rightarrow {}^4\text{He} + p + p, \tag{H.8}$$

producing a whopping 12.9 MeV of kinetic energy in the final particles. The ^{4}He nucleus has a much higher binding energy per nucleon (and therefore a lower mass per nucleon) than the ^{3}He nuclei, as can be seen from Fig. 12.2, which is the reason why this fusion reaction produces so much kinetic energy. Overall, to create the single ^{4}He nucleus, it takes two $p + p \rightarrow d + e^+ + \nu$ fusion reactions (to create two deuterons), two $p + d \rightarrow {}^3\text{He} + \gamma$ reactions (to create two He^3 nuclei), and finally a single $^3\text{He} + {}^3\text{He} \rightarrow {}^4\text{He} + p + p$ reaction. If we count the annihilation of the two positrons with two electrons in the Sun, which produces several photons with net energy $4m_ec^2 = 2.0$ MeV, the total kinetic energy released is a very respectable 26.7 MeV. The net result is that four protons and two electrons have been used up, leaving a single ^{4}He nucleus, two neutrinos, and several photons, with a net kinetic energy of 26.7 MeV. This entire set of reactions is called the *proton–proton chain,* and is the source of most of the Sun's energy.

Fusion reactions also took place in the very early universe, when it was so hot and dense that charged particles could overcome their electrical repulsion and merge together. Fusion reactions also produce much of the explosive energy in thermonuclear weapons ("hydrogen bombs"), and efforts have long been underway to control fusion reactions for the production of useful power. Such controlled reactions typically begin with the separation of heavy water (D_2O) from ordinary water (H_2O). In heavy water, the hydrogen atoms are actually deuterium atoms (D) with deuterons as the nuclei.

Natural water contains a small fraction of D_2O, so capturing the deuterons provides the fuel to initiate controlled fusion by means of the two reactions

$$d + d \to {}^3\text{He} + n + 3.27\,\text{MeV} \quad \text{(H.9a)}$$

$$\to t + p + 4.03\,\text{MeV} \quad \text{(H.9b)}$$

where the energies given are the net kinetic energies produced, shared by the final particles.[3]

With the ^{3}H (denoted by t) and ^{3}He nuclei now created, several other more energetic reactions become possible, including

$$d + t \to {}^4\text{He} + n + 17.59\,\text{MeV}, \quad \text{(H.10)}$$

$$t + t \to {}^4\text{He} + n + n + 11.27\,\text{MeV}, \quad \text{(H.11)}$$

and

$$d + {}^3\text{He} \to {}^4\text{He} + p + 18.35\,\text{MeV}. \quad \text{(H.12)}$$

Controlled fusion reactions have been achieved, in (for example) "Tokomak" plasma confinement machines, but so far they have not produced enough kinetic energy to make up for the energy needed to get the fusion reactions going in the first place. Strenuous efforts have been made, and although researchers have come ever closer to "breakeven," they have not yet arrived.

H.3 Nuclear Fission

In the nuclear fusion reactions just described, a couple of light nuclei fuse together into a heavier nucleus, so this final nucleus is closer to the middle of the binding-energy curve than are the initial nuclei. In nuclear *fission* reactions, the exact opposite takes place, but with a rather similar outcome. A nucleus at the far right of the curve of Fig. 12.2 is caused to split in two, producing two nuclei that are each closer to the middle of the curve. These final nuclei have larger binding energies per nucleon, so kinetic energy is released by this reduction in mass energy. An example is the nucleus ^{238}U. Left to itself, ^{238}U eventually undergoes α decay, but this usually takes billions of years. If instead we shoot a neutron at a ^{238}U isotope, the neutron is sometimes absorbed to form the slightly heavier nucleus ^{239}U in an excited energy state. The ^{239}U nucleus is quite unstable, however; often it quickly splits into two lighter nuclei, spilling out two or three neutrons as well. An example: Sometimes the ^{239}U fissions into barium and krypton nuclei,

$$n + {}^{238}\text{U} \to {}^{239}\text{U} \to {}^{138}\text{B} + {}^{98}\text{Kr} + n + n + n, \quad \text{(H.13)}$$

---

3. Some deuteron pairs undergo the reaction given in Eq. H.9a, while others react via H.9b. The relative probability of one reaction over the other depends upon the deuteron energies.

one of many possible sets of decay products. In other cases, the two final nuclei and the number of emitted neutrons may be different; however, the total binding energy is always greater at the end than at the beginning, so that kinetic energy is released.

The kinetic energy released in a fission reaction is huge. In uranium fission it is typically about 200 MeV per reaction, so the final particles have *considerably* more kinetic energy than the initial neutron and ^{238}U nucleus. Soon after the discovery of nuclear fission, physicists all over the world realized a stunning possibility. The initial neutron causes a uranium nucleus to fission, but additional neutrons are set free by the fissioning itself, as in Eq. H.13. So some of these secondary neutrons might cause a subsequent fission reaction, and so on. A single fission reaction might lead to a *chain reaction,* causing the ultimate fissioning of an enormous number of uranium nuclei. If the chain reaction were fast enough, the result would be a huge explosion, with a total energy released of about 200 MeV times the number of fissioning uranium nuclei.

Suppose we had 238 grams of ^{238}U, an Avogadro's number of 6.002×10^{23} nuclei. If every nucleus in the sample were to undergo fission, the energy released would be 1.9×10^{13} joules, which is the energy equivalent of the explosion of 4500 tons of TNT! Needless to say, numbers like that got everyone's attention.[4]

Fortunately, a ^{238}U bomb doesn't work, because to cause ^{238}U to fission, the initial neutron has to have a kinetic energy of at least 1 MeV, which is the *threshold energy* of the reaction. Although many neutrons produced by fission have an energy greater than that, their kinetic energies are quickly degraded by bouncing around in the sample, gradually giving up their energies to the nuclei they hit.

There is a different isotope of uranium,[5] ^{235}U, the second-most common after ^{238}U. Atoms with a ^{235}U nucleus are chemically nearly identical to atoms with a ^{238}U nucleus. There is a crucial difference between them, however; *any* neutron can cause fission of ^{235}U, with no "threshold energy" required. So if we could somehow prepare a sample of uranium highly enriched in ^{235}U and cause one of these nuclei to undergo fission, the neutrons emitted in the reaction would slow down, just like those from ^{238}U fissioning, but instead of rapidly becoming impotent due to their degraded energy, their likelihood of causing fission reactions in other ^{235}U isotopes actually *increases.* This would be a true chain reaction, where a modest quantity of ^{235}U could blow up with an energy comparable to thousands of tons of TNT.

Obtaining a highly enriched sample of uranium is difficult. It was first accomplished during World War II at a huge facility at Oak Ridge, Tennessee, where sufficient highly enriched uranium to make a bomb was obtained by the summer of 1945. It was delivered to Tinian Island in the Pacific, and the uranium bomb was then dropped over

---

4. It was also soon realized that there is a minimum mass, called the *critical mass,* for the chain reaction to be successful. The reason is that in a small sample of fissioning material, too many neutrons escape through the surface before they can cause fission in another nucleus. For larger samples, proportional losses become less. The critical mass of a bare sphere of ^{235}U was calculated to be about 50 kg.

5. About 0.7% of natural uranium is ^{235}U.

Hiroshima, Japan, on August 6, 1945. Seventy thousand people died instantly, and seventy thousand more by the end of one year.

Another kind of nuclear bomb was also envisioned and developed during World War II, which is easier to make in one way (no separation of ^{235}U from ^{238}U is required) but harder in others. Recall that an incoming neutron can induce fission in ^{235}U. However, that does not *always* happen. Sometimes the reaction $n + {}^{238}\text{U} \to {}^{239}\text{U}^* \to {}^{239}\text{U} + \gamma$ takes place instead, where the excited ^{239}U nucleus gamma decays, reducing its energy below what is needed to fission. The resulting lower energy ^{239}U nucleus then beta decays into a nucleus of a neptunium atom, the first "transuranic" element:[6] ${}^{239}\text{U} \to {}^{239}\text{Np} + e^- + \nu$. The neptunium nucleus in turn soon beta decays into the next transuranic element, a plutonium nucleus: ${}^{239}\text{Np} \to {}^{239}\text{Pu} + e^- + \nu$. The nucleus ^{239}Pu is also radioactive, but long-lived (half-life: 24,360 yrs). It has the distinction of being as fissionable as ^{235}U, in that a neutron of *any* energy can cause the fission of ^{239}Pu. So if enough ^{239}Pu can be manufactured in the above series of reactions, a nuclear weapon can be made from it. This nucleus was first made in quantity during World War II in nuclear reactors at Hanford, Washington.[7] In fact, the first nuclear test explosion, at Alamogordo, N.M., and the bomb dropped on Nagasaki on Aug. 9, 1945, both used ^{239}Pu. Forty thousand people died instantly in the Nagasaki explosion, and thirty thousand more by the end of one year.

The *yield* of a nuclear explosion is measured in terms of the number of tons of TNT required to produce the same explosive energy, where 1 ton of TNT explodes with energy 4.2×10^9 joules. The yield of the Nagasaki explosion was roughly 20 kilotons, so the amount of mass that suddenly disappeared in that explosion was

$$\Delta m = \frac{\Delta E}{c^2} = \frac{(20 \times 10^3\ \text{tons})(4.2 \times 10^9\ \text{joules/ton})}{(3 \times 10^8\ \text{m/s})^2} \cong 9 \times 10^{-4}\ \text{kg} = 0.9\ \text{g}.$$

The yield of the Hiroshima explosion was roughly 15 kilotons, so the mass used up then was about 0.7 grams. When mass energy is suddenly converted into kinetic energy, it takes very little mass reduction to produce horribly destructive explosions.

Problems

H–1. The nucleus ^{233}U undergoes α decay, releasing 4.909 MeV of kinetic energy. (a) Identify the daughter nucleus. (b) Protons and neutrons have mass energies

6. The nuclei of transuranic elements contain more than 92 protons.

7. A nuclear reactor is a facility in which many fission reactions can take place in a controlled way, with toxic materials contained by large amounts of shielding. Commercial reactors make electricity from steam produced from water heated by the reactor. "Breeder" reactors, such as those at Hanford, breed plutonium from uranium.

938.3 MeV and 939.6 MeV, respectively. The four nucleons in the α particle have an average binding energy per nucleon of 7.1 MeV. Find the difference in the total binding energies of the ^{233}U and daughter nuclei.

H–2. The power of sunlight striking a perpendicular square meter at the location of Earth is 1300 W/m^2, where 1 W (watt) $= 1$ J/s. Earth is 150×10^6 km from the Sun. (a) How much mass per second is lost by the Sun in radiation? (b) How long could the Sun burn at this rate, if the efficiency of mass to energy conversion in the Sun is 0.1%? ($M_{\text{Sun}} = 2 \times 10^{30}$ kg.) (c) The Sun burns by nuclear fusion reactions, in which four protons are converted into a helium nucleus, two positrons (which have the same mass as electrons), two neutrinos (of negligible mass here), and numerous photons. Every time this net reaction takes place, 26.7 MeV of energy is converted from mass energy into kinetic energy, nearly all of which ultimately shows up in the observed radiated power. How many protons per second are burned in the Sun?

H–3. One way to produce energy by fusion reactions in the Sun's core is the proton–proton chain described in the text. A different set of fusion reactions is the so-called CNO cycle, where a ^{12}C nucleus (the nucleus of the most common carbon isotope) is needed as a catalyst. The reactions are as follows:

$$(1)\ ^{12}\text{C} + {}^{1}\text{H} \rightarrow {}^{13}\text{N} + \gamma \qquad (2)\ ^{13}\text{N} \rightarrow {}^{13}\text{C} + e^{+} + \nu$$

$$(3)\ ^{13}\text{C} + {}^{1}\text{H} \rightarrow {}^{14}\text{N} + \gamma \qquad (4)\ ^{14}\text{N} + {}^{1}\text{H} \rightarrow {}^{15}\text{O} + \gamma$$

$$(5)\ ^{15}\text{O} \rightarrow {}^{15}\text{N} + e^{+} + \nu \qquad (6)\ ^{15}\text{N} + {}^{1}\text{H} \rightarrow {}^{12}\text{C} + {}^{4}\text{He}$$

Note that the ^{12}C nucleus used up in the first reaction is regenerated in the final reaction. A positron created in step (2) or (5) soon finds an electron in the Sun; these two particles then annihilate into photons. (The CNO cycle produces only a minor part of the Sun's energy, but it is much more important in very hot stars.) In terms of the masses of the proton ^{1}H, helium nucleus ^{4}He, and electron (or positron) e, find the net kinetic energy released in a single CNO cycle.

H–4. A hypothetical thermonuclear weapon uses the fusion reaction

$$d + t \rightarrow {}^{4}\text{He} + n$$

to power the explosion. That is, a deuteron and triton come together to form a ^{4}He nucleus and a neutron. The mass energies of these particles are

$$d\text{: 1875.6 MeV} \quad t\text{: 2808.9 MeV} \quad {}^{4}\text{He: 3727.4 MeV} \quad n\text{: 939.6 MeV}$$

(a) Find the kinetic energy (in MeV) released in one such reaction. That is, how much more kinetic energy is there after the reaction than before? (b) If the yield of the explosion is 10 Mt (that is, 10×10^6 tons of TNT equivalent), how many deuterons are used up in the explosion? (1 ton of TNT explodes with energy 4.2×10^9 joules; $1\,\text{eV} = 1.60 \times 10^{-19}$ J.)

H–5. Physicists are currently trying to produce useful quantities of energy from controlled thermonuclear reactions, including the "burning" of deuterons (d) in the reaction $d + d \to p + t$, where p is a proton and t is a triton (hydrogen nucleus ^{3}H). The mass energies are p: 938.3 MeV; d: 1875.7 MeV; t: 2809.1 MeV. In this reaction, how many deuterons per second must be burned in a reactor designed to provide 100 gigawatts ($= 10^{11}$ watts $= 10^{11}$ joules/second) of power, assuming 100% efficiency? (Note that $1\,\text{MeV} = 1.60 \times 10^{-13}$ joules.)

H–6. When 1 ton of TNT explodes, the kinetic energy released is 4.2×10^9 joules ($1\,\text{eV} = 1.60 \times 10^{-19}$ J.) In the following nuclear explosions, find how much mass (in grams or kilograms) is lost. (a) A 20 kiloton explosion of a ^{235}U bomb (that is, a kinetic energy equivalent to that of blowing up 20 kt of TNT). (b) A 100 megaton explosion of a thermonuclear (combined fission/fusion) weapon.

H–7. Assume that the critical mass of a sphere of ^{235}U is 50 kg. The density of uranium metal is 19 g/cm^3. (a) Find the radius of the critical-radius sphere. (b) We consider surrounding a sphere of pure ^{235}U by a shell of natural uranium, which is 99.3% ^{238}U. Would this tend to increase or decrease the critical mass of the ^{235}U sphere? (*Hint:* The shell of natural uranium would reflect some escaping neutrons back into the sphere.)

H–8. An exploding nuclear bomb creates a rapidly expanding shock wave in the air surrounding the blast. Within the shock wave the air glows brightly, because it has been strongly heated, giving the appearance of a "fireball." The radius R

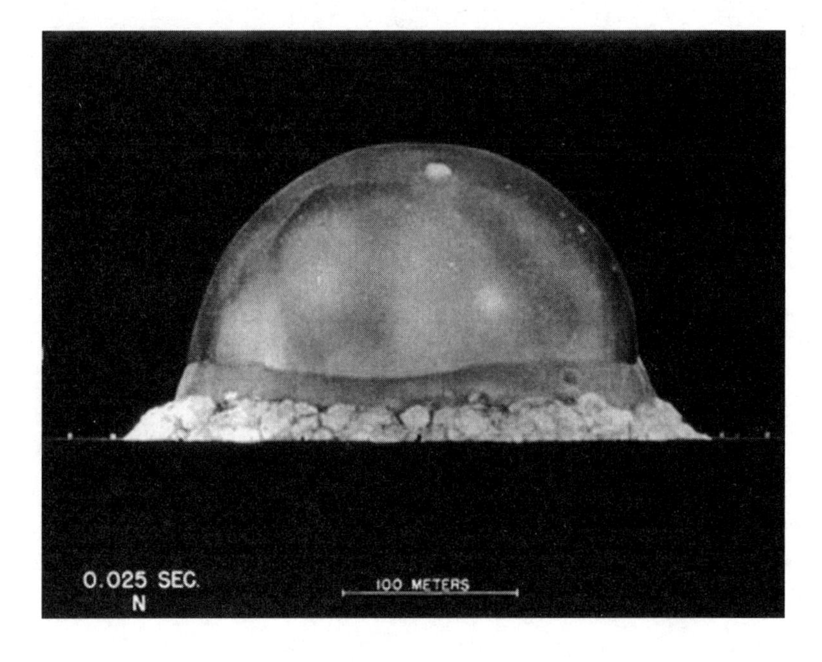

of the expanding ball of hot air depends upon time t, the energy E of the blast, and the density ρ of the air. (It might depend also upon the ambient air pressure, but to a good approximation it does not, because the ambient pressure is so much smaller than the pressures created by the blast.) (a) Using dimensional reasoning, find out how R depends upon t, E, and ρ. Dimensional reasoning simply means that an equation should have the same factors of mass, length, and time (the three basic dimensions of mechanics) on both sides. (b) If the shock wave has radius R_0 at time 0.01 s, what is its radius at 0.1 s? (c) The picture below shows the fireball of the Trinity test, the first nuclear explosion, at 05.30 hours, 16 July 1945, Alamogordo, N. M., at time 0.025 s after detonation. The diameter is about 250 m, as shown. Estimate the energy of the blast in joules and in equivalent tons of TNT, assuming the dimensionless coefficient in the expression for $R(t, E, \rho)$ is unity (the coefficient has been calculated to be 1.003). The explosive energy of 1 ton of TNT is 4.2×10^9 J. [Pictures of the first explosion were published in *Life Magazine.* Using dimensional reasoning, several physicists around the world deduced the yield of the explosion.]

APPENDIX I

Some Particles

Family	Mass Energy (MeV)	Mean Lifetime (seconds)
photon γ	0	stable
leptons		
$e^{\pm}$	0.511	stable
$\mu^{\pm}$	105.7	2.2×10^{-6}
$\tau^{\pm}$	1777	2.9×10^{-13}
ν_e, $\bar{\nu}_e$	(unknown, probably	stable, although neutrinos
ν_μ, $\bar{\nu}_\mu$	of order 10^{-3} eV)	can change from one kind to
ν_τ, $\bar{\nu}_\tau$		another (flavor oscillation)
mesons		
π^0	135.0	8.4×10^{-17}
$\pi^{\pm}$	139.6	2.6×10^{-8}
$K^{\pm}$	493.7	1.2×10^{-8}
K^0	497.7	K_1^0: 8.9×10^{-11}
		K_2^0: 5.2×10^{-8}
baryons		
p	938.3	stable
n	939.6	15 minutes
Λ	1115.7	2.6×10^{-10} s
Σ^+	1189.4	0.8×10^{-10}
Σ^0	1192.6	7.4×10^{-20}
Σ^-	1197.4	1.5×10^{-10}
Ξ^0	1314.8	2.9×10^{-10}
Ξ^-	1321.3	1.6×10^{-10}

APPENDIX J

Relativity and Electromagnetism

IT WAS EINSTEIN'S THINKING about electricity and magnetism in different frames of reference that led him to special relativity in the first place. The connections between them are fundamental. Here we explore just one of these connections for readers who already have some familiarity with electromagnetism, because it is an excellent illustration of the relativity of electric and magnetic fields and also of Lorentz-contraction effects, even when the velocities involved are small compared with the velocity of light.

The illustration involves the electric field **E** and magnetic field **B** outside an infinite straight wire, and the forces they exert on a hypothetical electric charge q placed outside the wire.[1] The sum of the electric and magnetic forces on q is

$$\mathbf{F} = \mathbf{F}_{\mathrm{E}} + \mathbf{F}_{\mathrm{B}} = q(\mathbf{E} + \mathbf{v} \times \mathbf{B}), \tag{J.1}$$

where $\mathbf{v}$ is the velocity of q and $\mathbf{v} \times \mathbf{B}$ is the cross product[2] of the vectors $\mathbf{v}$ and $\mathbf{B}$.

A wire contains both positive and negative charges, as shown in Fig. J.1, where the positive charges are the atomic nuclei together with many bound electrons, together forming positive ions, which are at rest relative to the wire, and the negative charges are the conduction electrons, which are free to move along the wire. We suppose that the ions have effectively uniform positive charge per unit length λ_+, and the conduction electrons have uniform negative charge per unit length λ_-, so the net charge per unit length of the wire is $\lambda = \lambda_+ + \lambda_-$, which may be positive, negative, or zero. The

---

1. Real wires are obviously not infinite, but we can imagine a very long wire and find the fields sufficiently close to its middle that its finiteness makes little difference.

2. The cross product $\mathbf{v} \times \mathbf{B}$ has the magnitude $vB \sin\theta$, where θ is the smaller angle between the vectors $\mathbf{v}$ and $\mathbf{B}$, and the direction of $\mathbf{v} \times \mathbf{B}$ is perpendicular to both $\mathbf{v}$ and $\mathbf{B}$ and governed by the right-hand rule, in which you begin by extending the thumb, forefinger, and middle finger of your right hand so they are perpendicular to one another. Then, placing your thumb in the direction of $\mathbf{v}$ and forefinger in the direction of $\mathbf{B}$, your middle finger will point in the direction of $\mathbf{v} \times \mathbf{B}$.

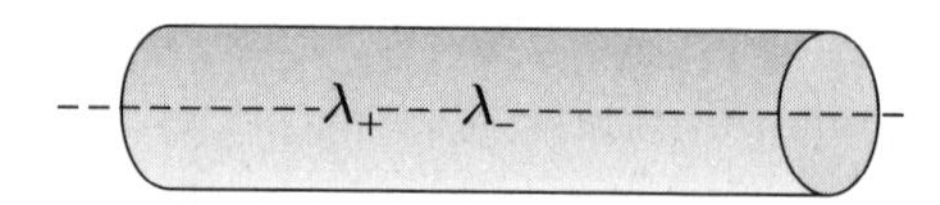

FIGURE J.1
An infinite wire with positive and negative charges per unit length λ_+ and λ_-.

overall charge density λ produces an electric field **E** at any point P outside the wire, of magnitude

$$E = \frac{\lambda}{2\pi\varepsilon_0 r}, \tag{J.2}$$

where ε_0 is a universal constant and r is the distance of P from the center of the wire. The direction of **E** is *outward,* away from the wire, if λ is positive, as illustrated in Fig. J.2(a), and *inward,* toward the center of the wire, if λ is negative.

If the conduction electrons are flowing, the wire is carrying a current. But since electrons are negatively charged, if they flow to the right the conventional current is to the left, and if they flow to the left the conventional current is to the right. The current causes a magnetic field to form outside the wire, whose magnitude is

$$B = \frac{\mu_0 i}{2\pi r}, \tag{J.3}$$

where i is the current and μ_0 is a universal constant, related to ε_0 by $\mu_0\varepsilon_0 = 1/c^2$, where c is the speed of light. The field chases its tail in circles around the wire, as shown in

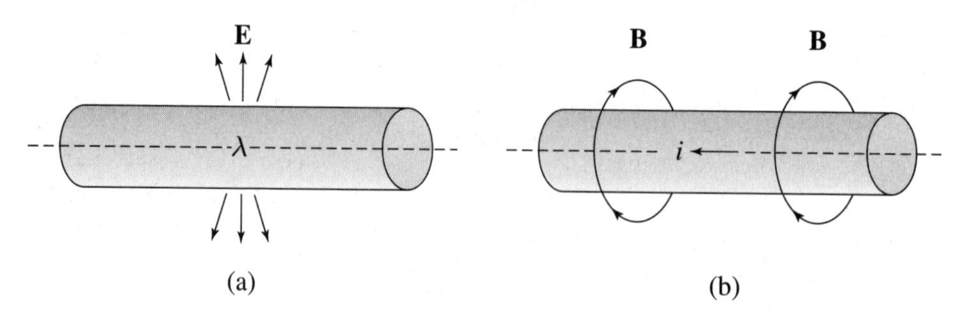

FIGURE J.2
(a) If the charge density in a wire is positive, there is an electric field surrounding the wire that is radially outward. (b) If the wire carries a current, there is a magnetic field surrounding the wire, which chases itself in circles. In the figure, the current is toward the left.

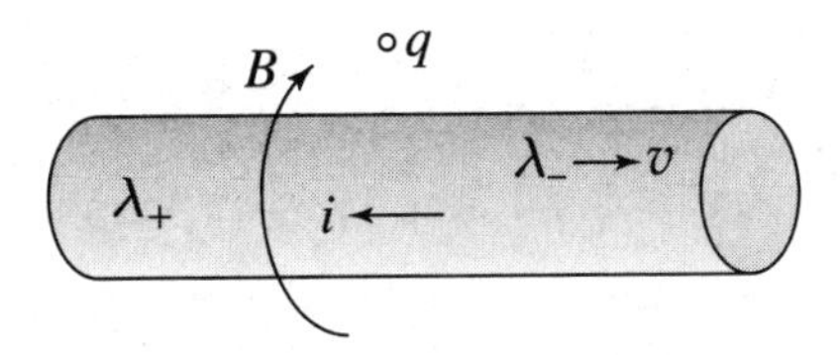

FIGURE J.3
A charge q placed at rest outside an electrically neutral current-carrying wire, in the rest frame of the wire. There is a magnetic field in this frame but no electric field. The charge q feels no force, so stays put.

Fig. J.2(b), where the direction of the circles can be determined by a variation of the right-hand rule.[3]

Now suppose that in the rest frame of our wire, which we will call the unprimed frame, the conduction electrons are in fact flowing to the right, so the conventional current is to the left. Also suppose that the net electric charge per unit length in the wire is $\lambda = \lambda_+ + \lambda_- = 0$, with a charge per unit length λ_- of the conduction electrons equal but opposite to the charge per unit length λ_+ of the positive ions. Then there is no electric field outside the wire. There is a magnetic field, however, because there is a nonzero current. The current i is the charge per unit time crossing through any cross section of the wire, which in this case is

$$i = \frac{\text{charge}}{\text{time}} = \frac{\text{charge}}{\text{length}} \times \frac{\text{length}}{\text{time}} = \lambda_- v \tag{J.4}$$

where λ_- is the charge per unit length of the flowing electrons, and v is their flow velocity. The magnetic field outside the wire is therefore

$$B = \frac{\mu_0 \lambda_- v}{2\pi r}, \tag{J.5}$$

circling around the wire.

Now place a charge q outside the wire, at rest in the rest frame of the wire, as shown in Fig. J.3. The charge feels no electric force because there is no electric field, and it also feels no magnetic force because, even though there is a magnetic field, the velocity of the charge q is zero.

Now view this exact same wire and external charge q in a primed frame moving to the right at speed v along the direction of the wire, where v is the speed of the conduction electrons themselves. The conduction electrons are therefore at rest in this frame, while the wire and its positive ions are together moving to the *left* at speed v. The external charge q is *also* moving to the left at speed v. What now is the force on q?

3. Point the thumb of your right hand along the direction of the conventional current (that is, opposite to the electron-flow direction.) Then the fingers of your right hand will naturally curl in the direction of **B**.

We already know the answer. The external charge remains at rest in the unprimed frame, and moves at constant velocity v to the left in the primed frame. The momentum of the charge stays constant in either frame, so there can be no net force on it in *either* frame!

Let us verify that there is no net force in the primed frame, by calculating the electric and magnetic fields in that frame, and the forces they cause. First of all, even though the net charge per unit length on the wire was zero in the unprimed frame, this is no longer true in the primed frame. The reason is Lorentz contraction. The ions were at rest in the unprimed frame, but in the primed frame they are moving, so those ions that were within some given length of wire in the wire's rest frame are now squeezed into a length that is shorter by the factor $\sqrt{1-v^2/c^2}$ in the primed frame. Therefore the positive charge per unit length has *increased* to $\lambda'_+ = \lambda_+/\sqrt{1-v^2/c^2}$. Also, in the original unprimed frame the conduction electrons were moving with speed v, but in the primed frame they are not moving at all. Therefore the squeezing is in the opposite direction for them: The same number of conduction electrons will occupy a larger length in the primed frame than they did in the unprimed frame, so their density is *decreased* in the primed frame to $\lambda'_- = \lambda_-\sqrt{1-v^2/c^2}$. The overall charge per unit length of the wire is therefore

$$\lambda' = \lambda'_+ + \lambda'_- = \frac{\lambda_+}{\sqrt{1-v^2/c^2}} + \lambda_-\sqrt{1-v^2/c^2} = \lambda_+\left(\frac{1}{\sqrt{1-v^2/c^2}} - \sqrt{1-v^2/c^2}\right)$$
$$= \frac{\lambda_+}{\sqrt{1-v^2/c^2}}\left[1-(1-v^2/c^2)\right] = \frac{\lambda_+(v^2/c^2)}{\sqrt{1-v^2/c^2}}. \tag{J.6}$$

This net charge density causes an electric field in the primed frame given by

$$E' = \frac{\lambda'}{2\pi\varepsilon_0 r} \tag{J.7}$$

and directed outward, away from the wire, since λ' is positive, as shown in Fig. J.2(a).

There is also an electrical force outward in the primed frame, given by $F'_E = qE'$, directed outward since the charge q is positive. Its magnitude is

$$F'_E = qE' = \frac{q\lambda'}{2\pi\varepsilon_0 r} = \frac{q\lambda_+(v^2/c^2)}{2\pi\varepsilon_0 r\sqrt{1-v^2/c^2}}. \tag{J.8}$$

Now we need to calculate the magnetic field in the primed frame. The conduction electrons are at rest in this frame, so their current is zero. However, the ions are now moving to the left, and so they produce a current to the left in the primed frame equal to $i' = \lambda'_+ v$, their (positive) charge per unit length multiplied by their velocity. The magnetic field they produce is therefore

$$B' = \frac{\mu_0 \lambda'_+ v}{2\pi r} \tag{J.9}$$

directed in circles around the wire, as shown in Fig. J.2(b). The magnetic force on the external charge q is not zero in this primed frame, because q is now moving with velocity $\mathbf{v}$ to the left. The magnetic force $\mathbf{F}_B = q\mathbf{v} \times \mathbf{B}$ has the magnitude qvB' (note that the vectors $\mathbf{v}$ and $\mathbf{B}$ are perpendicular to one another, so $\sin\theta = 1$, where θ is the angle between them.) Therefore the magnetic force has the magnitude

$$F'_B = \frac{q\mu_0 \lambda'_+ v^2}{2\pi r} = \frac{q\lambda_+ (v^2/c^2)}{2\pi \varepsilon_0 r\sqrt{1 - v^2/c^2}}, \tag{J.10}$$

where in the final equality we used $\mu_0 \varepsilon_0 = 1/c^2$ and $\lambda'_+ = \lambda_+/\sqrt{1 - v^2/c^2}$. The right-hand rule shows that the magnetic force is directed *inward,* toward the center of the wire. The total force on q is therefore (taking the outward direction to be positive)

$$F' = F'_E + F'_B = \frac{q\lambda_+ (v^2/c^2)}{2\pi \varepsilon_0 r\sqrt{1 - v^2/c^2}} - \frac{q\lambda_+ (v^2/c^2)}{2\pi \varepsilon_0 r\sqrt{1 - v^2/c^2}} = 0, \tag{J.11}$$

as expected. *If the net force on q is zero in one inertial frame, it must be zero in any other inertial frame as well, because the momentum stays constant in each frame.* However, the *reason* for the zero seems quite different from one frame to the other.

> In the unprimed frame of the wire there is no electric field, and even though there is a magnetic field, it causes no force on the external charge q, because q is not moving.
>
> In the primed frame moving along with the conduction electrons, there is both an electric field, which causes an outward force on q, and a magnetic field, which causes an inward force on q. The two forces exactly cancel one another.

We could recalculate the force in *any* frame. The electric and magnetic fields would differ from frame to frame, but whatever they are, the net force on q must always be zero. Note that even though the concept of electric and magnetic field *lines* is very useful, helping us visualize what is going on, they are really not there at all! They are really just very useful abstractions. In the rest frame of our wire there were no electric field lines at all, but if we simply viewed the exact same physical situation in a frame moving along the wire, electric field lines popped into existence.

Clearly the traditional electric and magnetic fields are not independent of one another. A purely magnetic field in one frame becomes both an electric and a magnetic field in a different frame. Similarly, a purely electric field in one frame, such as the frame in which a point charge is at rest, becomes both an electric and magnetic field in a frame in which the charge is moving. Einstein saw that **E** and **B** must be just aspects of a more general *electromagnetic* field that has a unified description. And this realization was so important to his thinking that he gave the title "On the electrodynamics of moving bodies,"[4] to his original 1905 paper introducing special relativity, and began this magnificent, revolutionary paper on the nature of time and space as follows:

> It is known that Maxwell's electrodynamics—as usually understood at the present time—when applied to moving bodies, leads to asymmetries which do not appear to be inherent in the phenomena. Take, for example, the reciprocal electrodynamic action of a magnet and a conductor. The observable phenomenon here depends only on the relative motion of the conductor and magnet, whereas the customary view draws a sharp distinction between the two cases in which either the one or the other of these bodies is in motion. For if the magnet is in motion and conductor at rest, there arises in the neighborhood of the magnet an electric field with a certain definite energy, producing a current at the places where parts of the conductor are situated. But if the magnet is stationary and conductor in motion, no electric field arises in the neighborhood of the magnet. In the conductor, however, we find an electromotive force, to which in itself there is no corresponding energy, but which gives rise—assuming equality of relative motion in the two cases discussed—to electric currents of the same path and intensity as those produced by the electric forces in the former case.
>
> Examples of this sort, together with the unsuccessful attempts to discover any motion of the earth relatively to the "light medium," suggest that the phenomena of electrodynamics as well as of mechanics possess no properties corresponding to the idea of absolute rest. . . .

---

4. From *The Principle of Relativity* by Einstein and others, Dover Publications, 1952. Translated from "Zur Elektrodynamik bewegter Körper," *Annalen der Physik,* 17, 1905.

Answers to Some Odd-Numbered Problems

1–1. $v' = (v_0 - V) + a_0 t$

1–3. (c) $a = -g$, $v = -gt$, $y = h - (1/2)gt^2$.

1–5. 10 m/s *KE* is not conserved.

1–7. Beforehand, the car moved at 60 mi/hr. Afterward, they move together at $20\sqrt{2}$ mi/hr $\cong$ 28 mi/hr.

1–9. (a) $(2/\sqrt{3})$ m/s $\cong$ 1.15 m/s (b) $(\sqrt{3} + 1/\sqrt{3})$ m/s $\cong$ 2.31 m/s (c) yes

1–11. (a) $v_0 = \sqrt{k/m}A$ (b) $v = \sqrt{3k/m}(A/2)$

1–13. (a) $v_{\text{He}} = 0.4v_0$, $v_n = -0.6v_0$ (b) $v_{\text{He}} = -0.6v_0$, $v_n = -1.6v_0$

2–1. 16.6 seconds of arc

2–3. (a) 190 mi/hr, at 39° (b) 90 mi/hr, at 53°

2–5. 13 m/s, at angle 63° to the vertical

2–7. Passengers will see more stars toward the front.

2–9. $t_{\text{A}} = 2Dv_{\text{s}}/\left(v_{\text{s}}^2 - V_0^2\right)$, $t_{\text{B}} = 2D/\sqrt{v_{\text{s}}^2 - V_0^2}$

3–1. (a) yes (b) no

3–5. (a) c and v_{s}, respectively (b) c and $v_{\text{s}} - v_{\text{w}}$ (c) c and $v_{\text{s}} - v_{\text{w}} + v_0$

4–1. 40 s

4–3. $v = 0.789\,c$

4–5. $\varepsilon = 5 \times 10^{-13}$

4–7. (a) 100/3 yrs (b) 80/3 yrs (c) 64/3 yrs

4–9. 2.6×10^{-10} s

4–11. (b) 20 days (c) 12 $c \cdot$ days (d) 5 days (e) 14 days

4–13. 1.04×10^{-7} s

5–1. (a) 1.01×10^5 yrs (b) 1.42×10^4 yrs (c) 1.41×10^4 $c \cdot$ yrs

5–3. $(3/5)c$

5–5. $v/c = 0.996$, 8.7×10^{-5} m

5–7. (a) $v/c = 1 - 3.0 \times 10^{-9}$ (b) 7.8 m

5–9. (a) $\varepsilon = 1 \times 10^{-30}$ (b) 5×10^{-11} m, about the radius of a hydrogen atom

5–11. (b) 36 $c \cdot$ yrs (c) 45 yrs (d) 27 yrs

6–1. (a) -10^{-7} s (b) 5×10^{-15} s

6–3. (a) 5/12 hrs (b) 5/13 $c \cdot$ hrs (c) 12/13 hrs

6–7. (a) 600 m/c (b) $1333\frac{1}{3}$ m/c (c) $1666\frac{2}{3}$ m/c (d) $1066\frac{2}{3}$ m/c

6–11. (c) 8/5 $c \cdot$ yrs (d) 10/9 yrs (e) 2/3 yrs (f) 8/9 $c \cdot$ yrs (g) 8/15 $c \cdot$ yrs

6–13. (a) 1080 hrs (b) 864 hrs; They are rescued!

6–15. (a) $(4/\sqrt{3})$ hrs (b) just before 3:20 a.m.

7–1. (a) A is 52 yrs old, B only 20. (b) 24 $c \cdot$ yrs (c) B is not always at rest in an inertial frame, so cannot always use the rules of special relativity.

7–3. The compatriots are right.

7–5. In the stick's frame, the plate and hole are tilted. The stick gets through.

7–7. In the original stick frame, the stick bends. The right side falls through first, and the left side later. There are no rigid bodies in relativity.

8–1. 11.5 s

8–3. (a) -1.2 m (b) 1.6 m/c

8–5. (a) $(15/17)c$ (b) (*i*) 80 m (*ii*) 100 m (*iii*) 80 m
(c) (*i*) 166.7 m/c (*ii*) 283.3 m/c

8–7. (a) $v = \sqrt{c^2 + 3v_0^2}\Big/2$ (b) $\theta = \tan^{-1}\left(c/\sqrt{3}v_0\right)$

8–9. speed $= (1-\varepsilon)c/(1-\varepsilon+\varepsilon^2/2)$

8–11. $\cos\theta' = (\cos\theta - V/c)/(1-(V/c)\cos\theta)$

8–13. $v\sqrt{2 - v^2/c^2}$

8–15. (b) $a'_x = a_x$ (c) $a_x = a'_x(1 - V^2/c^2)^{3/2}/(1 + v'_x V/c^2)^3$

9–1. (a) timelike (b) points on the past light cone

9–3. (a) spacelike

9–7. (c) S's clock reads 9 days when the tachyon arrives. (d) In S's frame, the tachyon arrives at S before it was sent by E!

9–9. $V = (4/5)c$

9–11. (a) negative (b) Al (c) Al

10–1. (a) 1.005 (b) 1.15 (c) 224 (d) 7.07×10^4

10–3. 2.25×10^5 m/s

10–5. (b) For A, $(v'_x)_{\text{before}} = 0$ and $(v'_x)_{\text{after}} = (-3v_0/2)\,/\,(1 + v_0^2/2c^2)$.
For B, $(v'_x)_{\text{before}} = -v_0$ and $(v'_x)_{\text{after}} = (-v_0/2)\,/\,(1 - v_0^2/2c^2)$.

10–7. $M/m = 9/16$

10–9. $v/c = 0.14$

11–1. $KE = 29mc^2$, $p = 29.983mc$, $v/c = 0.9994$

11–3. $KE = 8.1 \times 10^{25}$ joules $= 9mc^2$

11–5. (a) 4.265 MeV (b) The thorium nucleus also has kinetic energy.

11–7. 30 m

11–9. (a) $E = 0.511$ MeV $\lambda = 0.00243$ nm (b) 1.02 MeV

11–11. 2.6×10^{-10} s

11–13. (a) 5000 MeV/c (b) 12,000 MeV/c^2 (c) 8000 MeV/c^2

11–15. $\varepsilon = 5 \times 10^{-7}$

11–17. $\varepsilon = 4.9 \times 10^{-24}$; 25 m/s

11–19. $v'_x = -(4/5)c$ $v'_y = -(9/25)c$

12–1. $M = 2m + k(\Delta x)^2/2c^2$

12–3. 2.9 MeV

12–5. (a) 109.8 MeV (b) 4.1 MeV (c) 29.7 MeV/c (d) 0.27 c

12–7. $v_n = 0.11\,c$, $v_\pi = 0.61\,c$

12–11. (a) The ^{239}U must have KE. (b) $E_\gamma = (M^{*2} - M^2)c^4/2M^*c^2$

12–13. 0.60 MeV

12–15. yes, 146.3 MeV

12–17. $7mc^2$

12–19. $\Delta\lambda = 0.0024$ nm, $KE_e = 8.6 \times 10^{-6}$ eV

12–23. 1.66 m/s

12–25. (b) $(3/5)c$ (c) $M/M_0 = \sqrt{(1 - v/c)/(1 + v/c)}$ (d) 1/3

12–27. (a) (*i*) 0.728 (*ii*) 0.866 (*iii*) 0.950

13–1. (a) $p = 0.383$ MeV$/c$, $E = 0.639$ MeV
(b) $p' = -0.214$ MeV$/c$, $E' = 0.554$ MeV

13–3. $p_x = 0$, $p_y = 704.7$ MeV$/c$, $E = 1175$ MeV
$p'_x = -881$ MeV$/c$, $p'_y = 704.7$ MeV$/c$, $E' = 1469$ MeV

13–5. 1.0×10^{-3}

13–7. $v/c = 4/5$

13–9. (b) $v/c = 0.162$, 0.212, 0.258, and 0.317

13–11. $\nu_{\text{observed}}/\nu_{\text{emitted}} = M/M_0$

13–13. $7m_pc^2$

13–15. (a) greater than (b) $4m_ec^2$

13–17. (a) 3.10 GeV (b) $\varepsilon = 5.4 \times 10^{-8}$ (c) 9.4×10^3 GeV

13–19. (c) $144mc^2$ in colliding-beam expt., $12mc^2$ in stationary-target experiment.

13–21. $\pm(12/25)c$

14–1. (a) stays put (b) drifts toward the front

14–3. (a) tip backward (b) accelerate the car forward

14–5. (a) $y = (1/2)g(L/c)^2$ (b) 1.4 km

14–7. *Hint:* View the flights from an inertial frame in which Earth rotates eastward.

14–13. (a) 1.3×10^7 m (b) 1.8×10^3 m/s (c) ~ 3 ns (d) $\sim 10^{-13}$ s/s
(e) fractional altitude effect $\sim 3.6 \times 10^{-11}$, fractional time dilation effect $\sim 1.8 \times 10^{-11}$

A–1. (a), (b) 1.015 (c) 0.9850

A–3. (a) 3.00006 (b) $3 + 6 \times 10^{-17}$

A–5. (a) $(1/2)mv(v^2/c^2)$ (b) $v/c = 0.1414$

C–1. (a) $v_{\text{apparent}} = v/(1 - v/c)$ (b) $v_{\text{apparent}} = v/(1 + v/c)$

C–3. (a) $\Delta t'/\Delta t = \sqrt{(1 - v/c)/(1 + v/c)}$
(b) $\nu_{\text{observed}}/\nu_{\text{emitted}} = \sqrt{(1 - v/c)/(1 + v/c)}$
(c) same, with $v \to -v$

E–1. (a) $\beta = 1/\alpha$ (c) $V = c/2$

E–3. (c) The blades would not remain rigidly straight, because signals produced at the handles can propagate along the blades only at speeds less than c.

F–1. 4 s

F–3. (a) $2\sqrt{30}mc/qE_0$ (b) 37 μs

F–7. (b) for small t, $a \to F/m$; for large t, $a \to 0$

G–3. (a) 30,000 yrs (b) 10.5 yrs (c) 42 yrs

G–5. (b) 2.2 yrs

G–7. 4.0 yrs

G–9. (a) 11.7 years (b) 462 years

G–11.

$$V = \tanh T_{\text{ship}} = \frac{T}{\sqrt{1+T^2}} = \frac{\sqrt{X(X+2)}}{X+1}$$

$$T = \sinh T_{\text{ship}} = \frac{V}{\sqrt{1-V^2}} = \sqrt{X(X+2)}$$

$$T_{\text{ship}} = \ln\left(T + \sqrt{1+T^2}\right) = \ln\left(\frac{1+V}{1-V}\right)^{1/2} = \ln\left(X + 1 + \sqrt{X(X+2)}\right)$$

$$X = \sqrt{1+T^2} - 1 = \frac{1}{\sqrt{1-V^2}} - 1 = 2\sinh^2(T_{\text{ship}}/2)$$

H–1. (a) ^{229}Th (b) 23.5 MeV

H–3. $KE_{\text{released}} = 4M_{^1\text{H}}c^2 + 2m_e c^2 - M_{^4\text{He}}c^2$

H–5. 3.1×10^{23} deuterons/second

H–7. (a) 8.6 cm (b) decrease

Index